BOSTON STUDIES IN THE PHILOSOPHY OF SCIENCE
VOLUME TWO: IN HONOR OF PHILIPP FRANK

PHILIPP FRANK

BOSTON STUDIES IN THE PHILOSOPHY OF SCIENCE

Volume Two: IN HONOR OF PHILIPP FRANK

Edited by

ROBERT S. COHEN and MARX W. WARTOFSKY

PROCEEDINGS OF THE BOSTON COLLOQUIUM FOR THE PHILOSOPHY OF SCIENCE, 1962-1964

HUMANITIES PRESS
New York
1965

Library of Congress Catalog Card Number: 65-28459

Printed in the United States of America

TABLE OF CONTENTS

PREFACE

Professor Philipp Frank has lived in the Boston and Cambridge community for more than 25 years. He brought here his philosophical love of science, his extraordinary way of teaching both the sophisticated and the beginner, his willing tolerance of quite different interpretations of the world, his standards of personal modesty and intellectual toughness, his progressive sympathies, his great intelligence, his humor and wisdom, and his lovely wife, Hania.

He also brought the discipline of philosophy of science into the education and mature thought of many scientists as well as philosophers. Gratitude for these many gifts impels us to offer to Philipp Frank this volume of papers presented to the Boston Colloquium for the Philosophy of Science in the years 1962-1964, introduced by greetings from some of his students and colleagues.

Our Colloquium construes the philosophy of science broadly, as he has advised us to do. We try to discuss open problems in the foundations of science, and, wherever relevant, to bring material from the history and cultural relations of science to bear upon such problems. We try also to talk with each other across all boundaries of discipline, to include scholars from philosophy, logic and mathematics, the physical and biological sciences, history and the social sciences, and the humanities as well. Our contributions are in the first instance working papers, presented in the authors' hopes of mutually helpful criticism. In this second volume derived from our discussions, we have attempted to report to our colleagues in Boston and elsewhere the principal contributions of the years represented.

This volume is also the substantial part of our report to the National Science Foundation, which has again kindly and generously supported our Colloquium by a grant to Boston University. Boston University acts on behalf of those philosophers and philosophically-minded scientists from the colleges and universities, industrial research laboratories and government laboratories of metropolitan Boston who join together in this Colloquium.

ROBERT S. COHEN and MARX W. WARTOFSKY
Boston Colloquium for the Philosophy of Science

Departments of Physics and Philosophy
Boston University
August 25, 1965

HOMAGE TO PROFESSOR PHILIPP G. FRANK

Peter G. Bergmann

It is now almost thirty years that I left the Institute of Theoretical Physics of the German University of Prague, of which you were then the Director, to take up postdoctorate work under Einstein at Princeton. The Institute was physically located on the top floor of a nineteenth-century academic building, Vinicna 3, conveniently across the street from the psychiatric hospital, whose grounds were protected by a high masonry wall. The patients would look at us, and we at the patients, often wondering who was "in" and who was "out." I had my headquarters at the library, which was operated for the joint benefit of theoretical and experimental physics, right next to your office. More often than not, you would stop by at the library, not only to look at recent journals but also to talk with the students there, mostly Richard Wallauschek and myself, about the latest gossip in politics, about physics, and about epistemology.

As one of the three academic institutions operated by the Czechoslovak Republik of Masaryk and Benes for its German minority population, the German University at Prague was somewhat isolated from the scientific traffic that streamed through the larger scientific communities. But because it was independent of the political control of the Nazis though using German as its official language of instruction, it provided a haven for refugees from Germany and from the Balkans, victims of political, racial, or national oppression. And the faculty of the physical and mathematical sciences was politically liberal as it was dedicated to the harmonious enterprise of delving into the secrets of nature and of logical structures. There were five full professors close to each other, L. Berwald and K. Löwner in mathematics, you in theoretical physics and R. Fürth in experimental physics, and R. Carnap in philosophy of science. There were frequent joint seminars and other common undertakings. But the spirit of the unity of science and of scientific inquiry was taken so much for granted that I, for one, did not fully appreciate it until later, when I found compartmentalization the rule in many places elsewhere.

You, Professor Frank, were as much interested in the problems of philosophy, and of epistemology in particular, as in those of theoretical physics. I remember your own lectures, at our seminars and at the Urania, as well as those by distinguished visitors, such as Neurath. Concurrently, you were putting the finishing touches on the second edition of the "Frank-and-Mises," the standard encyclopedia of mathematical physics of the twenties and thirties. At the time I studied at Prague you were turning the introductory lectures in theoretical physics, a two-year cycle, over to Walter Glaser, who subsequently became the tool of the Protectorate (Nazi) government but who was then a promising young theorist. You concentrated on frontier subjects, and I recall your lectures on Dirac's relativistic theory of the electron, and on special and general relativity. Our attitudes toward science and research were formed by the attitude that pervaded all your teaching and writing activities: the quest for understanding, and the search for unifying principles.

Two years after I left Europe, Czechoslovakia was sacrificed on the altar of "peace in our time." Those who were fortunate enough to have contacts abroad left, you after almost thirty years of professional service at Prague. At Harvard University you were able to translate into contemporary American two aspects of your lecturing at Prague, your concern for the philosophical foundations of science, and your efforts to acquaint non-scientists with the beauty and the charm of scientific inquiry. Along with many others of your former students I followed your work in this country from the distance, being happy that your ever-active mind and your eagerness to teach did not go to waste because of the catastrophe engulfing Europe.

We all, students and friends, unite in saying thanks for the help we have received from you, our teacher, and in wishing you happiness in the years to come.

A FEW WORDS TO PHILIPP FRANK, FOR THE SECOND VOLUME OF THE *BOSTON STUDIES*

RUDOLF CARNAP

Looking backward over the many years of friendship and collaboration, I should like to recall to our memory two of the main aims which Frank has pursued throughout his long years of work as a teacher and writer. His first aim was to bring together science and philosophy, to bridge the gap that had separated them in the last century, because traditional philosophy, especially in the form of German idealism, used a language that was no longer acceptable for men accustomed to scientific thinking. The reform of philosophy which Frank welcomed and at which he himself actively cooperated, was initiated and developed by men in two different fields. On the one hand, scientists like Duhem, Poincaré and Einstein tried to come to a better understanding of the foundations of their own scientific activity. On the other hand, in the field of modern logic more exact conceptual tools were developed which helped scientists and philosophers to improve their methods of concept formation and theory formation.

To bring science and philosophy into closer contact is useful but not sufficient. Frank recognized more clearly, I think, than most other philosophers and scientists that it is of greatest importance that those who work in theoretical fields be aware of the role of their work in the wider context of life, of the whole of society and culture. Therefore Frank, both in his own thought and in his teaching activities, paid close attention to the historical development in order to show how currents of thinking are motivated not only by striving for knowledge but to a great extent also by practical or emotional needs and social situations. He showed that this holds for theoretical work just as much as for work in other fields like art or religion. And further he regarded it as important that those working in theoretical fields, even in fields which seem quite remote from practical life, like theoretical physics or analytical philosophy, should be aware of their responsibility for the development of humanity. He often pointed out that ways and tendencies of apparently purely intellectual thinking, either true insights or errors and misinterpretations, are far from being purely academic matters, but are rather significant and effective, either helping or hindering the development towards better forms of life in human society, forms more conducive to social justice and to harmonious and creative life than those existing today.

I, as one of the small number of surviving old friends, from the time of Vienna and Prague, wish to join the friends whom Frank has found during the long years of his life in this country in our common expression of appreciation and affection, and of cordial wishes for many more years of creative activity and satisfying life.

REMINISCENCES OF PHILIPP FRANK AT PRAGUE

R. FÜRTH

I have known Philipp Frank since 1912 when I started studying physics and mathematics at the former German University of Prague and he had just been appointed to the Chair of Theoretical Physics as successor of Einstein. At 28 years of age Frank was by far the youngest professor in the University, a fact unknown to his students who were much impressed by the mastery of his teaching and his full beard.

The first lecture course he gave was on Maxwell's theory of electromagnetism, and some of us, including Charles Loewner and the late Arthur Winternitz, had selected this course although we were only beginners. At first Frank did not realize that the course would be attended by first year students, who had no knowledge of partial derivatives, differential equations and vectors; but on being told so he took great trouble in helping us to overcome these difficulties.

His style of lecturing was, on the face of it, rather easy going and slow, and he normally devoted the first quarter hour of each lecture to a repetition of the contents of the preceding lecture. But in this way he made us understand the subject thoroughly, in spite of our inexperience, and to arouse our enthusiasm for it.

During the time of my undergraduate studies Frank never attempted to cover the field of theoretical physics systematically and apparently selected his lecture courses according to his inclinations. I remember in particular the courses on probability theory, statistical mechanics and kinetic theory, subjects that I found so fascinating that they have occupied my main interest ever since.

Frank also liked to teach specialised and topical subjects by way of seminars, where the more senior students were given the task of reporting in turn on chapters of books or papers. He did, of course, most of the talking himself, interrupting the students frequently in order to emphasize salient points and to make comments. I especially remember a seminar on relativity theory, Frank's main field of interest at that time. We found these seminars most enjoyable because they were conducted in an informal and friendly atmosphere and brought us in closer personal contact with Frank.

Once during my second year he asked me, knowing of my preference for experimental physics, whether I would like to do some experimental research work in his Department on some problem related to his lectures. Einstein's predecessor F. Lippich was a distinguished research worker in the field of wave optics and had accumulated a remarkably good stock of optical and other apparatus. He had also fitted out one of the large rooms of the Department as an optical laboratory; we called it the "black room" since walls and ceiling were painted completely black.

Frank himself had no inclination to undertake experimental work but was anxious to put the facilities of his Department to some good use. I was overjoyed by his offer and proposed to investigate the phenomenon of critical opalescence of binary liquid mixtures of which I had read in Perrin's famous book "The Atoms." Frank gave his assent, and naively unaware of the difficulties of this problem and my limited experience in experimental work, I started cheerfully to assemble the apparatus under Frank's admiring eyes.

The outbreak of the first world war in 1914 brought about great changes in the Departments of Experimental and Theoretical Physics which were housed in the same building. The former became completely depleted of scientific staff and, although still an undergraduate, I took on a great deal of teaching and laboratory work there. I also transferred my optical research to the experimental Department and so spent most of my time in it. Nevertheless Frank offered me, some time later, the post of "Assistant" in his Department which I held for twelve years until I was appointed to a Professorship.

Naturally during this long time I became more and more involved in the life of Frank's Department and that of the adjoining Mathematics Department, especially after I had once again transferred my experimental work to the "black room," which Frank had put at my disposal. I did not actually collaborate with him in his theoretical research work as he preferred to work on his own and never had a "research school." But he was always interested in my work and gave me every support and complete freedom to do what I liked within the narrow limits of the Department's facilities.

From time to time Frank used to give colloquia reports on the results of his investigations and on papers of interest to him, mainly from the fields of relativity and quantum theory. These talks were always lucid and presented in his own inimitable style. I learned to admire his great power of analysis of fundamental scientific problems which, in the course of time, led him to turn his interest more and more towards the philosophy of science.

Privately Frank's wide interests and acute observations made him a lively and witty companion. After his lecture in the morning he invariably went to the room of George Pick, the well known mathematician, who was Head of the Mathematics Department (and, by the way, together with Einstein and the experimental physicist Lampa, had made the proposal for Frank's appointment). He would sit there for a while, in later years also to be joined by Pick's colleague Ludwig Berwald, the successor of Kowalewski. Judging from the occasions when I was present myself, these talks were far from being confined to academic matters. Who would have thought at that time that Pick and Berwald, like so many of our friends, would perish in Nazi concentration camps!

With his students Frank was on easy and friendly terms and on many occasions joined them in departmental outings and parties. He was completely free of any racial or national prejudices; when after the end of the first world war the relations between the German and the Czech Universities in the new Czechoslovak state were in general rather tense he was always on the best terms with his Czech colleagues, and we used to meet them frequently on scientific and social occasions. Often we also met some of the Russian emigré scientists, who had come to Prague after the war and were made to feel at home by Hania and Philipp Frank. I remember in particular the physicists Eichenwald and Rashevsky.

For a short time in 1922 the "black room" became a centre of attraction for colleagues and friends when, together with my fiancée, who too was one of Frank's students, I had built a radio receiver, and there was great excitement in the evenings, when we were able to pick up fragments of the experimental broadcasts of some European stations. As time went by the "black room" became more and more crowded with my experimental research students and their equipment, and Frank was always interested in their work although he never took an active part in it.

Nevertheless I suspect that he was not too sorry when, upon my appointment to the Chair of Experimental Physics in 1931, this part of the activity of his Department came to an end. He was particularly pleased when, on my suggestion, we got permission to exchange the experimental equipment of the theoretical Department with the books and journals of the experimental Department. This arrangement put a very good library at his disposal which was accommodated in the vacated black room (after it had changed its colour!), and from then on his department became once again entirely devoted to theoretical physics.

I was Frank's colleague in the Faculty of Science for eleven years. In contrast to many of my other colleagues he did not like to be involved in academic affairs, and when I happened to sit next to him at Board meetings, he used to make amusing and sarcastic asides to me about some of the proceedings. On the other hand he never approved of the conduct of some other colleagues, who preferred to keep silent for the sake of peace, but spoke up courageously whenever anything seemed wrong to him, which sometimes did not make him particularly popular. He strongly resisted any attempts of Nazi sympathisers amongst staff and students of the University to apply racial doctrines to the admission of students or the appointment of staff, and he vigorously supported the appointment of the late E. Finlay-Freundlich to the Chair of Astronomy. He was also responsible for the creation of a Chair of Philosophy of Science within the Faculty of Science, against considerable opposition, and for the appointment of Rudolf Carnap to this Chair.

During the last years before the invasion of Czechoslovakia, Frank, who seemed to understand much better than most of us what was going on, had become very pessimistic about our future prospects. When in the autumn of 1938 he went with his wife to the U.S.A. to give lectures at various Universities I had the feeling that we would not see them again for a long time to come. Indeed, twelve years had to elapse before we had a happy re-union in London.

TO PHILIPP FRANK

GERALD J. HOLTON

Professor Philipp Frank's collection of essays, *Between Physics and Philosophy,* was published just before World War II. I was a student when it came into my hands that year. The effect upon me, as upon so many of Professor Frank's readers and students, was profound and lasting. Here was a link in the charismatic chain leading to such predecessors as Ernst Mach and Henri Poincaré.

Some years later, I had the chance to take a course in the philosophy of science which Professor Frank was giving at Harvard, and shortly after that to help out in the same course as an assistant. In this way was started a treasured association which brought me a little closer to the person whose scientific and philosophical writings I had already been studying, whose power and even whose style comes through on almost every page that he has written. Thus, the introduction, entitled "Historical Background–Discussions in a Vienna Coffee House," to the second edition of Professor Frank's collection of essays* begins with these lines:

> "At the time when the first chapter of this book was written (1907) I had just graduated from the University of Vienna as a doctor of philosophy in physics. But the domain of my most intensive interest was the philosophy of science. I used to associate with a group of students who assembled every Thursday night in one of the old Viennese coffee houses. We stayed until midnight and even later, discussing problems of science and philosophy. Our interest was spread widely over many fields, but we returned again and again to our central problem: How can we avoid the traditional ambiguity and obscurity of philosophy? How can we bring about the closest possible *rapprochement* between philosophy and science? By 'science' we did not mean 'natural science' only, but we included always social studies and the humanities."

* Philipp Frank, *Modern Science and Its Philosophy* (New York, 1955 reprinting of *Between Physics and Philosophy,* 2nd edition, Cambridge, Mass. 1949).

This paragraph illustrates what has been the lasting formula: deepest involvement with central intellectual issues; reconstruction of intellectual unities rather than merely iconoclastic revolt; and—equally important—pursuits of these discussions in a humane and deceptively informal setting rather than in a stuffy ex-cathedra manner.

This apparently paradoxical combination of serious power and effortless style has made Professor Frank the effective and beloved teacher of unnumbered readers and students all over the world. Each new essay, lecture, or conversation has left his audience with new recognitions. That this should be felt equally by Professor Frank's beginning students and long-term colleagues is the measure not only of a quality of mind but equally of spirit; for there seems to be in and around him a sort of calm equilibrium that allows problems and answers to be seen with new clarity.

And in this, as much as in his writings stretching over nearly sixty years, Professor Frank teaches us what is perhaps the best of the many meanings of the word "philosophical."

TO PHILIPP FRANK

EDWIN C. KEMBLE

We of the Boston-Cambridge academic community made the acquaintance of Philipp Frank and his wife in 1939, when they sought refuge at Harvard in the days of the Hitler terror. Frank was at that time 56 years of age, a distinguished physicist and mathematician and an international leader in the philosophy of science. In those days there was no proper niche for the philosophy of science in the Harvard departmental scheme and no appointment befitting his stature was available. However, with the warm support of Percy Bridgman and Harlow Shapley the Harvard Corporation appointed Professor Frank a research associate in physics and philosophy. Two years later he was granted an appointment on a half-time basis with tenure as Lecturer on Physics and Mathematics, a post which he held until his retirement in 1954. The Harvard courses which he taught in the philosophy of science and in elementary thermodynamics were extremely successful. The arrangement gave him the opportunity to accept alternate-term appointments as visiting lecturer at other universities.

Frank's position and economic circumstances, which forced him to devote so much of his time and energy to teaching and writing on a rather elementary level, have been a boon to the cause of clear thinking wherever he has gone. His extraordinary lucidity, his urbanity, and delightful sense of humor, have made him a popular speaker at scientific and philosophical gatherings. He has not been a victim of the common delusion that ideas expressed in technical jargon are in some way more important and profound than those expressed in plain English. Nor did he ever subscribe to the notion that the difficulty which a lecture course presents to its students is a measure of its value. His influence on the development of the philosophy of science in America and on the spread of the philosophical point of view among teachers of science has been far-reaching and always on the side of the angels. All who have been fortunate enough to come within the range of his influence are in debt to this remarkable man.

TO PHILIPP FRANK

HENRY MARGENAU

Although only a few hours intervene between receipt of the invitation to contribute to the Philipp Frank Volume and my departure for the Seneca Conference in Cordoba, I cannot allow this opportunity to pass without a brief acknowledgment of my personal debt to Philipp Frank as a philosopher of science, as a teacher and as a man. His writings on the philosophical implications of the theory of relativity, on the humanism of its creator, his later publications dealing with interpretations of the quantum theory are landmarks in our quest for deepened understanding of the spirit of modern science. He is unique in his generation as a philosopher of science who combines an expert knowledge of modern physics with professional competence in philosophy.

For a long time I admired these qualities from afar. When acquaintance with Professor Frank transformed him from an ideal into a living person he invaded my thinking to an extent that is difficult to measure and to acknowledge. The themes of his publications have reverberated in my own writings, and this is perhaps the best testimony I can give to his stimulating qualities as a teacher.

I am happy to join the contributors to this book in expressing my indebtedness to Philipp Frank, my high regard for him as a person, and my sincere wishes for more years of success and happiness. And to the editors I give a word of thanks for willingness to include this inadequate note of appreciation in this distinguished volume.

TO PHILIPP FRANK

HILDA VON MISES

I greet you, dear Philipp, also in the name of your old friend Richard von Mises. You and Mises had much in common, in your life and in your thinking. Both of you were Viennese, of the same age, having had the same classical education, and the same love of mathematics, physics and philosophy. Some of your friends may not know of your mathematical interests, of your collaboration with such men as G. Pick, K. Loewner, W. Blaschke. Your interest in the "slow changing middle-ground" that connects the system of physical hypotheses with experience, led you to work with Mises and the editing of the "Frank-Mises" (1925-27; expanded edition 1931-34; reprinted 1961). Rather than to discuss the character and influence of this widely-known work, I prefer to recall at this point your own great chapter on classical mechanics and ray optics, from which I, as so many others, learned analytical mechanics, and where, by "taking seriously" Hamilton's analogy between optics and mechanics, you prepared the understanding of the Schroedinger wave equation.

However, if the name of Philipp Frank is mentioned the primary associations are "philosophy of science," "Vienna Circle," "logical empiricism," "positivism." In his last address (at a meeting of the Institute of the Unity of Science, May 1953) von Mises said: "When I am asked by a student or by a colleague, what positivism is, I answer: 'He is a positivist who when confronted with any problem in life reacts in a manner in which a typical contemporary scientist deals with his problem of research.' " There is little doubt that Frank would accept this statement, he who points out again and again that there is but one method of reasoning which applies to the problems of the sciences (in the wide sense of "Wissenschaft") as well as to philosophy and to daily life. Frank's aim is to conceive science and philosophy as *one* coherent system of thought; not as a finished existing system which we merely have to discover but rather as one 'in statu nascendi,' subject to continuous modification. Of no scientific proposition, of no basic statement or axiom, do we know what change it will experience in the future. Certainty can be found only in tautological statements—which do not tell us anything about the world of experience.

Frank, a great personal friend of Einstein, had a deep understanding of relativity theory. He explains its physical meaning, its philosophical implications, their influence on each other and their related gradual change. The restricted theory of relativity nearly fulfilled Mach's "positivistic requirement" according to which all principles of physics should be formulated in terms of observable quantities. In the general theory, however, the fundamental principles fulfill these requirements only in an indirect way: Einstein merely required that from these principles conclusions could be drawn that are connected with observable experiences by semantic rules, the general principles themselves being "the product of mathematical and logical imagination."

Other physicists experienced a similar development. In 1928, P. W. Bridgman, a close friend of Frank during Frank's life in Cambridge, wrote: "We should now make it our business to understand so thoroughly the character of our permanent relation to nature that another change in attitude such as that due to Einstein shall be forever impossible." But 30 years later, Bridgman wrote—and this is a strong and beautiful expression of ideas he had in common with Frank—"To me it seems now incomprehensible that I should ever have thought it within my power, or within the power of the human race for this matter, to analyze so thoroughly the functioning of our thinking apparatus that I should confidently expect to exhaust the subject and to eliminate the possibility of a bright new idea against which I would be defenseless."

In the later years of his life Frank became more and more interested in the sociological aspects of science and philosophy—in their historical, cultural and psychological background. These studies led him to the conviction that scientific work, no matter how seemingly remote from practical life, had great influence on human behaviour. This does not relate to the much-quoted "role of science in an atomic age." Frank's idea is that the "relativism" of science slowly penetrates human conduct and helps "to get rid of empty slogans and to formulate the goals of human life sincerely and unambiguously." Clear thinking and unceasing logical criticism of language are in an inherent contrast to intolerance, fanaticism and injustice.

It is difficult to review in so many words what is new and characteristic in Frank's thoughts. His greatness lies in his clarity and in his thorough understanding of science and of philosophy. His life-work and its influence seem to be exceedingly well characterized by words which Mises wrote with respect to Ernst Mach: "Die Geschichte des intellektuellen Lebens kennt an vielen Stellen das Auftauchen von Gedankengaengen, die sich allmaehlich, fast unmerkbar durchsetzen . . . und die Denkweise der Menschen langsam umformen. Solche Ideen sind oft lange vorbereitet, finden dann einmal den richtigen Ausdruck durch den Mund eines Berufenen und werden schliesslich anonymes Gemeingut."*

I myself have known Philipp Frank for almost 50 years and I would like to finish with a happy recollection. Mises and I often visited the Franks, Philipp and the gentle and beautiful Hania. Mises knew the great men of his time: the poets, the great physicists, mathematicians and philosophers. When after such a visit with the Franks, after talks on many things and many people, we left, Mises would say to me: "Der Gescheiteste von allen ist doch der Philipp Frank."

*In the history of intellectual life we often encounter the emergence of ideas which prevail gradually, almost unnoticed, and slowly transform man's way of thinking. Such ideas are often long in preparation; they are, at last, adequately expressed by the right person and become eventually part of the anonymous acquisition.

PHILIPP FRANK–AN APPRECIATION

ERNEST NAGEL

The philosophy of science is currently cultivated and taught in America mainly by professional philosophers, for relatively few American scientists when compared with European ones have been seriously interested in the subject. However, the major contributions since classical antiquity to the analysis of scientific ideas and methods have for the most part been the work of men who have also contributed to the development of some branch of substantive science; and it is to the company of philosophically minded scientists, which includes such distinguished physicists like Helmholtz, Mach, Poincaré, and Einstein, that Philipp Frank belongs. His publications on questions in the philosophy of science, unlike those of most professional philosophers, are marked not only by a firm mastery of scientific ideas that can be achieved only by long experience in using them. They also display, again unlike much writing on these questions by professional philosophers, a rare ability to illuminate classical as well as contemporary physics, by exhibiting the uses of theoretical constructions in the various contexts in which they have a genuine scientific significance. Frank's philosophical contributions deal with issues that are intimately related to the primary materials and the social import of science; and although the questions he examines often require him to engage in subtle analyses, his analyses are never scholastic exercises whose only ascertainable aim is to multiply irrelevant formal distinctions. The task of the philosophy of science, as practiced by Frank, is to clarify the nature of scientific ideas and to dispel the misconceptions that so frequently envelop them. Accordingly, far from being a subject that can be of genuine interest only to a limited group of specialists, the philosophy of science at the hands of Frank is a profoundly liberal discipline from which all serious students of human achievements can profit.

It has been my privilege to know Philipp Frank personally as well as through his publications; and I can testify that the qualities of mind embodied in his writings appeared in heightened form in all my numerous conversations with him. His unusually quick intelligence and good sense, his remarkable gift for penetrating to the heart of a scientific idea and revealing its structure, and his steadfastly critical but humane intellectual temper are certainly manifested in his publications, but I think rarely with the dramatic force and good-natured wit that were so pervasive in his talk. In any event, since his very first published contribution to the philosophy of science, Frank has been concerned with fundamental but difficult problems, to many of which there are perhaps no definitive solutions. In this respect he is a thinker of the calibre described in an anecdote he once recalled from a conversation he had with Einstein. The latter distinguished two kinds of physicists, by comparing them with workers whose assigned task is to drill holes through planks of wood. According to Einstein, many physicists are like workers who busy themselves with putting innumerable holes through the softest and thinnest planks they can find; but there also physicists who resemble workers attempting to drill through the knottiest and thickest boards that are available, and for whom Einstein expressed boundless admiration despite the fact that their efforts may be only rarely or partly successful. The sense for what are the truly important questions and the uncompromising devotion to their pursuit which Einstein so greatly admired in physicists of the second type, are embodied to an eminent degree in the work of Philipp Frank in the philosophy of science.

PHILIPP G. FRANK, PHYSICIST EXTRAORDINARY

RAYMOND J. SEEGER, *National Science Foundation*

Philipp Frank, a physicist himself, is the physicist's philosopher of physics. His whole life has been devoted to seeking an understanding of physical phenomena from an experiential viewpoint, coupled with the panoramic outlook of a theorist. He has disclosed in high relief fundamental concepts embedded in the matrix of sense impressions; he has traced both their historical development and their indefinite boundaries, outlined both their cultural dependence and their philosophical significance. He has emphasized that a scientific idea cannot be resolved into particular events in a space–time continuum, but rather represents a diffuse area covering associated disciplines over a period of time. In order to appreciate the whole evolutionary pattern, therefore, one must have at least a "critical spread" for any field.

Although not a professional historian, Philipp Frank has contributed much to our understanding of the history of science by continuously reviewing science in the perspective from the past. The instantaneous photograph of current science is merely a phase of a sequential development, which affords a sense of direction. Too often modern historians are fascinated by their narcissistic view of the past and fettered by the recorded data of their own selection.

Although not a professional philosopher, Philipp Frank has made significant contributions to the "analysis and synthesis of knowledge," i.e., to the philosophy of science, in his insistence upon the phrasing of questions relative to an experiential framework of reference and upon the abandonment of meaningless questions. What is more he has indicated suggestive clues for discovering meaningful answers. Too often modern philosophers have become lost in the maze of their linguistic analysis and restricted by the narrow abstractness of their own methodology.

Although not a professional sociologist, Philipp Frank has been a pioneer in exploring science as a social phenomenon involving the growth and decay of living ideas. He has focused attention upon the role of scientists, both individually and socially, in selecting empirical data according to operational criteria, in devising creative concepts to unify the "scientific facts," in seeking generalized relations among them, in providing a simplified, comprehensive, imaginative theory for describing and predicting events, in establishing axiomatic foundations for logically consistent deductions. Too often modern sociologists, and even natural scientists, neglect science as a field of human behavior in either theoretical make-up or its technological consequences.

Although not a professional educationalist, Philipp Frank has pointed clearly to a communication link between the two apparently diverging academic cultures, viz., the great chain of thinking. He has cultivated the humanism of science in its historical roots, its philosophical branches, and sociological fruits. Too often sophisticated scholars are nowadays content to dwell in isolated ivory towers with little desire to communicate with the common people below. Professor Frank has shown himself a teacher par excellence in his ability and interest in explaining the meaning and significance of important scientific ideas to nonscientists. He has demonstrated how to bridge successfully the ever widening gap between the humanists and the scientists.

Philipp Frank is a living symbol of the very unity of science which he has long urged as a goal toward which all scientists should strive. I have always found him an encouraging leader in cooperative enterprises in which I have had a share. In the early 1940's, when I undertook to transform Section L of AAAS from a group interested strictly in the Historical and Philological Sciences to one concerned more broadly with the History and Philosophy of Science, he was largely responsible for securing the interest and cooperation of the professional philosophers. The Section soon became the primary meeting ground for historians, philosophers, and scientists in their common endeavor to understand natural phenomena in a social world. He conceived and organized a successful "Conference on The Validation of Scientific Theories," sponsored by the Institute for the Unity of Science, the Philosophy of Science Association, Section L of AAAS, the American Academy of Arts and Sciences, and the National Science Foundation. Ten years later, when I initiated the National Science Foundation Program on the History and Philosophy of Science, I relied heavily upon his stimulating suggestions and wise counsel.

Shortly thereafter he formulated the "Summarizing Remarks" of the "Conference on the History, Philosophy and Sociology of Science," sponsored jointly by the National Science Foundation and the American Philosophical Society—the guideline of the NSF program to this very day. One of the first grants by the NSF for research in this field was for his own project on "The Reasons for the Acceptance of Scientific Theories." His eager participation in the summer Institute on the History and Philosophy of Science for collegiate science teachers has been a major attraction ever since its inception about seven years ago at American University. He has given of himself unstintingly and graciously in every effort to promote a better understanding of science in its philosophical and social implications. Together with many others, I gratefully and joyfully regard him as Almus Pater in this area of learning.

"Gladly wolde he lerne and gladly teche."

TO PHILIPP FRANK

KURT SITTE

The pleasure of contributing a few lines to this volume dedicated to Philipp Frank is mingled only by a sad feeling of incompetence: How can I, in this short space, give proper expression to my gratitude for all I owe him, and to my admiration for the scientist and the man?

When I came to Prague in 1928, an eager but raw student, there were on the walls of the old Institute of Theoretical Physics forbidding-looking lists of publications by its members, and mostly by its Head, Professor Philipp Frank. There were a few photographs, too,— old ones, as I soon discovered—, for they showed him still bearded, every inch a 'Herr Professor,' one of those Lords of Creation we lesser beings could scarcely dare to approach. It did not take long to find out that this impression was quite wrong, that even in those days and in the stuffy atmosphere of a pre-war German university, rank and merit counted nothing for Philipp Frank but interest and enthusiasm for science did.

Yet we thought him lazy at first, until a little of his wisdom, so generously and in good humor shared out to us, had rubbed off to give us the rudiments of maturity. He used to come very late to his lectures, and then to present the cream only while leaving all the dirty details for us to work out. Again in his seminars on the then modern quantum theory he seemed content to learn new material through our poor efforts—though when we got stuck, it was always a sharp remark of his which pointed the way to us. By that time Philipp Frank had become deeply interested in the philosophical problems of science, and this, of course, accounted for what we philistines had deemed his 'laziness.' But we learned from him in this field, too, almost without knowing it. I still recall my amazement when by his words suddenly my eyes were opened to perceive interrelations which I had never even vaguely suspected—and without his help, might not have realized for a long time.

A test for what I had learned came a few years later when a colleague was given the hard task of typing out Philipp Frank's manuscript of his 'Kausalität und ihre Grenzen.' Half the time she was in despair, for she could not read his handwriting; and she came to me for succour. I fared no better in reading, but with a few cues deciphered I managed to dictate 'what Frank would have said' on the subject matter. And I am still proud that when he got our typewritten sheets, there were no major mistakes for him to correct. (Though there were a few occasions when we both failed, and had to go to Philipp Frank for elucidation: in these cases we found that he, too, could not read his manuscript but quickly formulated a new version.)

Yes, we were very fortunate, we students of Philipp Frank in those years. Perhaps more so than those of any other generation of his students, for at that time the brilliant theoretician, and the author of that Bible of ours, the 'Frank-Mises,' had not yet yielded so largely to the philosopher. We had them both in our Philipp Frank.

With all my heart I send my best wishes to him for many more years in good health, and my apologies for this inadequate expression of my sincere appreciation and my gratitude.

A BIBLIOGRAPHY OF PHILIPP FRANK

Selected Writings on the Philosophy of Science

1907 *Kausalgesetz und Erfahrung.* Ostwald's *Annalen der Naturphilosophie 6,* 445-450 (Eng. tr. in 1941b and 1949a).

1908 *Mechanismus oder Vitalismus? Versuch einer präzisen Formulierung der Fragestellung.* Ostwald's *Annalen der Naturphilosophie 7,* 393-409.

1910 *Gibt es eine absolute Bewegung?* in *Wiss. Beilage z. 23. Jahresbericht d. Philos. Gesell. a. d. Univ. Wien.*

1911 *Das Relativitätsprinzip und die Darstellung der physikalischen Erscheinungen im vierdimensionalen Raum.* Ostwald's *Annalen der Naturphilosophie 10,* 129-161.

1917 *Die Bedeutung der physikalischen Erkenntnistheorie Machs für das Geistesleben der Gegenwart. Naturwissenschaften 5,* 65-72 (Eng. tr. in 1941b and 1949a).

1918 *Josef Popper-Lynkeus zu seinem achtzigsten Geburtstag. Physikal. Zeit. 19,* 57-59.

1919 *Die statistische Betrachtungsweise in der Physik. Naturwissenschaften 7,* 701-740.

1920 *Wissenschaft und Theologie. Freier Gedanke,* Nr. 40.

1925 *Die Differentialgleichungen der Mechanik und Physik,* with Richard von Mises (Braunschweig; last revised edition, 1935; Russian tr., Moscow, 1937).

1926 *Theologische Elemente in den exacten Wissenschaften. Prager Tagblatt* for December 8th.

1928a *Über die Anschaulichkeit physikalischer Theorien. Naturwissenschaften 16,* 122-128.

1928b *Relativitätsmechanik* in *Handbuch für physikalische und technische Mechanik,* vol. 2, pp. 52ff. (Leipzig).

1928c *Gibt es ein irrationales Moment in den Theorien der moderne Physik? N. Zurich Zeitung* for December 17th.

1929 *Was bedeuten die gegenwärtigen physikalischen Theorien für die allgemeine Erkenntnislehre? Naturwissenschaften 17* (Eng. tr. in 1941b and 1949a; revised French tr. in 1934b).

1930 Exact reprint of (1929) in *Erkenntnis 1,* 126ff.

1931 *Der Charakter der heutigen physikalischen Theorien. Scientia.* (Expanded text of paper delivered at the 1928 Congress of Soviet physics at Saratov).

1932 *Das Kausalgesetz und seine Grenzen* (Vienna; French tr. in 1937c).

1934a *La Physique contemporaine manifeste-t-elle une tendence a réintégrer un élément psychique? Revue de Synthèse 8,* 133 (Eng. tr. in 1941b and 1949a).

1934b *Théorie de la connaisance et physique moderne.* (Paris; expanded version of (1929)).

1935a *Positivistische oder metaphysische Auffassung der Physik? Scientia.*

1935b *Das Ende der mechanistischen Physik. Einheitswissenschaft 5* (Vienna) (French tr. in 1936a; revised Czech tr. in 1938a).

1936a *La Fin de la physique mécanique* (Paris; revision of 1935b).

1936b *Logisierender Empirismus in der Philosophie der U.S.S.R.* in *Actes du congrès international de philosophie scientifique Paris 1935.* vol. 8 pp. 68-76 (Paris; Eng. tr. in 1941b and 1949a).

1936c *Philosophische Deutungen und Missdeutungen der Quantentheorie. Erkenntnis 6* (Eng. tr. in 1941b and 1949a; also revised as Part I of 1938b).

1937a *The mechanical versus the mathematical conception of nature. Philosophy of Science 4,* 41-74 (Reprinted in 1949a).

1937b *Was versteht der Physiker unter der "Grösse" eines Körpers? Theoria 3.*

1937c *Le Principe de causalité et ses limites* (Paris; tr. of (1932)).

1938a *Rozvrat mechanisticke fysiky* (Prague; revised tr. of 1935b).

1938b *Interpretations and Misinterpretations of Modern Physics* (Paris).

1938c *Ernst Mach—the Centenary of his Birth. Erkenntnis 7* (Reprinted in 1941b and 1949a).

1938d *Physik und Logischer Empirismus. Erkenntnis 7.*

1938e *Bemerkungen zu E. Cassirer: Determinismus und Indeterminismus in der modernen Physik. Theoria 4* (Reprinted in 1941b and 1949a).

1939 *Modern physics and common sense. Scripta Mathematica 6* (reprinted in 1949a).

1941a *Why do scientists and philosophers so often disagree about the merits of a new theory? Rev. Mod. Phys. 13,* 171-175.

1941b *Between Physics and Philosophy* (Cambridge, Mass.)

1941c *Is the universe infinite? Popular Astronomy 49,* No. 8.

1942 *The influence of an "uneven" anisotropy on the path of light rays. Phys. Rev. 62,* 241-243.

1944 *The philosophical meaning of the Copernican revolution. Proc. Amer. Philos. Soc. 87,* 381-386 (Reprinted in 1949a).

1946a *Science teaching and the humanities. ETC: A Review of General Semantics 4,* 3-24 (Reprinted in 1949a).

1946b *Foundations of Physics.* Vol. I No. 7 of the *International Encyclopedia of Unified Science* (Chicago).

1946c *The relativity of truth and the advance of knowledge. Analisi.* (Included, in expanded form, in 1950c).

1947a *The place of the philosophy of science in the curriculum of the physics student. Amer. J. Phys. 15,* 202-218 (Reprinted in 1949a).

1947b *Einstein: His Life and Times* (New York; revised, 1953; first German ed., Munich 1949).

1947c *The Institute for the Unity of Science, Its Background and Purpose. Synthèse 6.*

1948a *The place of logic and metaphysics in the advancement of modern science. Philosophy of Science 15,* 275-286 (Reprinted in 1949a).

1948b *Logical Empiricism I: The Problem of Physical Reality. Synthèse 7,* 457-466.

1949a *Modern Science and Its Philosophy.* (Cambridge, Mass.; enlarged revision of 1941b).

1949b *Einstein, Mach, and logical positivism* in *Albert Einstein: Philosopher-Scientist,* ed. P. A. Schilpp, pp. 271-286 (Chicago).

1949c *Einstein's philosophy of science. Rev. Mod. Phys. 21,* 349-355.

1950a *Comments on realistic versus phenomenalistic interpretations. Philosophy of Science 17,* 166-169.

1950b *Metaphysical interpretations of science. Brit. J. Phil. Sci. 1,* 60-91.

1950c *Introduction to the philosophy of physical science on the basis of logical empiricism. Synthèse 8,* 28-45.

1950d *Relativity: A Richer Truth* (Boston).

1951 *The logical and sociological aspects of science. Proc. Am. Acad. Arts Sci. 80,* 16-30.

1952 *The origin of the separation between science and philosophy. Proc. Am. Acad. Arts Sci. 80,* 115-139.

1953 *The role of authority in the interpretation of science* in *Freedom and Authority in our Time,* ed. L. Bryson *et al.* pp. 361-363 (New York).

1954 *Non-scientific symbols in science* in *Symbols and Values: an Initial Study,* ed. L. Bryson *et al.* (New York).

1956 *The variety of reasons for the acceptance of scientific theories* in *The Validation of Scientific Theories,* ed. Philipp Frank, pp. 3-17 (Boston; first published in *Science, 79* 1954).

1957 *Philosophy of Science: The Link Between Science and Philosophy* (Englewood Cliffs, New Jersey).

1958 *Contemporary science and the contemporary world view. Daedalus 87,* 57-66.

1963 *The pragmatic component in Carnap's "Elimination of metaphysics"* in *The Philosophy of Rudolf Carnap,* ed. P. A. Schilpp, pp. 159-164 (Chicago).

PROCEEDINGS OF THE BOSTON COLLOQUIUM FOR THE PHILOSOPHY OF SCIENCE, 1962-1964

1

G. SCHLESINGER

INSTANTIATION AND CONFIRMATION*

Presented February 7, 1963

I

The paradoxes which plague the prevailing theory of confirmation are the symptoms of its basic infirmities. Hempel's is a widely known one, but there are other paradoxes as well. First we shall consider Hempel's paradox which is briefly this: it is commonly agreed that the observation of a black raven serves to confirm the generalization 'All ravens are black,' but that it would be absurd to claim that the observation of an object that was neither a raven nor black, e.g., a white shoe confirmed the same, to any degree. But 'All ravens are black' ≡ 'All non-black things are non-ravens' and a white shoe instantiates a non-black non-raven and its observation confirms 'All non-black things are non-ravens' and therefore to the same extent its logical equivalent 'All ravens are black.' Thus logical analysis forces us to maintain something commonsense rejects as absurd.

II

Let us consider some of the better known solutions to the paradox. G. H. von Wright, for example, tries in *The Logical Problem of Induction*[1] to meet Hempel's paradox by pointing out that not under all circumstances is it against commonsense to regard the observation of a white shoe as a confirmation of the general law that all ravens are black. It all depends on how we

* I wish to thank David Stove for his useful criticism.

come to know that the object in front of us is a white shoe. If the first thing that came to our notice was its color, the possibility of its being a white raven was not ruled out for us right from the beginning. Upon subsequent verification, finding that it was not a raven, we may truly say that we have a genuine confirming instance of the law, since through the observed object's turning out to be a shoe, the law has been saved from falsification after it has stood the risk. It is only when we first found that it was a shoe but that its color was white we discovered afterwards, that it would be paradoxical to regard it as a confirming instance of 'All ravens are black,' since the object in question has never presented a possible threat to this law. But then, von Wright points out there is complete symmetry here, because the observation of a black raven too confirms the law only if we first knew of its ravenness before we knew of its blackness but not when its blackness was the first thing we discovered about it, since in that case it has never constituted a possible source of falsification.

Some may find it objectionable that confirmation should be made dependent on subjective factors like the order in which one happened to observe that a given object is both black and a raven. It should however be pointed out that the fact on which the generalization 'All ravens are black' is based is not the fact that black ravens and black ravens only *exist* but the fact that black ravens and black ravens only have been *observed.* When the observation of blackness has preceded that of ravenness then we have an objectively different observation from the one in which the order is reversed. Thus von Wright does not lay himself open to the accusation that he has introduced psychological considerations into the logic of confirmation.

Others[2] attempt to tackle Hempel's paradox by arguing that the ratio of the number of ravens to the number of all other things in the universe is very small. On the other hand, black is just one of many possible colors. Consequently the prior probability of a raven not being black is much larger than the probability of a non-black thing being a raven. Hence even though a non-black thing constitutes a potential threat to the statement 'All ravens are black' it is a very small threat compared to the threat a raven represents in case it may not be black. It follows therefore that the observation of a non-black non-raven lends the statement confirmation to a much lesser degree than the observation of a black raven. It is reasonable therefore to treat the former as virtually irrelevant and look only for black ravens in our efforts to confirm the statement.

But what if we stood in the zoo in front of a sign 'Section 3: Woodcocks, Ravens, Magpies' where if we pointed at a bird the prior probability of its being a raven is quite considerable? The answer is, yes, in such a situation the observation of a brown woodcock would constitute a confirmation that all ravens are black since the prior probability of a brown bird turning out a raven is not negligible. But then in such a situation it does not strike us as paradoxical that a brown woodcock should be instrumental to the confirmation of our generalization. Indeed Alexander himself points out that in the case of a generalization like 'All deciduous trees are less than 200 feet in height' it is not merely not paradoxical to regard trees which are more than 200 feet in height and non-deciduous as relevant instances but as actually preferable in this respect to trees that are less than 200 feet in height and deciduous. The reason is that in the light of our background knowledge that trees taller than 200 feet are scarce the prior probability of a tree's being taller than 200 feet is very small while that of its being deciduous is considerable.

Finally we should mention Hempel's own radical solution suggested in the same paper in which he posed the paradox. Hempel insists that we have to abide by logical analysis no matter how odd its results may seem to untutored commonsense. He declares " . . . the impression of paradoxicality arises from a misguided intuition in the matter."[3] We must therefore try and get used to the idea that there is nothing really strange about the observation of a white shoe counting as a confirming instance of the proposition 'All ravens are black.'

III

Let me now propose another paradox not mentioned in the literature. This will help to impress upon us the need for fundamental rethinking to be done on the whole subject of confirmation since none of the solutions considered in the last section can cope with this new paradox. The new paradox may also be regarded as more destructive than Hempel's because his paradox consists only in turning, by logical analysis, an instance which commonsense treats as neutral, into a confirming instance. The new paradox however shows that an obviously confirming instance may not only be turned by logical analysis into a neutral but even into a disconfirming instance. And this is how it runs:

What would confirm the more general proposition 'All birds are black'? Again the observation of a black raven which is an instance of a black bird. At the same time, a black raven is a non-white non-swan and by Hempel's analysis its observation confirms 'All non-white things are non-swans' and its logical equivalent 'All swans are white' from which it logically follows that not all birds are black.[4] Hence the selfsame observation supports 'All birds are black' and its contradictory 'Not all birds are black.'

Now of course it is not objectionable that an object (e.g., a black raven) should instantiate a generalization which is empirically false (e.g., 'All birds are black') because it is known to be false through independent observation (e.g., the observation that there are grey sparrows). But what does seem absurd is that an object should be instrumental to the confirmation of a proposition when it is at one and the same time instrumental to its disconfirmation as well.

Let us see how this affects von Wright. Suppose we have observed a thing to be a bird but have not yet established its species or color. For argument's sake suppose we are walking in the field and observe the shadow of a bird and the ground, thus we can be reasonably sure that a bird is flying above us. Are we faced with a potential falsifier of the proposition 'All birds are black'? Yes, because the bird may turn out to be non-black. Are we also faced with a potential falsifier of 'All swans are white'? Yes, because the bird may turn out to be a non-white swan. Consequently when we look up and see that the bird is a black raven we obtain, according to von Wright, a confirming instantiation of both of these propositions. In other words, von Wright would have to agree that the same object on one and the same occasion may be instrumental to the confirmation of 'All birds are black' and of its contradictory.

It may seem at first as if with the kind of solution suggested by Alexander one could meet the new paradox as well, for it is still true that swans represent a very small fraction of birds and hence the apriori probability of a non-white bird being a swan is much smaller than that of a swan being non-white and consequently it is reasonable not to regard a non-white non-swan bird as a confirmation of the statement 'All swans are white.' However, it is possible to state the new paradox in a different form which will turn the observation of a black raven not merely into a support for 'Not all birds are black' but for 'Not all ravens are black' and then we shall immediately witness the ineffectiveness of the suggested solution.

A raven is either male or female. Suppose the observed black raven is a male raven, i.e., a non-white non-female raven; then it instantiates 'All non-white things are not female ravens' which is logically equivalent to 'All female ravens are white' which entails that not all ravens are black.[5] But from what we know about the distribution of sexes among other animals it is unreasonable to assume that females constitute a very small fraction of the raven population and the plausible assumption to make is that they constitute roughly half of it. Consequently the prior probability of a black raven turning out female is by no means negligible. Therefore there is no reason why the observation of a black male raven should not be regarded as a good confirmation that all female ravens are white comparable to the one obtained through the observation of a white female raven. Of course Alexander would say here as in the case of the deciduous trees generalization that because of the high ratio of females amongst ravens it is not at all absurd that the observation of a black male raven should confirm that all female ravens are white. However this is precisely why the new paradox arises. For after all it is also undeniably true that the observation of a black male raven confirms that all ravens are black and now we must admit that it confirms its contradictory as well.

Hempel's own solution to his paradox based on the idea that if the results of intuitive commonsense clash with those of logical analysis, so much worse for commonsense which will have to yield to logic, cannot apply here either, as the clash in the new paradox is between two results of logical analysis coupled with objective facts.

IV

Now we shall consider a very simple suggestion for solving all the paradoxes. In the next section we shall see how, in spite of its initial plausibility, it fails as indeed, I believe, will any solution advanced before our fundamental notions on the process of confirmation have undergone complete revision. These two sections will, it is hoped impress further upon us the urgency of such revision.

To begin, we must distinguish between mere instantiation and unique instantiation. As soon as we have become aware of this distinction it is most natural and reasonable to suggest that unique instantiation alone, but not mere instantiation, provides confirmation. The distinction is simply this: a generalization is uniquely instantiated by an object, if and only if, the same object cannot be construed as an instantiation of any of the contraries of the same generalization. But when an object is equally an instance of contrary generalizations it does not instantiate uniquely either of them.

There is an obvious difference between the way a black raven is related to the generalization 'All ravens are black' and the way a white shoe is related to it. True enough both instantiate it. However a white shoe does not instantiate it uniquely. For a white shoe is just as much a non-green non-raven as it is a non-black non-raven and its observation would tend therefore just as much to confirm 'All ravens are green' as it tends to confirm 'All ravens are black' and on the supposition that ravens exist, these two are contraries. But an object is not a unique instance of a universal generalization when there is no difference between the way it is related to it and to a generalization which is or can be construed as its contrary. For precisely the same reason, no support for 'All ravens are black' is provided by the observation of a blue book, green emerald, brown shirt or red herring. A black raven and a black raven only stands in an unique relationship to 'All ravens are black' since it cannot be construed as an instance of any of its contraries.

Now of course we can also easily see how this provides us with the suggestion that the new paradox may be dissolved on the same principle. The observation of a black raven is not to be regarded as a support for 'All swans are white' since a black raven is not an unique instance of this generalization. A black raven is just as much a non-green non-swan as it is a non-white non-swan and hence its observation could with equal cogency be construed as a support for 'All swans are green.' Similarly a black male raven does not constitute an unique instance of 'All female ravens are white' as it is just as much a non-green non-female raven as it is a non-white non-female raven.

To summarize, consider the following statements:

1) All books are blue
2) All female ravens are white
3) All animals are black
4) All birds are black
5) All ravens are black
6) All young slim male ravens are black
7) All black things are young slim male ravens
8) All non-black things are non-ravens

and the way the observation of a black young slim male raven is related to each one of them;

Since a black young slim male raven is an unique instance of each of the statements 3-7 the observation of it supports each one of them and there is nothing odd about it. However on observing a black cat for example we reject 7 as false. On observing a grey horse we reject 3 and settle for the more restricted generalization 4. On observing a grey robin we have to reject 4 and can maintain only the even more restricted 5. As long as we do not observe any female ravens or old ravens or fat ravens which are of a color other than black we need not abandon 5 for the more restricted 6. To statements 1 and 2, however, our observation is neutral, since it is symmetrically related to them and to some of their contraries. The object of our observation is not an unique instance of 1 or 2.

What about 8? It is important to point out that even though

this statement explicitly mentions non-black things and non-ravens it is not uniquely instantiated by an object which is both non-black and non-raven, e.g., a white shoe. For on the supposition that there are ravens, 'All non-black things are non-ravens' is a contrary of 'All non-green things are non-ravens' and a white shoe is symmetrically related to them both. Thus the present solution is not based on an attempt to drive a wedge between 'All ravens are black' and 'All non-black things are non-ravens.' The two are logically equivalent. A black raven uniquely instantiates them both. A white shoe does not uniquely instantiate either of them.

V

The distinction between mere instantiation and unique instantiation seems sensible and the suggestion that when an object instantiates a universal statement and its contrary, it should not be allowed to present itself as a candidate lending support to either one of such statements sounds a most reasonable one. However this does not lead to the dissolution of our paradoxes. There are at least two types of cases in which the paradoxes remain:

(i) Consider generalizations like 'All Britishers speak English.' Is it confirmed by the observation of a Frenchman who speaks no English? Commonsense says no, but how do we account for this? After all, such a person provides an instance to the generalization 'All non-English speakers are non-Britishers' which is the logical equivalent of 'All Britishers speak English.' True enough the same Frenchman (in case he speaks no German either) may be an object of instantiation to the generalization 'All Britishers speak German' but this leads us nowhere in the present case for the two generalizations are not contraries since the ability to speak English does not exclude the ability to speak German. Thus we have no answer to the question why the observation of a Britisher who speaks both English and German may support simultaneously 'All Britishers speak English' and 'All Britishers speak German' while the observation of a Frenchman unable to speak English or German may not lend support to either.[6]

This example indicates the nature of a whole group of cases where the proposed solution fails: cases in which the generalization is made about non-exclusive properties. Blackness is an exclusive property; once it is known to apply to a given object no other property of its kind may apply (i.e., no other color property) for nothing can be black and green all over. The ability to speak English on the other hand is not an exclusive property of its kind (it allows the ability to speak other languages). Further examples of exclusive properties are: being a raven, having temperature T, having mass M, being at position P, occurring at time *t;* examples of non-exclusive properties are: being a British subject, being a carpenter, being a music lover.

(ii) But even when the generalizations are made about exclusive properties the solution only functions as long as we work with cases in which they are not exhaustive, but when we consider cases in which the properties are exclusive and exhaustive the solution breaks down. Example: 'All bats are blind.' Why is it not confirmed by the observation of a sighted raven? The latter can after all not be construed as an instance of a contrary generalization since blindness and sightedness are exhaustive. Or a white shoe while admittedly does not uniquely instantiate 'All ravens are black' it does uniquely instantiate 'All ravens are not white' and its observation should at least lend support to the latter.

On the other hand the solution breaks down because of different considerations as well according to which even a black raven does not uniquely instantiate 'All ravens are black.'[7] For if the raven is female i.e. non-male then it instantiates 'All non-(male ravens) —e.g. shoe strings and sealing wax—are black' and given that there are non-black things these two statements are contraries.

VI

No remedy can be brought to the situation until a radical reorientation of our views on confirmation has taken place. The simple-minded theory which has tacitly been assumed throughout the whole of the previous discussion, a theory in which we dealt with nothing but generalizations and their instantiations, has to be replaced by a more elaborate one. As a first step toward such reorientation we should realize how distorting or oversimplification it was to maintain that the degree to which a generalization of the kind 'Everything that has A has B' has been established depends solely on how often we have found A and B conjoined in the past and that the observation of a very large number of things having both A and B necessarily confirms strongly the generalization, just as it is not true that when a few instances only of such conjunctions have been observed, then the degree to which the generalization has been confirmed is inevitably small.

The truth is that the degree to which the statement 'A and B always go together' is confirmed, is directly determined by the degree of confidence earned by the rational grounds upon which the constant conjunction of A and B is asserted. These grounds themselves become established in the course of what sometimes may be a rather complex process. The observation of instances in which A and B are conjoined in only one of the many factors which may play a role in this process. Its relative contribution–if any–to the preparation of the logical basis upon which the generalization 'Whenever A then B' rests, depends on other generalizations we want it to co-exist with in harmony; on other relevant observations we have made; on the various assumptions we already hold and the background theories which belong to the system of scientific knowledge to which we subscribe.

To show how the weight of instantiation varies with varying circumstances we may begin by pointing out that it is conceivable that the law 'Everything that has A has B' could become established without anyone ever making a single observation of a thing in which A and B occur conjointly. It is not very much of a practical possibility, yet it is conceivable, that the law about the blackness of all ravens be established without anyone ever having seen the color of even one raven. Imagine we have seen only dissected birds in which we discovered a certain mechanism, e.g., a certain gland, which according to well-established theories was indispensable for the production of some of the essential characteristics of ravens and which at the same time inevitable produces blackness of plumage, then this would once and for all place the law that all ravens are black on a firmer foundation than any amount of instantiation.

Of course such cases are the exception. However even in the general case where instantiation does constitute a factor, its relative role varies greatly with our other observations, assumptions and theories relevant to the situation. Compare the two generalizations 'All people with malaria have been bitten by mosquitoes' and 'All children with an allergy have an over-the-average intelligence.' The reason why we believe nowadays in the truth of the former is because we have discovered the mechanism which is essentially associated with the production of the disease, i.e., the malaria bacterium which cannot attack the victim unless injected into his blood and which at the same time constrains the mosquito to go with the illness because the only vehicle through which the organism may gain access to the victim's blood is the mosquito (or to be quite exact, a 'mosquito-like' contraption, e.g., a physician's syringe). But even before the direct discovery of this mechanism the constant conjunction of malaria and mosquito bite was believed to be a fact because of the numerous instances of such conjunctions observed. Numerous instances of the observed conjunction of allergy and intelligence have on the other hand not established confidence in the truth of the second generalization. Why? Because the belief that many diseases are carried about by animals was accepted quite early in the history of medicine and therefore the suggestion that malaria is carried by mosquitoes fitted in well with existing knowledge. On the other hand the second generalization does not fit in well with existing knowledge. The natures of intelligence and allergy were believed to be basically different and hence the existence of a direct causal connection between the two, unlikely. Consequently the numerous instantiations were believed to be explainable through some alternative hypothesis (a belief vindicated when it was found that the parents of intelligent children, who as a rule are themselves intelligent, are more concerned about such indispositions and consult the physician, while the allergy of the plain child whose parents do not care, remains unreported and inaccessible to public knowledge). In the case of malaria too of course many hypotheses were available to counter the suggestion that there is an intrinsic link between the disease and the mosquito (e.g. that dampness in the bones is the real cause of malaria and the ubiquitous presence of mosquitoes in malaria cases can be put down to the presence of stagnant water which is the source of dampness in the air and the breeding place of mosquitoes at the same time). However we were not motivated to search hard for alternative hypotheses because of the prima facie plausibility of the generalization.

After these remarks we may return to our ravens and attempt to give a general preliminary outline of the process of confirmation. When we begin our observations and have noticed the first few ravens encountered to be black, we have no support yet at all for the generalization that all ravens are black. We would be inclined to say even after observing several instances of black ravens that it is more likely than not that ravens will be found in all sorts of colors because of what we know already of the way individual birds of the same species usually differ in color from one another.

When however the blackness of ravens has been instantiated a great many times, the situation becomes a different one. Members of other species which were of no uniform color were not consistently, over a long period, observed to be of the same color. Instead of saying that the ravens observed just happened to be black, it is more plausible now to suggest the existence of a particular cause responsible for the uniform blackness of ravens.

Of course this unique causal influence may not necessarily reside in the anatomy of ravens. If ravens always nested in chimneys this would just as well explain their black appearance, or perhaps it might reasonably be suggested that the specific climatic conditions prevailing in the places from which all the observed ravens originated caused their darkness. To eliminate for instance this last alternative explanation and thus strengthen the credibility of our generalization we need to observe ravens hailing from different climatic conditions from those previously observed. This is just another point in illustration of the varying weight of instantiation in confirmation: the degree to which the credibility of 'All ravens are black' is strengthened depends not only on the amount but also on the kind of black raven observed.

It goes without saying that even if no one actually suggested that the cause of the blackness of the ravens lies in the common climatic conditions under which the hitherto observed ravens were living, the observation of black ravens originating from different climates eliminates the possibility of such a suggestion. Thus the rationale underlying the principle "observe the greatest possible variety of ravens" it that in this way we may eliminate best potential alternative hypotheses.

In general: we have a very well-vindicated conviction that if in a considerable number of objects manifesting A, B has invariably occurred as well, then there is likely to be a single cause linking the two properties. The function of observing more instances of the occurrence of A under diverse conditions, is to determine whether perhaps B's co-occurrence is due to some unique external set of circumstances and not to a mechanism rooted in the nature of the thing possessing A. Further instantiations then form a part of a process of elimination, in the course of which we eliminate the possibility that the cause of B resides outside A; in the end of which we have traced the cause of B to be inherent in A, simply by having shown that it cannot be inherent in anything else. We shall of course continue to hope that one day, with our increased knowledge the presence of the causal mechanism associated with A will also positively be demonstrated and we shall be able to learn its exact nature and way of functioning.

It is important to realize that even though our understanding of the process whereby birds receive their color, unlike our understanding of how malaria is transmitted by mosquitoes; may be incomplete and we may yet be unable to demonstrate positively the existence of a mechanism through which ravens become black, when it is asserted that all ravens are black what is meant is that all ravens have to be black; that a thing has to be black in order to be a raven for blackness is a physically necessary by-product of ravenness ensured by the existence of a causal mechanism eventually to be identified positively. By this all we mean is that the natural kind of ravens of which it is already known that they possess a number of common features do in addition possess an unique locatable physical complex, which according to higher order theories produce blackness.

The examination of more and more individual ravens is relevant to the generalization because it may be instrumental to the positive discovery of such a mechanism or failing this, to the indirect tracing of such a mechanism to the ravens by eliminating alternative hypotheses according to which the causal influence is embedded in circumstances residing outside the ravens themselves.

VII

By now we have gone a considerable way toward the dissolution of the paradoxes. By now it should be clear on the one hand that it is by no means necessarily paradoxical that an object neither possessing A nor B should serve as confirming instantiation to the generalization 'Whatever has A has B' and on the other that in cases when it does seem paradoxical it does so for good reasons. For everything depends on the possible rational grounds upon which the generalization may be asserted. The immediate thing that matters is what prior notions we have formed about the causal machinery, which in the context of our background knowledge may with plausibility be operative in any given case and whose presence could serve as an explanation why A and B are always conjoined.

If circumstances happen to be such that the presence of a particular mechanism ensuring that all objects with non-B have non-A as well, may reasonably be assumed, then naturally the observation of such objects will not be thought of as absurdly irrelevant to the law 'Everything that has A, has B'. A necessary condition for circumstances to favour such an assumption is, that things having non-B should be regarded as of one natural kind because of their possessing a number of other common properties and for being united by specific theories appertaining to them. In such cases upon finding that an object exhibiting non-B has also non-A, we assume—in the absence of reasons against doing so—that such objects are further united by all of them having non-A.

Non-black things are not regarded as forming a natural kind; beside being non-black we could name few properties shared by all such objects or theories which apply specifically to them. Had it been otherwise had we seriously entertained the idea at any stage that something uniquely common might be associated with white shoes, green emeralds, blue books, red herrings, etc., something which is uniformly associated with all non-black objects, objects forming a natural kind and which due to this singular thing associated with them have been constrained to become beings different from ravens, we would not have found it repugnant that objects like these present themselves as confirming instantiations of the proposition 'All ravens are black.'

Similarly all the things in the universe which are not male ravens are not thought of as belonging to a natural kind. Hence the fact that female ravens were found black provides no rational grounds whatever for the assertion that all non- (male ravens) are black.

Now we can see the error which started off all the paradoxes. We began by saying 'Everything that has A has B' is confirmed by the observation of a thing that has both A and B. But of course this is not correct. For the relevance of a thing to the generalization 'Everything that has A has B' is by no means guaranteed by the mere fact that it has A and B. Only when we know what precise property A and B stand for and understand what particular generalization we are faced with, can we form an idea of the kind of reasons which may serve to explain why the generalization is true. Only then can we decide whether the observation of a particular object may have an effect on either the strengthening or the weakening of the case for any of the possible justifications for asserting the generalization and hence whether that object is a suitable candidate for instantiation. Also then we may know which additional features besides having A and B or lacking both A and B an object ought to have so that its observation may be instrumental to the preparation of the rational grounds upon which the generalization rests.

VIII

We may now take one further step and consider for a moment generalizations concerning the conjunction of A and B which are made without the assumption of the existence of anything linking the two together. It will be illuminating to inquire what the situation was in cases of bare contexts of the kind where the generalization is so basic that no one is expecting to derive it from any more fundamental laws and its validity is not thought of as vouched for by the operation of any causal mechanism. It is generally accepted for instance that light just travels at velocity *c* because it does so, but not because 'something made' light travel at velocity *c*. We do not regard 'All light travels at velocity *c*' as being in the need of further explanation but as a basic irreducible fact of the universe. Thus statements like 'All light travels at velocity *c*' which are not asserted on the presupposition of a causal mechanism of independent existence and form therefore the basic or ultimate laws of the universe are to be regarded as paradigms of generalizations made in bare contexts.

We may notice at once that unlike in the cases of the blackness of ravens, whiteness of swans, greenness of emeralds or the connection between malaria and mosquito bites which were instantiated many times before they were accepted as facts, that light travels at velocity *c* was accepted as a fact upon the very first experiment. Admittedly for many years physicists kept on performing experiments determining the velocity of light but these were designed not to increase the credibility of the generalization 'All light travels at velocity *c*' by multiplying the observed instances of it, but to approximate more closely to the true value of *c* whose constancy never seemed in question.

Physicists did not ask themselves, perhaps on this occasion or under these circumstances only, but not always, does light travel at velocity *c*, because they adopted the policy not to assume unnecessarily, diversity and variations and expect uniformity and constancy in the absence of positive reasons for not doing so. In bare contexts then, no enumeration at all is done and the assertion that A and B go always together is not made on the basis of the extended observation of their repeated conjunction. Such an assertion is made on the grounds provided by the principle to assume the maximum possible stability and constancy in nature; this principle is the ultimate justification of all basic laws asserted in vacuous contexts.

It was of course hardly possible for the observation of the first black raven to have taken place in an entirely bare context for how can it be that we have absolutely no knowledge about any animals yet recognize a raven as belonging to a specific kind within the animal world. Irrespective of this we may assert that would it have been possible for the observation of the first black raven to occur in an empty context then here too we would have declared at once that all ravens are black. But the actual observation of the first raven took place when it was known that members of natural kinds similar to those formed by ravens do not in general have the same color. It was in fact therefore more in keeping with our desire to maximize uniformity to assume even after observing a few ravens and finding them black that ravens, just like most animals, come in many colors.

On the other hand it is also clear that had we observed light's velocity to vary under some sets of circumstances and to remain constant under other we would have no longer treated this an an ultimate fact, and would have searched for an explanation. That is, we would have assumed that other differences, correlated with this difference, also exist between the two sets of circumstances and would have tried to find out what they were.

In the last account if required to state what in general is the effect of observing an instance of the conjunction of A and B we can say no more than that in every case we shall make the assumption which maximises uniformity and shall try to accommodate our observation in a way it harmonizes best with our background knowledge and information. In the last three sections an attempt was made to indicate how this very general principle may work out when applied to some actual situations.

Dept. of Philosophy
Australian National University
Canberra, Australia

NOTES

1 (Oxford, 1957) 122-127.

2 e.g., H. G. Alexander in his 'Paradox of Confirmation', *British Journal for the Philosophy of Science 9* (1958), 227-233.

3 *Mind 55* (1946), 79.

4 To be more precise, the observation of a black raven confirms 'All swans are white' without existential import only, from which it does not follow that there are some non-black birds since it is compatible with the proposition 'Swans do not exist.' However if we have already been given that swans exist, then the observation of a black raven does lend support to the statement 'Not all birds are black.'

5 Here it is not even necessary to say that in addition to our present observation we have also observed that female ravens exist and that these two observations

jointly confirm that not all ravens are black. It may well be claimed that taking for granted elementary facts of life, the existence of a male raven in itself entails that female ravens exist and hence the observation of a black male raven on its own confirms that not all ravens are black.

[6] To be precise the generalization we have to consider has 'Britishers who speak some language' as its subject and not merely 'Britishers.' If the latter were the case one could object that a Frenchman who speaks no English (but speaks some language) does instantiate a contrary, namely, 'All Britishers speak no language.'

[7] Prof. I. Scheffler suggested, in correspondence, another reason why this may be so. For it instantiates $(x)\ (Rx \vee Bx)$, [R=raven, B=black] which is equivalent to $(x)\ (\sim Rx \supset Bx)$ which is a contrary of $(x)\ (Rx \supset Bx)$, given that there are non-black things.

However, against this, one might suggest that instantiation is restricted to a very particular form, namely: An object possessing properties A and B instantiates the universal $(x)\ (Ax \supset Bx)$ and nothing else.

Thus an object with properties A and B does not instantiate $(x)\ (Ax \vee Cx)$ nor $(x)\ (Bx \vee Cx)$. Consequently a black raven instantiates only $(x)\ (Rx \supset Bx)$ or $(x)\ (Bx \supset Rx)$ the first of which transforms into $(x)\ \sim)\ Rx \vee Bx)$ the second into $(x)\ (Rx \vee \sim Bx)$ but neither of course into $(x)\ (Rx \vee Bx)$.

In the light of the above restriction the observation of a black raven does not support $(x)\ (Rx \vee Bx)$ nor does it support therefore $(x)\ (\sim Rx \supset Bx)$.

COMMENTS

CARL G. HEMPEL

1. In his stimulating paper, Mr. Schlesinger refers extensively to the "paradoxes of confirmation", to which I called attention in earlier publications [1]; and I will therefore begin with a brief statement of the basic points at issue, which are essential to my subsequent comments.

The paradoxes concern what Carnap has called the qualitative concept of confirmation, which does not admit of degrees, and which would be used in statements to the effect that a certain object, or a certain piece of information, is confirmatory for a given hypothesis. The paradoxes are logical consequences of two assumptions concerning this concept of confirmation, each of which appears to be reasonable and sound.

One of these assumptions is suggested in a statement by Nicod, and I therefore called it *Nicod's criterion of confirmation.* It is to the effect that if h is a hypothesis of universal conditional form and contains one-place predicates only, then h is confirmed by any object that satisfies both its antecedent and its consequent; and h is disconfirmed by any object that satisfies its antecedent but not its consequent. For example, the hypothesis that all ravens are black, or

(h_1) $\quad (x)\ (Rx \supset Bx)$

is confirmed by any object i such that Ri and Bi, i.e., by anything that is a raven and black, and it is disconfirmed by any object j such that Rj but not Bj.

The other assumption is simply that whatever confirms one of two logically equivalent hypotheses also confirms the other. I called this the *equivalence condition.*

Now h_1 is logically equivalent to

(h_2) $\quad (x)(-Bx \supset -Rx)$, i.e., All non-black things are non-ravens and to

$h_3)$ $\quad (x)[(Rx \vee -Rx) \supset (-Rx \vee Bx)]$,

i.e., Anything that is or is not a raven is not a raven or is black.

Consequently, any object i such that $-Bi$ and $-Ri$ (for example, a white shoe), and indeed any object that is not R, and any object that is B, will confirm h_1, since, according to Nicod's criterion, such objects confirm the hypotheses h_2 or h_3, respectively. These and some related consequences of the two assumptions constitute what I called the paradoxes of confirmation.

2. Mr. Schlesinger proposes a "new paradox" of confirmation, according to which an obviously confirming instance may be turned into a positively disconfirming instance. One of his examples may be stated as follows. Let i be a bird (Bd) that is black (B) but is neither a swan (S) nor white (W); so that we have

$$Bdi, Bi, -Si, -Wi.$$

Then, by virtue of its first two characteristics, object i is a confirming instance of

$$(h_4) \qquad (x)(Bdx \supset Bx),$$

while by reason of its third and fourth characteristic, i confirms

$$(h_5) \qquad (x)(-Wx \supset -Sx),$$

or, equivalently,

$$(h_6) \qquad (x)(Sx \supset Wx).$$

And h_6 finally implies the contradictory of h_4, namely:

$$(h_7) \qquad -(x)(Bdx \supset Bx),$$

provided, as Mr. Schlesinger rightly notes, that "we have already been given that swans exist"—let us call this additional required premiss P_1, and provided, as we should add, that we are also given the further premisses P_2, that all swans are birds, and P_3, that no white things are black. Strictly, then, h_7 is inferred here from h_6 in conjunction with P_1, P_2, and P_3.

But this argument fails to show what it claims to show, namely, that i, which is indeed a confirmatory instance of h_4, is also a disconfirming instance for h_4. For the argument does not establish that i disconfirms h_4 in the sense of the only criterion here under discussion, namely, the Nicod criterion. Indeed, this would require that i satisfy the antecedent, but not the consequent of h_4, i.e., that i be a bird but not black; whereas, by hypothesis, i is a bird and black. Hence, by the Nicod criterion, i does not disconfirm h_4.

Now Mr. Schlesinger does not define his concept of disconfirmation and does not state how he takes it to be related to that of confirmation; but his reasoning in the example just considered is clearly to this effect: Object i does confirm h_6, and since h_7 "follows from" h_6, i must confirm also h_7 and disconfirm its contradictory h_4. But apart from the fact just noted that this reasoning is incorrect for the concepts of confirmation and disconfirmation determined by the Nicod criterion, the argument involves at least one assumption that seems to me unsound on *any* reasonable construal of confirmation. For h_7, as we noted, is inferred here not from h_6 alone, but from h_6 in conjunction with the further premisses P_1, P_2, and P_3. And even if we were to allow that P_2 and P_3 are analytic (and that accordingly the conjunction of h_6 with P_2 and P_3 is logically equivalent to h_6 alone), this surely is not true of the premiss P_1, that there are swans. Thus, Mr. Schlesinger's argument presupposes that if an object i confirms h_6 and if h_6 in conjunction with some further empirical statement P_1 (which is taken to be true) implies h_7, then i must also confirm h_7. But this cannot generally be the case, for h_6 in conjunction with P_1 also implies P_1 and it would follow therefore that an object that confirms a given hypothesis (in our case, h_6) also confirms any other empirical statement that is taken to be true (in our case, P_1); and this is clearly absurd.

An analogous remark applies to Mr. Schlesinger's example of the male and the female ravens; for "fact of life" or not, the assertion that there are female ravens, which is needed as an additional premiss here, is an empirical statement that is *not* entailed by the information that i is a male raven: otherwise i could not be observed to be a black raven unless a female raven were observed as well; and on this strange construal of 'male,' Mr. Schlesinger's example would simply lose its point.

3. But while Mr. Schlesinger's argument does not establish his new paradox of confirmation, it is actually true that if the Nicod criterion is generalized in a perfectly natural fashion to cover universal hypotheses containing also relational predicates, then the resulting concept of confirmation has the peculiarity envisaged by Mr. Schlesinger. This fact, which might indeed be referred to as a further paradox of Nicod-confirmation, was pointed out in

my earlier article[2]. Briefly, this result can be established as follows: Consider the hypothesis

$$(h_8) \qquad (x)\,(y)\,[-(Rxy \cdot Ryx) \supset (Rxy \cdot -Ryx)]$$

The natural generalization of the Nicod criterion would stipulate that an ordered pair of objects, *a, b* counts as confirming or as disconfirming the hypothesis according as it satisfies its antecedent and its consequent, or its antecedent but not its consequent. Suppose now that *a,b* are two objects such that *Rab* but not *Rba*. Then the ordered pair *(a,b)* confirms h_8, but it disconfirms the following hypothesis, which is equivalent to h_8 by contraposition:

$$(h_9) \qquad (x)(y)[(-Rxy \vee Ryx) \supset (Rxy \cdot Ryx)]$$

Thus, the "new paradox" does indeed arise in this case.

It is for this reason among others that in my earlier article I rejected the Nicod criterion and replaced it by a different definition of confirmation, which is applicable not only to universally quantified hypotheses but also to hypotheses of existential and of mixed quantificational form. The definition construes confirming and disconfirming cases for a hypothesis not as objects, but as sentences that report on observational findings.[3] Confirmation thus defined is demonstrably immune to the "new paradox": No observation report can both confirm and disconfirm one and the same hypothesis; in our last example, for instance, the report that $Rab \cdot -Rba$ disconfirms both h_8 and h_9. But the modified definition preserves those consequences of the Nicod criterion and the equivalence condition which were indicated in section 1 above. In regard to these, I argued that their appearance of paradoxicality is seen to be deceptive[4] when one considers the intent of the proposed definition, and in particular its objective of characterizing the relation of confirmation for the case where no additional, "background," information is available in judging the confirmatory relevance of a particular item for a given hypothesis.[5]

The concept of confirmation, then, that gives rise to the paradoxes is meant to apply specifically to the austere kind of situation in which there is no antecedent information and no background theory that might be considered in deciding whether a given individual case with certain observed characteristics is to count as confirmatory, as disconfirmatory, or as neutral for a contemplated hypothesis. Admittedly, this is an idealized kind of situation which will rarely be encountered in practice; but in an attempt to deal rigorously with problems in the logic of confirmation it might well be advisable to tackle simple cases first.

Mr. Schlesinger deals very suggestively with some of the intricacies of the case where a background of empirical information and of theory is available; as a result of his discussion he purports to identify "the error which started off all the paradoxes," namely, in effect, a failure to realize that in dealing with the generalization that all A are B, it is only when we know precisely what properties A and B are and by what causal mechanism they might be connected, that we can "form an idea of the kind of reasons which may serve to explain why the generalization is true." But surely, in the context of confirmation we are not concerned with *reasons explaining why* a given generalization is true, but with *grounds for believing that* it is true. Nonetheless, the considerations Mr. Schlesinger adduces are pertinent also to the matter of confirmation. In particular, if the previously accepted background assumptions include a well substantiated theory which logically implies, and thus explains, a given generalization, then of course the generalization is well confirmed even if no particular instance of it has been observed. Mr. Schlesinger rightly mentions this possibility, though as an exceptional one. But perhaps it is not quite so rare as he seems to suggest. For example, when on the basis of his periodic table Mendeleev predicted various characteristics of elements that had never been observed, and of some of their compounds, he was surely establishing generalizations in this way, using the periodic system as part of his background theory. A similar remark applies to recent and current predictions of uniformities that would be exhibited by trans-uranic elements not as yet produced; the theory on which these predictions are based is, of course, considerably advanced over Mendeleev's schema. Again, it might be argued that the technology of artificial satellites and of space travel not infrequently relies on laws of which no instances have been observed, but which have the support of well-corroborated theories from which they are derived.

Dept. of Philosophy
Princeton University
Princeton, New Jersey

NOTES

1 The paradoxes were adumbrated on p. 222 of my article "Le Problème de la Vérité, *Theoria 3* (1937), 206-246; and they were made explicit and analyzed in detail on pp. 9-21 of my essay "Studies in the Logic of Confirmation", *Mind 54* (1945), 1-26 and 97-121.

2 "Studies in the Logic of Confirmation", p. 13, fn. 1. There is an erratum in the formulation, on line 5 of that note, of the hypothesis that serves to illustrate the paradox: the clause '$\cdot R(y,x)$' was omitted in the antecedent. The correct version, which is given by h_8 in the present text, was stated in Hempel, "A Note on the Paradoxes of confirmation", *Mind 55* (1946), 79-82.

3 Inasmuch as such sentences may be viewed as expressing presumptive facts, this construal conforms well to Nicod's own conception; for he speaks of "facts," such as "the presence of *B* in a case of *A*" as confirming a universal proposition, such as that all *A* are *B*. *Cf.* J. Nicod, *Foundations of Geometry and Induction* (London 1930), p. 219.

4 See also N. Goodman's argument to this effect in his book *Fact, Fiction, and Forecast* (Cambridge, 1955), 71-73.

5 For a detailed critical discussion of my definition of confirmation see R. Carnap, *Logical Foundations of Probability,* (Chicago, 1950), 223-224, 468-482.

2

ERNAN MCMULLIN

FROM MATTER TO MASS

Presented March 19, 1963

In this paper*, I would like to explore the genesis of two historically related concepts: the concept of *matter,* which developed first in Ionian natural philosophy in the sixth century B.C., and the concept of *mass,* which came to focus in the mechanics of Newton more than two thousand years later. I want to ask of each of these: to what intellectual needs did they respond? And to inquire into the sort of historical link, if any, that joins them. The matter-mass transition is possibly the clearest instance of a philosophic concept's both suggesting, and eventually (at least in some domains) giving way to a scientific one. Our inquiry may, then, be expected to throw some light on the thorny question of the conceptual relations between philosophy and natural science.

§ 1 *Ionian "Stuff":*

Speculative thinking appears to have begun in Ionia from the question: "how can the multiplicity of physical things, and the changes they undergo, be brought into some intelligible pattern?" To most minds of the time, this must have seemed like a hopeless quest, unless one were to appeal to supra-human sources of knowledge for help. Small pieces of pattern were known: in astronomy, in meteorology, and so forth. But the existence of a single over-all pattern must have seemed extremely dubious. The difficult search for unity beneath all the multiplicity of becoming took the most obvious path: the unity is a simple physical one of a "stuff" that underlies, whether water, fire, air, seeds, atoms. In one stroke, the

* Most of the research for this article was done while the author was working under a National Science Foundation grant at Yale University in 1957-59. I would like to express my appreciation of the help afforded both by the Foundation and by several members of the Yale Department of Philosophy, notably Irwin Lieb, John Smith and Rulon Wells.

multiplicity of different sorts of change is reduced to a single one: if one knows the properties of the primary "stuff", and the ways in which it normally undergoes change, all others can (hopefully) be understood in terms of these.

The paradigm of Ionian cosmological explanation is, therefore, the discovery of one or a small number of underlying "stuffs", whose behavior is (1) either already familiar and sufficiently complex to provide analogies for many different sorts of physical change, or (2) is postulated precisely to meet the needs of diverse sorts of explanations of change (the local motion of Democritus' atoms), or (3) is left deliberately unspecified and open (that of Anaximander's Indefinite). The first of these alternatives is the most empirically-oriented, for it would require for its support the observation of changes (like those where water appears to become mud or air) and the drawing of inductive analogies between them. The doctrines of Thales, Anaximenes, Empedocles, insofar as they are supportable by evidence, would rank (in modern terminology) as insecurely-based inductive generalizations. Whereas Democritus' view draws its strength entirely from the re-orientation forced upon Parmenides' original conceptual analysis of *Being* and *Change* by the identification of Non-Being with Space. It does not rely on *specific* observations, even indirectly, and so might be classed as a philosophic doctrine rather than a primitive form of natural science.

The claim made by these doctrines to *explain* rests on the reductive character of the model of physical change they offered. Whether they relied upon conceptual analysis or on crude observations, they resembled one another by postulating a fundamental matter-stuff of some kind. This does not mean that they rejected other models of explanation (by formal aspect, by *telos,* by efficient cause), as Aristotle was later to charge. Rather, they incorporated *all* of these in a confused way: most of them endowed their matter-stuff with life (and thus *telos* and efficiency); some of them stressed the structural aspect (as Democritus did the shape and relative motions of his atoms); the clashing contraries must obviously be taken as efficient cause . . . But the dominant aspect of their explanations, the one from which the intelligibility of the explanation was thought mainly to spring, was undoubtedly the matter-stuff which underlay, and which was simpler to comprehend, both in its nature and in its behavior, than the complex entities and changes of everyday experience.

§ 2 *Aristotle's "Matter":*

How far the Ionians meant their cosmological theories to be pressed is not clear. The ideal of an *"epistémé"*, a set of necessary principles whose truth was guaranteed by some sort of special insight, had not yet matured sufficiently to make them ask the more sophisticated questions about evidence and certitude that were to be the commonplace of post-Socratic thinkers. But when Aristotle came along and looked at their mode of procedure, an implication of it (to which few of them had ever adverted), seemed to him a quite decisive reason for its rejection. If all changes are not simply likened to the changes of a single stuff, but are asserted to *be* changes of that stuff, then everything ultimately *consists* of that stuff, and there is only a single kind of world-substance. The apparent coming-to-be of different substances is deceptive. All that happens is simple a re-formation of what was already there; it is the same substance, before and after (though a change in the *number* of individual similar substances is possible, except in Democritus' theory).

For Aristotle, the existence of "genesis", i.e. the coming-to-be of completely new beings that are in nowise re-shuffled versions of older ones, was so obvious as to need no proof. One had only to think of the coming-to-be of humans. Hence, any schema of natural explanation which ignored or, worse, implicitly denied, this fact, had itself to be called in question. Yet this schema had an inherent attractiveness that could not be altogether set aside. Aristotle's proposal is to retain a matter-category, but to alter the whole structure of analysis leading to it, and thus alter the category itself. Unfortunately, his proposal is really two quite different proposals, and the consequent ambiguity is one of the most serious hindrances to the understanding of Aristotle's cosmology.

In the *Physics,* Book One, he is concerned to affirm the number and nature of the "principles" of physical things in a way that will transcend by its self-evidence and necessity the unsure and unclear sayings of the Ionians, and yet will retain *some* continuity with their scientific effort. He achieves this by transforming the analysis into a purely conceptual one, while retaining something of the old terminology of contraries. By analyzing the concept of change, he shows without any difficulty that three conditions are necessary and sufficient in order that there may be a change. There must be a subject which changes; it must first lack a certain characteristic; it must later possess this characteristic. The first of these is called the "matter" of the change; in the example he gives, where a particular non-musical man becomes musical, the "matter" would be the man. By calling the man "matter" relative to this change, Aristotle does not intend to say *anything* of the man other than that he could come to possess a characteristic not previously possessed. In other words, 'matter' here conveys no properties (watery, fiery . . .) ; it is simply a word which indicates, in the most unspecific way possible, that its referent has a particular function vis-à-vis a designated change. This matter-form-privation analysis applies, therefore, to *any* context of which the word 'change' can be used, to the changes of immaterial Intelligences as well as to those of terrestrial bodies. In his attempt to reach an *"epistémé"* of the desired Platonic sort, Aristotle has thus sacrificed the empirical character of the Ionian analysis entirely.

But elsewhere (and in one brief passage in Book One), Aristotle's thought turns in a fundamentally different direction. If instead of stressing *"epistémé"*, one stresses *genesis,* an unexpected difficulty arises for the matter-form-privation analysis. In *genesis,* there is no predicate which is properly predicable throughout the change, and thus the change (it would seem) lacks a "matter" of the sort that the prior conceptual analysis of predications about change required. For it was essential to the validity of this analysis that there should be a *term* (like 'man') which is predicable before and after; it was on the ground (and *only* on the ground) of such predicability that the analysis yielded a "matter-subject" which would be the referent of the term.

In analyzing *genesis,* the entire argument shifts in a subtle, yet very definite, way. The *evidence* for the claim that there *is* a matter-subject in substantial change can no longer be the simple continued predicability of a substance-term before and after the change, but rather the *empirical* affirmation that in those events commonly regarded as substantial changes, *"we find in every case* something that underlies, from which proceeds that which comes to be; for instance, animals and plants from seed" (*Phys.* I, 190b 2-4). In other words, Aristotle is saying that the label, 'change', is properly applied to such cases, and that a matter-form-privation analysis similar to the one he has already given applies, therefore, to them, despite appearances. In particular, even though there is no *nameable* matter-subject which perseveres, there is, he claims, a matter-subject of an analogous sort here, one which is not nameable, however. Since, by hypothesis, no substantial predicate remains applicable throughout the change, the matter-subject (he calls it "primary matter") cannot be characterized by any substantial predicate.

The question immediately arises: if it cannot be so characterized, then how can Aristotle know that it is there in the first place? Or to put this in a different way, how can he be so sure that those events normally regarded as instances of coming-to-be are really changes at all? His analysis has previously shown beyond any doubt that for 'change' to be correctly predicated, there *must* be a matter-subject of some sort (otherwise the apparent "change" will really be a replacement). Other than saying that "we find something that underlies" in all such cases, without specifying how we find it, Aristotle does not elaborate on the sort of evidence his statement requires. Since the "something" here has (apparently) no continuous substantial properties to identify it, this is a serious problem. He appears to be relying on an "inductive insight" of the sort that in his *Posterior Analytics* he mentions as grounding the indemonstrable premisses from which *epistémé* must spring.

§ 3 *Matter and Continuity:*

What has to be shown here is the *continuity* between the before and after in instances of coming-to-be; the continuity can then be assumed to have as its ontological basis some sort of real "principle" in the thing itself. (The analysis of, and justification for, this last statement will have to be left aside in this essay).[1] If one is not simply to rely on a "hunch" that *A* has *changed* into *B* (rather than that *B* has replaced *A*), one would use as concrete indicators such factors as continuous identifiability at a particular spatial location; no outside "materials" brought in to any notable extent; the following of some intelligible causal pattern of a kind that would suggest that *A* has been transformed into *B;* a partial continuity of perceptible properties like color and feel. Among other things, one would expect some sort of *quantitative* continuity, though an index of quantity may be difficult to decide on. A pebble cannot *change* into a boulder. This claim rests not so much on an assumption that "stuff" cannot be created, as on the fact that if a pebble *did* appear to "change" into a boulder, some outside "stuff" would have to appear (either by way of creation or replacement), and then the event would no longer be properly described as a *change* but as an *addition.* On the other hand, if an acorn grows into an oak, that is called *change,* even though a great deal of "stuff" has had to come in from the outside. The difference would seem to be that in the case of the acorn, there is a central unifying identifiable principle governing the process continuously throughout, whereas in the other instance, there is only the brute addition of material.

Could a table disappear from this room and re-appear more or less instantaneously next door, without following a localizable continuous spatial path between the two places? The answer is: no, unless there were some repeatable physical process of "projection" involved, where each part of table-at-*B* had a unique and law-specified physical relation with the corresponding part of table-at-*A*. The reason for this is not some sort of physical difficulty (e.g. a relativistic embargo on instantaneous transfer). It is much simpler than that: there would be no way of knowing that it was the same table. Or more accurately, to say it was the "same" table in these circumstances would either have no meaning or would have a meaning remote from the everyday one. (If a being with a *memory* were involved in our "instantaneous transfer", its memory might allow us to apply to it the notion of "sameness" meaningfully. But this special case can be disregarded here, since our basic interests is in *substantial* changes where such individual memory-continuity would be excluded).

When we look at instances of coming-to-be, then, and affirm that there is "something that underlies", we mean that there are certain types of quantitative and causal continuity manifested that would not be manifested were it to be merely replacement. By speaking of the substratum as indefinite, Aristotle wished to stress that it is not a *stuff,* i.e. a material with recognizable uniform qualities. He has already repudiated the Ionian attempt to define such a stuff. But by making it *entirely* indefinite, he seems to leave no room for the continuities we have seen to be necessary. His reasons for doing this have, therefore, to be carefully scrutinized.

First, in the well-known passage where primary matter is spoken of as "neither a particular thing nor of a certain quantity nor otherwise positively characterized" (*Meta.* VII, 1029a 23-5), it is defined as the ultimate subject of *predication,* not as the ultimate subject of change. And it is by no means evident that the two are equivalent. Second, in true coming-to-be it is necessary (he implies) that no substantial predicate be applicable throughout (or else the substance it denotes will have remained unchanged). But this leaves all kinds of loopholes. Where *B* comes to be from *A,* both *A* and *B* might, in a given instance, be transparent (to use one of his own examples). This would not, however, entitle us to say that "that which underlies" is transparent. But this is rather because the substratum is not an ordinary substantial thing and so one cannot apply predicates to it in the normal way, no more than one could to the *form* of the thing, taken separately. The substratum is in some way the *ground* for the continued applicability of the predicate, *transparent,* in this case, even though transparency may not be attributed to the substratum itself directly. So that lack of applicable predicates is no argument for *total* ontological indeterminacy.

Thirdly, Aristotle is trying to find a schema that will apply indifferently to *any* substantial coming-to-be (just as the triadic schema of the *Physics* did to changes of *any* sort). If he is to find a substratum-principle whose description would fit *all* instances, then he must abstract from those ontological differences that distinguish the substratum of one substantial change from that of another. And so, once again, he is forced to speak in terms of "indeterminacy", but an indeterminacy now of the *concept* relative to its instances (akin to that of a genus relative to its subspecies). It is perfectly obvious that "that which underlies" the change whereby a particular group of chemicals become a baby elephant is ontologically *not* identical with that which underlies the change that occurs when a particular tree dies. So that the "indeterminacy" attributed to the primary matter by Aristotle cannot be a literal ontological one of each instance of substratum, of a sort that would make the primary matter of one change ontologically equivalent to that of every other.

§ 4 *Matter and Quantity:*

Let us now return to a discussion of the historical legacy left by Aristotle regarding the matter-substratum of things. It is not a *stuff,* strictly speaking. In the context of change, the word, 'matter', will be used either for the individual subject of *any* change (with no connotation of a stuff or material that would be common to different instances), or else for the substratum disclosed only in substantial change (though presumably present at all times). The latter *sounds* as thought it were somehow the same from one instance to another (though lacking any nameable property), but as we have seen, this position is a difficult one to maintain. One can say, therefore, that there were *some* apparent grounds in the work of Aristotle for retaining the old Ionian metaphor of an underlying common stuff, now a rather mysterious propertyless "principle", but because this doctrine had to be qualified in so many ways, it would be hazardous to speak of a straightforward "Aristotelian position" on the question.

Matter and quantity were only distantly related in the Aristotelian cosmology. Matter pertains to the *essence* of a natural thing; it is a substantial principle. Whereas quantity (thought of principally as an answer to the question: "how much?") is an accident, outside essence, in nowise revelatory of the nature of things. It is true that quantity is necessary for a physical essence even to exist, but this does not make it any the less accidental, and peripheral to the search for *epistémé*. Furthermore, since the primary matter is spoken of as "indeterminate", it will usually be assumed that it cannot be regarded as quantified in any way. Thus, though the matter is "conserved" through change, this conservation cannot, it would seem, be represented as a quantitative conservation (to Aristotle, a typical atomist error).

So what does "conservation" here mean? It is a permanence in the category of substance, not that of quantity or quality. Yet it does not mean that substances may not come to be or pass away: this is the very point Aristotle wished to safeguard. What is rejected is radical *"de nihilo"* creation or annihilation. In Book One of the *De Generatione,* primary matter is defined as that which makes both the perpetuity and the unbroken continuity of coming-to-be possible. By requiring a matter-continuant, radical creation is rejected; by making it indeterminate, the genuine novelty of the beings that come-to-be is secured. The emphasis here is not so much on conservation as on something that remains the same, on an ontological basis for continuous univocal predication.

Nevertheless, the exclusion of radical creation implicit in the whole Greek tradition (Aristotle remarks that this exclusion "pre-occupied earlier philosophers more than any other theme", *De Gen.* I, 317b 29) does suggest that there is *something* that remains the same, a cosmological measure whose constancy is an index of the non-occurrence of radical creation. What could this be? The atomists had an easy answer: the total number of beings (atoms) in the universe remains the same. But, as we have seen, this meant the denial of the substantial unity of living beings made up of a host of atoms; in Aristotle's view, the passing-away of a single such being would ordinarily involve the coming-to-be of a plurality of simpler substances. So that number is not an adequate conservation-measure.

Earlier thinkers undoubtedly had in mind some vague sort of constancy of the quantity of "stuff" in the universe when excluding creation from their account of process. If all changes are to be explained by successive transformations of an underlying stuff, that stuff must not itself be allowed to come-to-be. But this was not open to Aristotle, although he does occasionally talk about "all the matter in the universe" as though it had some sort of quantitative measure. If, *per impossibile,* creation were to occur, if, for example, in the dissolution of a living body one small piece of matter were to come to be *ex nihilo,* the only way open to Aristotle to identify it as creation would be through the category of substance: a new thing has appeared that has no traceable roots in the past. But this will be very difficult to prove, if the overall process within which the new being appears is sufficiently complex.

Without the use of *some* quantitative measure, how is one to know whether the process should yield *A, B* and *C* as end-products, or only *A* and *B?* And which quantitative measure ought one choose as an index for so fundamental an occurrence as radical creation? Is it possible to justify one such measure rather than another in such a highly metaphysical context? These latter questions were to arise only later, in the Christian framework of creation, nature and miracle. But the first question must have posed itself from the first for followers of Aristotle.

Our conclusion is that the Greek notion of a common matter-stuff was subject to two contrary pulls as regards quantification. Maintaining the reality of substantial change seemed to imply that primary matter had no intrinsic measure whereby it could be said to be quantified. While the rejection of the possibility of radical creation implicitly suggested the conservation, in some sense, of primary matter. Unless a non-quantitative index of conservation could be found, it would seem that one or other of these theses would have to be abandoned. This, then, was the categorial scheme that the medievals inherited. We have left aside in this account other important aspects of the Greek matter-concept (individuation, contingency, relation with space . . .), in order to concentrate on that aspect which is most clearly linked with the later notion of mass.

§ 5 *Quantity of Matter:*

In medieval natural philosophy, the relationship between primary matter and the category of quantity was extensively discussed in the context of the problems of individuation, by then regarded as of central importance. Distinctions between different types of quantity (e.g. *dimensiones interminatae* and *dimensiones terminatae*) were introduced with a view to explaining the respective roles played by matter and by quantity in the individuating of physical objects.[2] The central question was whether *any* sort of actuality or determinacy could be assigned to primary matter in itself, independently of the forms that successively qualified it. Writers in the Platonist and Augustinian traditions were usually willing to admit some "form of corporeity" or "beginnings of form" in primary matter, whereas the "stricter" Aristotelians (like

Aquinas) insisted on its complete separation from all determinate categories, including that of quantity.

There was one important domain of process where the Aristotelian schema ran into special difficulties. In modern physics, density changes are not considered especially basic. But because notable density changes were associated with the change of any one of the four elements into another (water into "air", i.e. water vapor, for example), and because elemental changes were the most fundamental sort of substantial change, the nature of condensation and rarefaction was of particular interest. The atomists had the simplest explanation: the changes in density were simply caused by spatial changes among the atoms. But if one refused to admit the possibility of void, matters were not so simple.

Aristotle remarks that "dense differs from rare in containing more matter in the same volume" (*De Caelo,* III, 299b 9), which suggests that density is a quantitative characteristic. In the *Categories,* he emphasizes that although

> 'rarity' and 'density' . . . seem to be terms indicating quality, yet it would appear that they really belong to a class other than quality. For it is rather a certain relative position of the parts composing the thing thus qualified which, it appears, is indicated by each of these terms. A thing is *dense* if its parts are closely combined with one another, *rare* if there are interstices among them. (10a 17-22)

This is quite a striking passage: not only does he exclude density from the category of quality (despite the latitude of this category, which includes e.g. figure and shape and everything else that "makes something be such-and-such"), but the reason for which he does this is the Galilean-sounding one that density is reducible to (explainable in) quantitative terms, and *thus* is not a quality.

In the *Physics,* however, he reverts to a somewhat different position. He is bothered by a puzzle inherited from the Pythagoreans: when "air has been made out of a cupful of water", either the universe as a whole must change its volume (must "bulge"), or "out of an equal amount of air elsewhere a cupful of water must be made", or else one must admit the existence of void (216b 23-30). Instead of opting for one of these possibilities or showing that another exists, he prefers to avoid the issue by reformulating the problem in terms of contraries and thus in the category of quality. Rarity and density are likened to lightness and heaviness; when they succeed one another in the "same matter", condensation or rarefaction takes place.

> When air is produced from water, the same matter has become something different, not by acquiring an addition to it, but by becoming actually what it was potentially . . . It is the matter, being potentially both, that comes to be each of the two (217a 27-34).

When he speaks here of the "same matter", it is clearly not a quantitative sort of sameness that he has in mind. Rather, it is the substantial sameness of a substratum which, independently of quantity, is informed successively by different qualities. Quantity (and the Pythagorean puzzle that went with it) is simply left out of account as irrelevant.

A quite different approach to questions of density is found in the tradition of hydrostatics going back to Archimedes. In the latter's own work, *On Floating Bodies,* bodies are identified and distinguished from one another in terms (equivalently) of their specific weight. Alloy problems were very popular in early medieval natural science, mainly, one suspects, as textbook problems in simple arithmetic. Many methods of determining specific weight were described; the combining of two contingent measures (volume and weight) to produce a number which served at all times to identify a particular "nature" undoubtedly brought the notions of quantity and of matter nearer together. But because hydrostatic problems were not concerned with density *changes,* as such, the concept of matter did not appear in their formulation.

The relationship between weight and density is still not entirely clear in the earlier medieval treatises; the first "operational" definition of density in terms of weight and volume is found in the *De insidentibus in humidum* (thirteenth century; author unknown). The term, 'specific weight', is discovered here for the first time, and is carefully distinguished from 'numerical (gross) weight'. The use of 'specific' is significant:

> The relation of one heavy body to another can be considered . . . according to *species,* as when we seek to compare the specific weight of gold to that of silver; and this should be done on the basis of equal volumes of gold and silver.[3]

The choice of term suggests that the relation between two natural species can be described in terms of a number.

Fourteenth century Aristotelianism was somewhat affected by this Archimedean tradition; we find, for instance, in Albert of Saxony a discussion of hydrostatics employing a curious mixture of Aristotelian and Archimedean arguments. A topic of growing importance at this time was that of the "latitude of forms", i.e. the changes in intensity or degree of qualitative forms. Scotus had suggested that such *changes* of quality might themselves be estimated in a quantitative way. This device was fully explored by the Merton School (1328-1350); though density was regarded by them as a quality, changes in it (they thought) could be estimated by numerical means. Their notion of an "intensive" measure is ambiguous: sometimes it refers to the ranking of degrees of a quality in terms of their "intensity"; at other times, it means a ratio between two bodies on the basis of some property (whether qualitative or quantitative). Thus specific weight is often called for this reason "intensive" weight.

Density is defined by the Mertonians as a relation between *materia* and quantity (volume). But they never explained how *materia* is to be taken here. In fact, there is a very curious dichotomy in the works of this period between discussions involving the relatively operational notion of specific weight in the context of hydrostatics, and those concerning the entirely theoretical notion of density in the context of the remission of forms. The putting together of specific weight and density would have furnished an operational way of estimating density in terms of weight measurement, and would thus have made weight the primary measure of the *materia* mentioned in the Mertonian definition of density.

But this did not happen, perhaps because weight for the Aristotelians was an accidental characteristic which varied with the distance of the body from its natural place, a variable force which acted downwards upon some bodies and not on others (fire had no gravity, only levity).

Thus when Swineshead in his *Book of Calculations* asserted that in condensation and rarefaction the *"quantitas materiae"* remains the same, the conservation referred to does not rest on weight measurement, but rather on an intuitive "amount-of-stuff" notion akin to that of the Ionians. It may also have been influenced by the recurrent suggestion (with something both of Plato and of Aristotle behind it) that density could be thought of in terms of the number of "units of matter" per unit volume. One finds this suggestion for example, in the *De Macrocosmo* of Marcus Trevisano (1378). It might seem, at first sight, that "quantity of matter" might be defined in this way via a counting process, without recourse to weight-measurement. But, of course, this is not so: we must know that the units of matter are of the same weight before this definition will work. If it be assumed that they are of the same size and homogeneous, the number of the "units of matter" will be a measure of the "quantity of matter". But the assumption of homogeneity here involves an implicit reference to a dynamic measurement of weight or to percussion properties.

The ambiguity about the status of density has unfortunate consequences here. If it be regarded as an irreducible quality, like color, the notion of "quantity of matter" will still be a basically qualitative one, incapable of being fitted into an operational mechanics. On the other hand, if it be taken as quantitative, it is no longer an independent variable, since it will ultimately be defined in terms of weight and volume. In this case, any attempt to define "quantity of matter" in terms of "density and bulk taken conjointly" to quote Newton's phrase, will risk either error or circularity.

For in order to know the density, one would already need to know the weight (or some equivalent dynamic property). Without a theory of gravitation, i.e. a way of separating the weight-vector into a scalar quantity (mass) characteristic of the body, and a vector quantity (gravitational acceleration) characteristic of the place, the "quantity of matter" will come out to be identical with the weight. And this is clearly wrong, because weight—as the Aristotelians well knew—is not conserved. With a theory of gravitation, on the other hand, density is definable as mass divided by volume, but now our definition of mass (or "quantity of matter") has evaporated into the truism that $m = v \cdot (m/v)$ which can tell us nothing about the nature of *m*, since any measure-number one pleases can be inserted in the equation. Newton inherited this conceptual tangle from his Mertonian predecessors, and he was never quite able to extricate himself from it. His definition of mass or "quantity of matter" as the product of density and volume is an empty one, unless one either assumes density to be an irreducible quality whose "value" is somehow known in advance (this is the direction in which Jammer suspects Newton was going), or else a numerical measure based on the counting of "units of matter" (defended by Crew, Burtt, Hoppe; but this, as we have seen, is ultimately circular).[4]

§ 6 *Matter and Motion:*

The "matter" we have been discussing so far is a substrate of change, an agent of continuity or conservation. The problems associated with its quantification arise from the difficulty in defining a factor that will be conserved through all forms of change. The *purpose* of the quantification has merely been to ensure some form of conservation in any given change. To know the "quantity of matter" is only to know how much stuff we have; it does not seem to help in distinguishing changes from one another. But now we must ask whether the "quantity of matter" will affect the motion itself. "Matter" is to be thought of as that which determines the kind of motion a body will have.

Aristotle had said that if two bodies of the same volume are allowed to fall, the heavier of the two will have the faster natural motion downwards (*De Caelo,* IV, 308a 31-3). Indeed, the theory of natural movement (and thus of nature) contained in the *De Caelo* entirely disregards the traditional pairs of contraries (hot-cold, dry-moist), and concentrates on the heavy and the light. As Solmsen points out, Aristotle's cosmology really only has two elements (fire which is light, earth which is heavy); the addition of the other two in the *De Caelo* is justified only in the most perfunctory way, as an obvious gesture to a quite different and older mode of defining 'element'.[5] Natural movement is determined entirely by these two contraries, and not in any way by the matter in which they inhere. Aristotle does not seriously try to quantify natural movements, as a means towards comparing them. In his discussion of void, he asserts, for example, that speed of fall is inversely proportional to the resistance of the medium, but as Solmsen remarks, this "law" of free fall is quite incidental to his theory of natural movement, and was clearly elicited by the needs of his polemic against Democritus and Plato.

In the later Aristotelian tradition, Philoponus, Avempace, Bradwardine and others pointed out the deficiencies of Aristotle's law, not by comparing it with experiment or by trying to find operational definitions of the three quantities concerned–speed, force (or weight), and resistance–but rather by simply exploring the relations between the three concepts, taking them to be intuitively understood from ordinary usage. The claim that the speed of free fall depended directly on the weight of the body had been challenged on the basis of experience as early as Philoponus. It led to obvious paradoxes where, for example, two equal weights were tied together so as to form one–which would then have to fall twice as fast. But, in general, it was assumed that the speed of fall *did* depend somehow on weight, and (because of the resistance of the medium) on volume.

It was only with the *impetus* theory of Buridan, however, that the notion of matter became involved in this discussion. "The greater quantity of matter a body contains, the more impetus it can receive and the greater the intensity with which it can receive it." [6] The impetus is an intrinsic quality of the moving body–by means of this he avoids the classic difficulties in Aristotle's theory of projectile motion–but it is defined in terms of the speed and the "quantity of matter" of the body. In a dense body, he says, there is more *materia prima,* and the *materia* here is to be regarded as the recipient of impetus; its own intrinsic *"quantitas"* will determine how much impetus it will receive, and how long, therefore, the motion will last, since the impetus will be gradually worn down by the resistance of the ever-present medium.

Buridan's innovation is a far-reaching one for the concept of matter. Matter had always been regarded as the "principle", the necessary condition, of mobility. But now it is related directly to the sort of motion that occurs, in the sense that a measure of the matter of a body will also provide a measure for the motion which is natural to the body. Through his use of the phrase, 'quantity of matter', Buridan suggested a connection both with the Aristotelian matter-substrate whose measure would be something intrinsic to the body, and with the Archimedean methods of weight-measurement. The relationship between the two is not yet clear: mass is not yet distinguished from weight, but Buridan has laid it down that there *is* an intrinsic matter-measure which is not identical with weight but is somehow to be measured through weight, and which determines the response of a body to outside moving causes. This is perhaps the crucial moment in the transformation of the concept of matter into a concept of mass. It is essential at this point that mass be taken as a quantitative measure of *materia:* it is this, plus the assumption that *materia* is the proper recipient of impetus, that tied the threads together. Later on, the notion of a "quantity of matter" will turn out to be unnecessary; the concept of *materia* will be dropped entirely from mechanics. Its historical role was to link the notions of weight and of response to impressed motion in such a way that a single intrinsic measure of gravitational and inertial mass would ultimately impose itself.

§ 7 *Matter and Space:*

What Buridan did was to suggest that there is a quantifiable factor intrinsic to every physical body whose measure is a measure also of the body's response to impressed forces. It is called *"materia"* because its measure is proportional to the product of the density and the volume of the body, just as though it were the measure of a kind of stuff, and also because it is an invariant through change, as Aristotle's primary matter was. The relations between weight (or "gravity") and "quantity of matter" are not clear. The former initiates natural motion and can impress an ever-increasing impetus on a falling body. The latter is an index of the amount of impetus a given extrinsic force can impart on a body by way of violent motion. The climate here is still thoroughly

Aristotelian; both the ontology and the methodology are traditional. Only the role played by the *materia* is new.

When Galileo rejected Aristotle's ontology and methodology, the mechanics he constructed sounded like Buridan's in some respects. But impetus is no longer a motive force; a body remains in inertial motion not because the impetus keeps it going but because a counter-force would be needed to stop it. Every body has an "intrinsic and natural quality which makes it resistant to motion", a point Kepler also was to dwell on; the measure of this resistance is to be the same as the measure of the body's tendency to move downwards. Galileo is groping here for the distinction between inertial mass and weight. Matter is seen by him more and more clearly as inertial, as having the property of continuing in the same state, whether of rest or of motion, unless disturbed. But weight is still an intrinsic property, place-dependent, and unanalyzable. The "quality of matter" that appears in his definition of impetus is thus sometimes equated with weight and sometimes (equivalently) with mass. Weight is the force propelling a falling body, and mass is its resistance to motion. If every body falls at the same speed *in vacuo,* it can only be that weight and mass are proportional, in such a way that an increase in one is counterbalanced by the other. But Galileo did not press his law of the free fall as far as this.

With Descartes there was a sharp break in the development of a concept of mass. His identification of matter with quantity allowed him to geometize physics in an attractively simple way, but it made the notion of a *quantitas materiae* meaningless. Matter itself is now a form of quantity, i.e. extension or space. The resistance to motion is measured by the volume of a body only, not by the product of density and volume, as heretofore. This means that geometrically similar bodies ought to react in the same way to the same extrinsic force, which is manifestly not the case. Further, it leaves no room for a coherent notion of density; the old Greek paradox about how a geometrically-defined *plenum* can be compressed is back again.

The utter failure of Descartes' system in the realm of mechanics came from his quantification of matter in terms of volume only. It is irretrievably incorrect to reduce matter to geometrical shapes, as Aristotle had long ago showed against Plato and the Pythagoreans.

This does not mean that matter cannot be quantized, only that it cannot be quantized in terms of Euclidean volumes. The distinction between mathematics and physics comes precisely from the irreducible element of "matter" (in the Aristotelian sense of that term) in the object of physics.

The claim of some recent quantum theorists to have reduced physics to a form of geometry (and thereby somehow "evaporated" matter) is rather misleading. What has happened here is rather that the geometry has been "physicalized", so to speak. It is no longer an account of 3-dimensional volumes; it contains dynamic elements of its own through its definition of world-lines, through the time-change in its curvatures and so on. If the notion of geometry be sufficiently broadened, physics *might* be reduced to "mathematics" and matter to "space", but it should be remembered that the "mathematics" and the "space" this would require could quite as easily and as accurately be described as "physics" and "matter". Not matter in the irreducible Newtonian sense of points endowed with a certain quality; if the field approach works here, it is because the "space" and the "geometrization" involved already incorporate the dynamic features that the Aristotelian matter was intended to safeguard.

§ 8 *The Concept of mass:*

Newton brought together these scattered threads stretching back to Thales, to Pythagoras, to Aristotle, to Archimedes, to Buridan and all the rest. He did it by means of a conceptual system as precarious and delicate in its empirical hold as it was breathtaking in its sweep. By analyzing force in terms of mass and acceleration produced, he was able to separate at last the intrinsic mass-factor from the position-dependent acceleration-factor in determinations of weight. Thus mass can be measured for the first time by means of a measurement of weight. From this point of view, mass is an intrinsic invariant quantitative measure of the response of a body to gravitational attraction. Its definition involves no explicit reference to the concept of matter, although many of its conceptual features are inherited from that source.

Newton himself thought of mass as "quantity of matter"; it was by drawing on the complex history of this latter concept that he succeeded in formulating the three conceptual relations we call Newton's Laws. But when he came to define 'quantity of matter', in the manner of Buridan, as the product of density and volume, the question of circularity to which we already alluded in § 5 immediately arose. It would seem that the density of a body cannot be discovered without calling in some way on a prior knowledge of its weight. If "quantity of matter" were unrelated to weight, as had formerly been the case, the circularity might still be avoided. But now that mass is directly derivative from weight, this is no longer so simple. Since a weight-determination and a volume-determination suffice (where the gravitational acceleration is known) to give both the mass *and* the density, it seems clear that mass cannot meaningfully be defined in terms of density and volume.

An inspection of the *Principia* shows that the "quantity of matter" definition is an isolated formula which plays no explicit role in the Newtonian system. If mass be defined in terms of the three "laws", taken together, there need not be any reference to "quantity of matter" at all. Indeed, it would seem as though the inclusion of this single reference in the *Principia* might be no more than a gesture to the older tradition of a quantified stuff. Newton may have wished to retain the connection between his new term, 'mass', and the older, intuitively intelligible, 'quantity of matter'. There can be no doubt that in the conceptual alchemy that went to the making of Newton's notion of mass, the various strands that we have seen associated with 'quantity of matter' played a definite role. He himself may even have felt that the notion of mass still derived part of its meaning from the matter-substrate tradition directly.

But the business of the *Principia* can be carried on without any allusion to matter or to its quantification. Mass is now on its own feet, able to carry on without its parent. Indeed, the only way matter can be introduced into Newtonian physics is to say that it is "that of which mass is the quantity". But what *is* mass the quantity of? Is it, in fact, a "quantity of" in this sense at all? A measure of the response of a body to moving causes is rooted in *materia* in some way, to be sure. But that it is *the* measure of *materia* is not so clear. There is a complex conceptual scheme at work here; mass is not defined in terms of an operation; the "laws" are not separately verifiable as observed regularities. It is the scheme *as* a whole which works, which is operational.

It is a far cry from the matter-stuff of Thales to the motion-factor of the Second Law. The two ideas have little or nothing in common; they respond to quite different norms of evidence and significance. But the audacious hope that there would *be* an invariant mass-measure intrinsic to every body finds its historic beginning in the equally audacious hope of the Ionians that there would be an underlying intelligible stuff.

Dept. of Philosophy
University of Notre Dame
Notre Dame, Indiana

NOTES

[1] See *The Concept of Matter,* ed. E. McMullin (Notre Dame), essay on "Matter As a Principle," pp. 169-208.

[2] See M. Jammer, *Concepts of Mass* (Cambridge. Mass., 1961), chap. 4, and J. A. Weisheipl, O.P., "Matter in fourteenth century science" in *The Concept of Matter,* 1963.

[3] See M. Clagett, *The Science of Mechanics in the Middle Ages* (Madison, Wisc., 1959), 94.

[4] See Jammer, *op. cit.,* pp. 64-72, for an account of the different attempts that have been made to defend Newton's definition of mass against the charge of circularity.

[5] F. Solmsen, *Aristotle's System of the Physical World* (Ithaca, 1960), chap. 13.

[6] See A. C. Crombie *Medieval and Early Modern Science* (New York, 1959), vol. 2, p. 68.

COMMENTS

Marx W. Wartofsky

The history of the concept of matter, insofar as it concerns the "transition from matter to mass", is a history of the successive attempts at the quantification of matter. Concurrently, it is a history of the conceptual reformulations of what it is that is being quantified, and also of alternative modes of quantification. The problems that McMullin properly raises with respect to the notion "Quantity of Matter" concern a clarification of the concept "quantity" as much as of the concept "matter".

The views of the Ionian *Physiologoi* already involve fundamental quantificational notions, the paradigm of which is the so-called many-one problem. One could say that this problem is at the conceptual heart of Ionian natural philosophy. The objection that the concern here is really a qualitative, not a quantitative one, i.e., a concern with unity and plurality, continuity and discontinuity, with qualitative change and becoming–already presupposes some favored sense of quantification; I take this to be an unnecessarily restrictive view. Parmenidean "quantification" develops as a two-valued logical scheme, with Being and non-Being as the values, and with the first clear (though only implicit) use of the "laws" of contradiction, identity, and excluded middle. The Pythagoreans founded quantification in terms of number, explicitly; the atomists emphasized quantification in terms of shape, size and motion. In short, Pre-Socratic philosophy already exhibits *alternative* modes of quantification, often fused and not separated from each other in any distinct way. In fact, the intuitive and speculative richness of Pre-Socratic philosophy owes much to this very indistinctness, for the heuristically significant free-play which it permitted, and for the range of conceptual problems which it raised for later philosophy and science.

A schematic reconstruction of the *sources* of these alternative modes of quantification might look something like this:

(1) *Natural Language,* with respect to concepts of substance, predication, and of a two-valued logical scheme.

(2) *Counting,* with respect to arithmetic concepts of number, element, and additivity.

(3) *Spatial Perception,* with respect to geometric concepts of shape, size, motion, extension, divisibility.

The logic of these alternative conceptual (and practical) frameworks delimits the quantificational scheme in each case. The evolution of quantificational concepts is, in effect, the discovery of rules by means of which transformations or reductions from one quantificational "language" to another may be carried out. This evolution, of course, is still proceeding. But the accomplishment of Greek thought was to begin the analysis of the indistinct, intuitive fusion of these alternative schemata and to distinguish them one from the other.

It seems to me that such a sketch clarifies some major questions concerning the quantification of matter. For example, if matter is taken to be the subject of predication, on the one hand (in terms of a logic derived from natural language), and at the same time is taken as the divisible, space-occupying stuff, whose parts can be counted, then there is bound to be a conceptual snarl. In the first case, "matter" is the semantic referent for talk about change, as McMullin points out; it is required for making sense, i.e., for talking about *what it is that changes.* In the second case, "matter" is not a "what" which changes, but a combinatorial concept, a sum of elements, and "change" in this framework can only be taken as the additive combination, and dissolution of elements. The point is this: that quantification of matter means quantification in *some framework* of quantification, which needs to be specified. For the notion "*the* (i.e., the one and only) framework of quantification" is otherwise ambiguous.

This is nowhere clearer than in the passage from the *Physics* in which Aristotle talks about two alternative senses of the term "matter".

> "We must distinguish a sense in which 'matter' is destroyed or is produced and a sense in which it is not. As that 'in which' [there is a privation], a material really ceases to be; what ceases to be is the privative aspect of it; but as a potentiality matter itself does not really pass away, but must be indestructible and unproducible. If matter were produced, some first constituent would have to be present out of which matter would arise; but to be such a constituent is matter's own nature. Were matter produced, matter would therefore have been before it arose! But 'matter' is by definition the 'first' persistent being out of which anything arises. . . . So too, if matter were destroyed, it would pass into matter in the end; hence, matter would have perished before it perished!" (*Physics,* I, 9, 192a 25 ff.)

In effect, "prime matter" is a logically and semantically necessary prerequisite for any talk about identity through change, for only an impredicable "matter", as the persisting subject of reference, would permit one to make sense of this identity. As McMullin has pointed out, in another place,[1] this "matter" is a conceptual and metaphysical *principle,* and not a *stuff*. The dilemma that McMullin alludes to concerning "conservation of matter" is resolved, if the alternative senses of "conservation" are made clear. The dilemma is this: "prime matter" is "conserved" (neither produced nor destroyed), but by the very definition of prime matter, it is impredicable, so that quantity may not be predicated of it. There is then no sense in which a "quantity of matter" can be conserved since there is no measure of a *how much*. I think the "tension", as McMullin puts it, rests on the ambiguity as to which of several alternative quantificational schemata is being appealed to, to make sense of this "conservation". That Aristotle does not mean conservation in terms of *number* is clear, since this presupposes a sum of elements which constitute prime matter; this is disallowed by the definition. Nor does he mean conservation with respect to *form,* or some *morphic* self-identity, since this is to predicate actualities of prime matter, which again is disallowed by the definition. The alternative, in my sketch, is the sense of quantification derived from natural language. "Conservation" here means the preservation of referential identity. And Aristotle's argument is in the form of a *reductio ad absurdum,* i.e., the conclusion is in the form of a contradiction: "matter would therefore have been before it arose". Matter is "conserved" (as "prime matter") neither in the arithmetic or geometric sense, but in the semantic sense of referential identity.

Thus, it is not "conservation of matter" *as such* which presents the dilemma of an unmeasurable conservation, but the attempt to impose the requirements of *one* quantificational sense of conservation on a different sense of conservation.

If we talk about conservation of *mass,* we are talking about conservation in the framework of *laws of conservation,* which are mathematical expressions, interpreted in terms of numerically measurable properties. We have abandoned the *principle* of conservation of matter, as a semantic requirement, or as a *philosophical* principle [2] for the *laws* of conservation, where these latter are interpreted in terms of operations of measurement, expressible in numerals (or in some simple order isomorphic with numerals).

This sort of quantification becomes preeminently the one represented as the classical Newtonian concept of mass. Its operational interpretation is a measurement operation, i.e., the empirical determination of quantity by means of fundamental measurements, e.g., of weight, of distance, of time, of direction of motion. What we are ostensibly "counting" here are *units* of weight, or of distance, etc. With respect to mass itself, we are not counting units, but our mass numbers are *ratios* (of a given mass to a unit mass) which are themselves determined in terms of the fundamental units of measurement. Thus, what underlies, or apparently underlies the concept of matter *as mass* is not a notion of quantity of matter, but of ratio-invariance among certain properties of a given physical body, i.e., with respect to measurements of the properties of weight, or acceleration, etc. In other words, the determination of a mass-number requires specification of a certain frame of reference, and of a scale of measurement with certain formal properties. A mass-number is thus *not* a count taken of entities of a certain ideally homogeneous unit-magnitude, but is rather a mathematical function of certain measurement-statements, systematically related in a complex theory. It is no surprise then that *mass* in this operational sense is not to be taken as itself a "measure of" anything else, e.g., of quantity of matter. McMullin is clearly correct in this conclusion, for once mass has been divorced (or to follow the "parent" metaphor, has had its *umbilicus* cut) from its conceptual origins in transcendental, non-operational notions of quantity of matter, its independent status is assured.

Now, having said all this, I would like to question it. For this represents a view of mass which is reductionist in the older positivistic, eliminative sense. It is a reduction of the concept of mass to its operational definition, in terms of measurements of weight, acceleration, etc. And this is the basis for the charge that Newton's identification of *mass* with *quantity of matter* is not only gratuitous, but involves circularity. It is true enough that the definition of mass does not require any notion of quantity of matter, and that "the business of the *Principia* can be carried on without any allusion to matter or its quantification", as McMullin says, *if* "the business of the *Principia*" is taken strictly in terms of its systemic coherence, its formal consistency and its empirical or operational interpretability; that is, if one is interested in no more than a system of laws, and not in their theoretical explanation.[3] The charge of circularity against Newton is traditionally the hallmark of positivist criticism, from Mach and Volkmann to the present.[4] Its basis is the rejection or elimination of the theoretical or conceptual model of mass, in favor of its operational definition, on the grounds that the former has no physical significance.

On this view, one is still permitted the use of "transcendental" terms like mass, if they are in turn intra-systemically defined in such a way that they reduce to operational concepts and their relations (the laws of the system). But Newton perversely persists in talking of *mass* as *quantity of matter* and Definition I of the *Principia* says that this quantity is "the measure of the same, arising from its density and bulk taken conjointly". The charge of circularity supposes that the "density" is the culprit here, since in order to establish density, one has two alternatives, apparently: either to explicitly measure a dynamic property, e.g., weight, and to define density in such a way that it is a derived measure proportional to this property; or to "count" presumably homogeneous entities per unit volume, so that the number n of such entities in specific bodies of unit volume gives us a simple order of specific densities. The second turns out, according to the criticism, to be a "hidden" version of the first, since the homogeneity of the entities itself needs to be established, and this in turn requires a measurement operation, e.g., weighing. Since weight is proportional to mass, the definition of mass as quantity of matter involves the circularity that one of the terms of the definition of quantity of matter itself involves the operational definition of mass in its determination. The circularity depends on the identification of quantity of matter and mass. If it can be shown, however, that "mass" has *two* senses here, then it can be shown that what appears as a circularity does so because of the confusion of two different quantificational schemes, and of two levels of quantification, one in terms of a theoretical or conceptual model, the other, in terms of empirical measurement.

That physical bodies have weight which is proportional to mass, and varies with the gravitational constant, that mass is an "intrinsic" invariant property of bodies, and that the mass-ratios of bodies can be established by measurements, e.g., of weight, acceleration, etc., is not in itself an *explanation of mass,* but an *explanation of variations in measured properties in terms of mass,* or more exactly, in terms of mass-invariance. But Newton's definition of quantity of matter is not an alternative, gratuitous and, what is worse, circular definition of mass, in terms of an operationally conceived density. Rather, it is a theoretical explanation of mass-invariance in terms of a *model* of this invariance. This involves "density", not as it is quantified in terms of derived empirical measurement, but rather in a way which explains mass-invariance, and mass-ratios in terms of theoretical entities which are *postulated* as ideally homogeneous units. (Units of "matter", or of anything else taken to be the "matter" referred to as the object of the transitive verb "to count".) It is not an operational model in terms of measurements made, but a theoretical model in terms of what it is that the measurements are made of. Thus, the postulation of homogeneity is certainly "circular" in an odd sense, but not circular in the sense that Newton is charged with. It is the same "circularity" which attends *any* measurement operation, and serves as its fundamental postulate: namely, that the units referred to in any measurement are ideally homogeneous, whether these are units of weight, or of time or of distance. In its simplest form, it is the semantic principle of referential identity once again: a way of keeping our meanings fixed, so that one gram is always one gram, although the same object may weigh one gram here, and more than a gram there. This ideal homogeneity of units is itself the formal requirement for any additive system. One cannot, in turn, require that these units themselves be *measurably* equivalent, on two grounds: first, that if they were operationally measurable, in order to measure their equivalence, the same assumption of homogeneity would have to be made with respect to the units of *such* a (second order) measurement, and so on *ad infinitum*. (The argument of using the *same* unit to measure others fails, of course, since "same" would also require an operation of measurement on two occasions, by some ideally homogeneous unit, etc.) The second, and more serious, is that *ideally* homogeneous units are not operationally measurable in the first place. They are *theoretical* entities, and are "measurable" only in a derived and interpreted sense (in Aristotelian usage, only in terms of their *actualities*).

Thus, the condition that the "parts of matter" be identical, and the definition that quantity of matter is "measured" by "counting" such entities constitute a conceptual or theoretical model which embodies a formal condition of any measurement whatever. The sense in which mass may be identified with quantity of matter is the theoretical sense in which mass is taken to be an additive property. What "density" represents in this model is the formal condition of additivity–i.e., the ideal homogeneity of all units. Thus "density", as number of units per unit volume, serves here as a *model,* not as a *measure.* The "counting" of such units is not an *operation* of counting, but rather a postulation of *countability,* i.e., of additivity, a central feature of classical mass. Thus, the "counting" of units, in the density-volume model of mass, as quantity of matter, is not to be confused with the measurement operation of weighing, in the operational definition of quantity of mass. If indeed, the concept of mass, in its operational sense, is independent of the concept of quantity of matter, in its conceptual sense, why then confuse them, in the charge of circularity? To confuse the two, as I think the circularity-charge does, is the same as to demand of a system of weights, that one *weigh* each of the units of weight, to make sure they are all identical, and thus additive. A little consideration will show that this apparently practical and honest-minded demand is "circular" in an instructive way.

Dept. of Philosophy
Boston University
Boston, Massachusetts

NOTES

[1] Ernan McMullin, "Matter as a Principle," in *The Concept of Matter,* ed. E. McMullin, (Notre Dame, 1963), 169-208.

[2] Cf. Ernst Cassirer, *Determinism and Indeterminism in Modern Physics,* (New Haven, 1956), Chapter 5, 45-57.

[3] Cf., on this distinction between "theoretical explanation" and "laws," Norman R. Campbell, *Physics: The Elements,* pp. 120 ff., reprinted in *Readings in the Philosophy of Science,* ed. H. Feigl and M. Brodbeck (New York, 1953), 288 ff.

[4] See Max Jammer s illuminating discussion on this in *Concepts of Mass,* (Cambridge, 1961) 69 ff.

3

WILFRID SELLARS

THE IDENTITY APPROACH TO THE MIND-BODY PROBLEM*

Presented April 10, 1963

1. My primary aim in this paper is to set the stage for a discussion of some of the central themes in the so-called "identity approach" to the mind-body problem. I have particularly in mind Herbert Feigl's elaborate statement and defense of this approach in Volume II of the *Minnesota Studies*.[1] A secondary, but more constructive, purpose is to bring out some of the reasons which incline me to think that the theory is either very exciting but false, or true but relatively uninteresting.

2. I shall begin with a preliminary formulation of the identity theory which will highlight the topics I propose to discuss. Roughly put, the theory claims that what it calls "raw feels"—a technical expression which is intended to cover impressions and images pertaining to the external senses, as well as bodily sensations and feelings in a more usual sense—are identical with "brain states." It hastens to add that in speaking of "raw feels" as identical with "brain states" it does not simply mean that the very same logical subjects which have "raw feel" characteristics also have "brain state" characteristics, or that "raw feel" characteristics do not occur apart from "brain state" characteristics, but rather that the very characteristics themselves are identical. As Feigl puts it, "raw feel" universals are identical with certain "brain state" universals.

* I am grateful to colleagues at the University of Pittsburgh who forced me to clarify key passages in the argument.

3. This rough and ready formulation of what is actually a highly sophisticated philosophical thesis blocks out three topics with which any attempt to assess the identity theory must come to grips. Each of these topics turns out on the most cursory inspection to involve highly controversial issues which are at the very center of the philosophical stage. I shall not attempt to resolve all or, indeed, any of these issues. My aim will rather be to thread my way through them in such a way as to bring out the common ground I share with the identity theory and thus make possible a meaningful joining of issues.

4. It will not have passed unnoticed in this particular climate of opinion that the identity theory as formulated above is committed to the idea that it makes sense to speak of the identity of attributes or universals. This is the first of the thorny topics on which *something* must be said. This may be the place, but it is not the time to develop a theory of abstract entities.[2] I shall simply mobilize some of the pre-analytic strands which any theory must take into account, and develop them in a way which gives the claim that "raw feel" universals are identical with certain "brain state" universals at least the appearance of being in keeping with the spirit of a scientifically oriented philosophy.

5. Universals, then, are a subset of abstract entities. Their distinctive feature is that they are expressed in language by predicates (e.g. 'red') or by predicative expressions (e.g. '3 feet long', 'between red and yellow in color'). I shall say that predicates (under which term I shall usually include predicative expressions) 'stand for' or 'express' universals. Universals may be referred to as well as stood for or expressed. But predicates do not *refer to* universals; indeed, they are not referring expressions at all. Among the expressions which refer to universals, a particularly important role is played by those which are formed from predicates or predicative expressions which stand for or express the universals to which reference is made; thus

Triangularity
Being three feet long
Being between red and yellow in color

6. Universals are public objects. They are identities not only with respect to their many instances, but also with respect to the many minds which think in terms of them, and the many languages which give expression to them. This intersubjective and inter-linguistic character must be accounted for by any adequate theory of abstract entities. Equally important, and even more "platonistic" in tone is the distinction which must be drawn between those universals which have been "discovered" or come to be "known" and those which have not, and, within the sphere of the former, between those which are effectively taken account of by our language, and those which are not. To unpack this a bit, I shall assume that a universal is "discovered" or comes to be "known" in the course of coming to know what *use* a predicate would have to have in order to stand for or express it. The universal is *effectively taken account of* by our language if our language contains a predicative expression which actually has this use.[3]

7. Notice, therefore, that while we can refer to unknown undiscovered universals (I drop the quotation marks from these metaphorically used terms) and to universals which are not effectively taken account of by our language, only universals which are effectively taken account of by our language can be referred to by referring expressions formed from predicates which stand for or express them. Thus, although we can *refer to* the as yet unknown property of persons which would explain their telekinetic powers (supposing them to have such), our language contains no predicate which stand for or expresses this property.

8. Against this background, the following criterion for the identity of two universals can be formulated:

> Two universals are identical if, were a language to contain predicative expressions which stand for or express them, these predicative expressions would either independently have the same use, or one would be a definitional abbreviation of the other.

9. Clearly, much of the burden of the above distinctions is borne by the word 'use' and the phrase 'the same use.' My general

strategy is clear. It is to connect "realistic" talk about universals with "nominalistic" talk about linguistic expressions. My further strategy would be to connect talk about the use of expressions with talk about uniformities in the occurrence of linguistic inscriptions, and, therefore, to build a bridge to "behavioral criteria of synonymy." But that is strategy for a war and not a battle. Here I shall limit myself to pointing out that the patterns of use I have primarily in mind are (1) the reporting or observation pattern; (2) the consequence pattern. The latter is, roughly, the pattern which would find its explicit formulation in what Carnap calls 'transformation rules,' L-transformation rules, P-transformation rules and others. I add "and others" to Carnap's list because it is not clear that it is an exhaustive classification. Other possible candidates are "bridge laws" and "correspondence rules."

10. I pointed out that we can refer to universals for which we have no corresponding predicates. There are two types of case, one of which is, for our purposes uninteresting. Thus there is a sense in which it can be said that there are color universals for which we have no predicates. We can imagine that we had no predicate for the color between red and yellow. It should be noticed, however, that while we might not have had the predicate 'orange' we might well have had the predicate expression 'between red and yellow in color.' And, indeed, for a person to be in the logical space of color is for him to know how to use predicate expressions adequate to the job of *introducing* predicates in the narrower sense, such as 'orange.'

11. The interpretation of statements asserting the identity of universals where the logical space of the universals is in this sense familiar is relatively straightforward. Consider, for example the schema:

The universal which. . . . = the universal which. . . .

We can distinguish two forms the descriptions might take: (a) Each locates a universal with respect to a point outside the logical space of the universal located. Thus,

The color of Plato's beard = the color of your father's moustache.

Here, if we have the relevant information we can go from one of the descriptions either directly to an illustrating name of the universal, i.e. a name formed appropriately from the predicative expression which stands for it, thus

The color of Plato's beard = orange (i.e. being orange)

or to a description which locates the universal with respect to the logical space to which it belongs, thus

The color of Plato's beard = the color between red and yellow

(b) This last identity statement gives us an example of the second form the descriptions might take. In it at least one of them locates a universal in a manner internal to the logical space to which it belongs. From an identity statement of this form, for example the above, we can move, given that we have a predicate—say, 'orange'—which expresses the universal located as between red and yellow, to an identity statement

The color of Plato's beard = orange

in which a *description* is identified with a *nominatum* as in the paradigm of identification.

The author of Waverly = Scott.

If we do not have such a predicate, we at least have the predicate expression 'between red and yellow in color,' and could *introduce* a predicate having the use of 'orange'—which does not mean that the latter would be "short for" 'between red and yellow in color' any more than proper names are "short for" their criteria.

12. But the important case of referring to universals for which we have no corresponding predicates is that in which we refer to what I have called unknown or undiscovered universals. Consider, thus,

(A) The property which an adequate theory of telekinesis—if we but had it—would ascribe to persons having this power.

Contrast this with

(B) The property which the theory (current) of chemical interactions assigns to catalysts.

In (B) it is implied that we have a predicate in our language which stands for or expresses the property in question. Not so in the case (A). There the property in question is referred to by relating it to properties expressed by the predicates of the science of telekinesis at its operational and instantially inductive level. The logical space of the latter empirical constructs is not that of the properties to which access would be gained by constructing a sound theory of telekinetic phenomena.

13. In general, then, the universals which it is the task of theoretical science to "discover" are referred to *via* a reference to the unborn or undeveloped theory, the predicates of which would stand for or express them, and, therefore, *via* a reference to the logical space of the empirical properties of the phenomena to be explained by the theory.

14. Yet the predicates of even sketchily developed theories express or stand for universals. Here it is essential to note that as a theory develops, its predicates cannot, in general, be said to continue to stand for or express the same universals. This brings me to a fundamental point which adds an element of symmetry to our previous classification of universals. To the classification (which highlights the temporal dimension);

(1) Not yet discovered.
(2) Discovered or known:
  (a) Not yet effectively taken account of by our language
  (b) Effectively taken account of by our language

we must now add a third heading under (2)

  (c) No longer effectively taken account of by our language

and a new major category:

(3) Lost, or, so to speak, dis-discovered universals

15. These considerations strongly suggest that the objective or "platonistic" status I am ascribing to universals might be construed in a Peircean way as relative to the continuing scientific community. Thus, if on hearing the above proliferation of universals, one is tempted to expostulate, 'which of these universals really exist!' I would reply by recalling Peirce's characterization of a true proposition as one that the continuing scientific community would ultimately accept—and then changing the subject.

II

16. Now if the claims of the identity theory are placed in the frameword of the above distinctions, it is clear that the theory can scarcely intend to assert the identity of "raw feel" universals with certain "brain state" universals which are effectively taken account of by existing language. For on the above analysis, this would involve that some "brain state" predicates currently have the same use as "raw feel" predicates. And this is obviously not the case. The claim must, surely, rather be that among the universals which would find expression in the predicates of a to be developed "brain state" theory, some are identical with "raw feel" universals.

17. At first sight, this is hardly much better. For, it might be urged, how could any predicates in a "brain state" theory have the same use as "raw feel" predicates? After all, the latter doesn't even presuppose the knowledge that there are such things as brains! But before we take up this and other objections, we must explore the notion of a "raw feel" universal.

III

18. The "rawness" of "raw feels" is their non-conceptual character. The sense in which "raw feels" (e.g. a feeling of warmth) are "of something" is not to be assimilated to the intentionality of thoughts. To say that they are non-conceptual is, of course, not to deny that they can be referred to and characterized by the use of concepts, or even directly responded to by concepts in direct self-knowledge. 'Non-conceptual' does not mean non-conceptualized.

19. The word "feel" in the expression "raw feel" is an extension to all sense modalities of a use of the word "feel" which has its ultimate source in such context as,

(1) He felt the hair on the back of his neck bristle.

In this primary context, 'to feel' is clearly a cousin of 'to see,' and feeling in this sense can properly be classified as a mode of perception. Notice that feeling in this sense *is* conceptual; a propositional attitude. One would, perhaps, be more comfortable about this remark if the example had been,

(2) He felt *that* the hair on the back of his neck was bristling.

The relation between (1) and (2) is an interesting and important topic in the philosophy of perception. I shall simply assume on the present occasion that (1) is a stronger form of (2) which emphasizes the non-inferential character of the experience.

20. Notice that to ascribe a perceptual propositional attitude to a person in the form,

(3) He perceived that–p

is to endorse the proposition involved in the attitude. We can, however, ascribe the same propositional attitude in a non-endorsing way by using such locutions as

(4) He thought he perceived that–p

(5) It seemed to him that he perceived that–p

(6) It appeared to him that–p

(7) He was under the (perceptual) impression that-p

None of these is *completely* neutral with respect to endorsement. They all tend to imply the falsity of the proposition involved in the attitude, and have other overtones which are irrelevant to my purpose. I shall make a technical use of (7) in which it will imply neither the truth nor the falsity of the proposition involved in the perceptual propositional attitude. In this usage the statements,

(8) He was under the tactual impression that the hair at the back of his neck was bristling

(9) He was under the visual impression that there was a red and triangular physical object in front of him

ascribe perceptual propositional attitudes while making no commitment concerning the truth or falsity of the proposition involved in the attitude.

21. Now a classical theme in the philosophy of perception is that

the truth of statements such as (9) implies the occurrence of something which is variously called a '(visual) sensation' (a sensation of a red triangle) or a '(visual) impression' (an impression of a red triangle), where this occurrence is understood to be a nonconceptual episode which *somehow* has the perceptible qualities which the corresponding conceptual episode, i.e. the propositional attitude, takes to be exemplified in the world of perceptible things. Thus, the fact that a person is under the visual impression that a certain stick in water is bent is taken to imply that he is having a visual impression of a bent object. I shall assume that this is true. This does not mean that I accept the "sense datum inference," for it should not be assumed that to have a visual impression is to sense a sense datum as these terms are used in classical sense-datum theories.

22. Notice that visual impressions are classified by the use of the word "of" followed by the phrase which would appear in the statement of the propositional attitudes which imply their occurrence, thus

> Impression of a red and triangular object in front of one

corresponds to

> Impression that there is a red and triangular object in front of one.

The idea that there are such non-conceptual episodes was put to use in explaining, for example, how a straight stick (in water) can look bent, and a red object (in green light) look black. It was postulated that the propositional attitude expressed by

> He is under the visual impression that there is a black object in a certain place

involves, among other things, (a) the occurrence of an impression of a black object, (b) the occurrence of the thought that there is a black object in a certain place, the thought (or perceptual judgment, as it was called) being evoked by the impression. Roughly speaking, *impressions that* were construed as conceptual responses to *impressions of*.) To this was added the idea that while in standard conditions viewing red objects results in an impression of a red object, and viewing bent objects results in an impression of a bent object, in non-standard conditions (e.g. viewing a straight stick in water) the viewing of an object that is not bent may result in an impression of a bent object, and the viewing of an object that is not red may result in an impression of a red object.

23. Although the examples I have been using come from vision, exactly the same distinctions were in the case of feeling. Here "feeling *of* . . .' are the counterparts of 'visual impression of . . .' We can therefore understand the philosophical use of the expression "raw feel" as an extension to all modes of perception of an expression which stands for the non-conceptual kind of episode which explains why a person can be under the impression that he is being pricked by something sharp when in fact this is not the case.

24. I pointed out in 22 that sense impressions or raw feels are classified according to the perceptible qualities which are ascribed to some parts of the world by the perceptual propositional attitudes which they evoke, and which characterize their causes. As I see it, the "of" phrases in

> Sense impression of a red triangle in front of one
> Raw feel of being pricked on the hand by a sharp object

are adjectives which, in addition to classifying raw feels *extrinsically* by their causes and effects, also classify them with reference to their intrinsic character.

25. How are we to understand the intrinsic character of raw feels? Obviously the sense impression of red triangle is not, in the literal sense, either red or triangular; nor is the raw feel of being pricked on the hand by a sharp object a being pricked on the hand by a sharp object. The most that can be said is that the families of qualities and relations which intrinsically characterize raw feels or sense impressions correspond in a certain way to the families of qualities and relations which characterize perceptible objects and processes.[4] I shall return to this point later. For the moment I shall simply say that the logical space of the qualities and relations which characterize raw feels is, in certain respects, isomorphic with the logical space of the perceptible qualities and relations of physical objects and

processes. It would be useful, therefore, to introduce predicates for raw feels which are formed from predicates which stand for perceptible qualities and relations by adding the subscript 's'. Thus a triangular$_s$ impression or raw feel would be one which in standard conditions is brought about by viewing a triangular object and which, *ceteris paribus,* results in being under the impression that a triangular object is before one.[5]

26. It will have been noticed that even my characterization of the intrinsic properties of raw feels has been, so to speak, extrinsic. For I characterized them in terms of their correspondence with the perceptible qualities and relations of physical objects and processes. It might be referred from this that I think of our access to the logical space of impressions as indirect, as based upon a prior access to the logical space of perceptible qualities and relations. I shall postpone taking a stand on my own on this matter, and limit myself for the moment to pointing out that the type of identity theory I am examining rejects this suggestion and insists that our access to the logical space of sense impressions or raw feels is direct and, indeed, is the presupposition of our access to the logical space of physical objects and processes. It insists, indeed, that the qualities and relations of "raw feels" are "directly given" and that physical objects and their properties are "existential hypotheses" whose reality is guaranteed by the fine job they do of saving the appearances.

27. Finally a categorial point about raw feels which is implicit in the preceding remarks. They are construed as "pure episodes" and are contrasted with dispositions and mongrel categorical-hypothetical states. It should be noticed that the fact that one has in some sense "privileged access" to a state of oneself doesn't by itself imply that this state is a pure episode. Children can be trained to respond linguistically to short-term dispositional states of their organism, e.g. anger. Nor, as this point in turn suggests, need "privileged access" be construed in terms of classical theories of the given. The identity theory we are examining, however, is committed to the idea that raw feels are pure episodes and that raw feel facts are "given" in something like the classical sense.

IV

28. Before taking the bull by the horns, a word or two about the other terms of the identities envisaged by the identity theory. It will be remembered that according to the theory, raw feel universals are identical with certain brain state universals. Which brain state universals? Indeed, which brains?

29. For there is, in the first place, the brain as an empirical object to which empirical properties definable in observation terms can be ascribed. Can raw feel universals be identical with universals which characterize the empirical brain? They cannot, of course, be identical with any universals expressed by empirical predicates defined in terms of the publicly observable features of the brain, for raw feels are pure episodes which are public only in the sense that others can infer that which is given to oneself. (What authorizes the inference is, of course, a classic question.) Nevertheless it is important to see that there is a sense in which it is perfectly legitimate to suppose that raw feels *are identical* with certain states of the empirical brain. This, for the simple reason that it makes sense to suppose that they *are* states of the empirical brain. Imagine a person who has been defleshed and deboned, but whose nervous system is alive, intact and in functioning order. Imagine its sensory nerves hooked up with input devices and its motor nerves hooked up with an electronic system which enables it to communicate. Without expanding on this familiar piece of science fiction, let me simply suggest that we think of what we ordinary call a person as a nervous system clothed in flesh and bones. In view of what we know, it makes perfectly good sense to introduce the term "core person" for the empirical nervous system, and to introduce a way of talking according to which raw feels and, for that matter, thoughts are in the first instance states of "core persons" and only derivatively of the clothed person.

30. I submit that in this sense most scientifically oriented philosophers think of raw feels and thoughts as brain states. But while the thesis that raw feel universals *are,* in this sense, brain state universals and therefore trivially[6] *identical with* certain brain state

universals is almost undoubtedly *true*, it is relatively non-controversial and unexciting. If the issue is put in these large scale terms, only those who demand really distinct logical subjects for the sense-impressions and the material states of persons—thus Cartesian dualists—would demur. Aristotelians and Strawsonians could take it in their stride.

31. For the claim that raw feels (or even thoughts) are in this sense identical with brain states simply transfers the episodes and dispositions initially attributed to persons to the central nervous system, now conceived of as a core person. All of the important philosophical problems pertaining to the relation of mental states remain.

32. These considerations give proper perspective to the fact that according to the identity theory in the more challenging form in which it is currently defended, the brain state universals which are identical with raw feel universals, are universals which would be expressed by certain predicates of *an as yet to be elaborated theory* of brain activity. Thus, instead of making the relatively unexciting claim that raw feel universals are identical with certain brain state universals, where this reduces to the claim that raw feel universals *are* brain state universals (i.e. ascribable to brains as core persons) the identity theory in question claims that raw feel universals are not only brain state universals in this unexciting sense, but are identical with certain complex micro-physical universals to be 'discovered' in the course of developing a scientific theory of brains.

33. Thus the question arises, "Is it reasonable to suppose that the scientific study of brains will lead to the discovery of brain state universals which are identical with raw feel universals?" And to this question we are strongly tempted to answer 'No!!' For interpreted along the lines sketched at the beginning of this paper it becomes, "Would an adequate theory of brains contain predicates which had the same use as raw feel predicates?" And the idea that this might be so has a most implausible ring. It will be useful to formulate some of the objections which this idea tends to arouse.

34. The first is that since predicates which would stand for the relevant brain state universals are *ex hypothesi theoretical* predicates, they would not have the avowal or reporting use which is characteristic of some, if not all, raw feel predicates. To this objection the identity theorist replies that once the theory was developed, people could be trained to respond to the brain states in question with the predicates of the theory—which would then gain an avowal use.

35. The second objection is that raw feel predicates do not have a theoretical use, or, to put it in the material mode, raw feels are not theoretical entities. Here the identity theorist might reply that the *other*-ascriptive use of raw feel predicates is, in effect, a theoretical use. The force of this reply will be explored subsequently.

36. The third objection is the challenge "How can a predicate (e.g. 'having an impression of a red and triangular surface') which applies to a single logical subject (a person) have the same use as a predicate which applies to a multiplicity of scientific objects (micro-physical entities)?" The effect of this challenge is to make the point that the identity theory involves not only the identity of raw feel universals with certain brain state universals, but of persons with systems of scientific objects. The identity theorist can be expected to reply that it is enough for his purposes if raw feel universals which differ only in this categorial respect from the raw feel universals expressed by predicates which apply to persons as single logical subjects are identical with certain brain state universals. We shall leave this reply untouched, although we shall return to something like it at the end of our argument.

37. The fourth objection, however, is the most familiar and goes to the heart of the matter. "How," it asks, "can a property which is in the logical space of neurophysiological states be identical with a property which is not?" Otherwise put, "How could a predicate defined in terms of neurophysiological primitives have the same use as (be synonymous with) a predicate which is not?" To *this* question the inevitable answer is "It could not."

V

38. It might seem, as it has to many, that this is the end of the matter. The identity theory is absurd, and that is all there is to it. And, indeed, the identity theory as we have so far described it has no obvious defense against this standard objection. Yet it is not difficult to discern the fundamental strategy of the identity theorist in the fact of this objection. It consists in an appeal to a supposed analogy between the speculatively entertained identity of raw feel universals with brain state universals, and the once speculative but now established identity of chemical universals with certain microphysical universals. The story is a familiar one, and I shall not bore you with the details.[7] The relevant points are quickly made. Suppose U_c is a certain universal which the predicate 'P_c' in the chemical theory current at time T stands for. And suppose that this chemical theory has a degree of sophistication essentially that of chemical theory today, but that micro-physics current at T is rudimentary. An 'identity theorist' puts forward at T the thesis that chemical universals will turn out to be identical with certain to-be-discovered micro-physical universals, i.e. universals which would be expressed by the predicates of a more sophisticated micro-physics. An opponent raises the following objections:

> (1) How can micro-physical predicates which are not tied to Chem Lab observables have the same use as chemical predicates which are?

The 'identity theorist' replies that once the theory is developed, these defined micro-physical predicates are given this new use, and therefore acquire a chemical-theoretical role.

> (2) How can predicates of current chemical theory, which have no definitional tie to micro-physical primitives have the same use as any predicates of future micro-physical theory which will have such a tie?

This objection corresponds to the fourth and most telling objection to the mind-body identity theorist. And once again the objection is, in a certain sense, decisive. But there the 'identity theorist' has available to him a move which is, at first sight, not available in the raw feel, brainstate case. He can argue that *both* of the universals involved in the identification are *to be discovered* universals, the chemical ones as well as the micro-physical ones. Roughly, the identity claim takes the form,

> The universals which will be expressed at T′ by the predicates of a more adequate theory of chemical processes are identical with the universals which would be expressed at T′ by the predicates of a more adequate micro-physical theory.

and while the universals which the predicates of chemical theory current at T express would not be identical with micro-physical universals, the universals which would be expressed by its more powerful successor might be.

39. For just as universals can be 'discovered' and 'given effective expression in our language' by our coming to use predicates in various ways, so universals can be 'abandoned' by no longer finding expression in our language, and even lost. A chemical predicate which at T did not stand for a micro-physical universal may come to do so at T′. And the chemical universal for which it stood at T may be left in the lurch for a more sophisticated face.

40. The situation can be represented as one in which chemical theoretic predicates cease to stand for universals which are merely constantly co-exemplified with micro-physical universals ("bridge laws") and come to stand for micro-physical universals. The identification is *made* rather than *discovered*–though the possibility of identification is discovered.

VI

41. Is anything like this move possible in the raw feel, brain state case? Can the identity of raw feel universals with brain state universals be assimilated to the identity of chemical and micro-physical universals? Can raw feel predicates and brain state predicates be regarded as on the move towards a possible synonymity as was correctly predicted for the predicates of chemical and micro-

physical theory? Summarily put, can raw feels be *reduced* to neurophysiological states?

42. This suggestion runs up against the obvious objection that according to typical identity theories, raw feel predicates, at least in their first person use, are as *untheoretical* as predicates can be. Unlike the predicates of chemical theory, they are not on the move towards a more adequate logical space which they might come to express. Like the Bostonian, they are *there*. This is often put by saying that they "label" directly given qualities and relations.

43. And even if the identity theorist were to hold that the *other*-ascriptive use of raw feel predicates is to be reconstructed as involving something like a common sense theory which postulates them as inner episodes to account for the perceptual behavior (including verbal behavior) of *others,* so that the identification of the raw feel of *others* with neurophysiological states would be immune to this objection, the latter would still apply to any attempt to identify the theoretical 'raw feels' of others with the (supposedly) *radically non-theoretical raw feels of first person discourse.* But no identity theorist worth his salt would restrict his thesis to the raw feels of others.

44. Suppose that at this point the identity theorist switches his tactics to conform to his reductionist strategy and abandons the thesis of givenness. In other words, he now argues that instead of the use of raw feel predicates being a confluence of two autonomous uses, a *self*-ascriptive use in which they "label given universals" and an *other*-ascriptive use in which they can be compared to theoretical predicates, there is one basic use in which they can be compared to predicates in a common sense theory of perceptual behavior; that this is an *anybody*-ascriptive use, and that such avowal or reporting use as raw feel predicates have is a dimension of use which is built on and presupposes this anybody-ascriptive common sense theoretical use. Would not this complete the parallel with the chemistry-physics model? For if both raw feel and brain state predicates are in a suitably broad sense *theoretical* predicates, can we not conceive of a reduction of raw feel theory to brain state theory? Would not this complete the parallel with the chemistry-

physics model? For if both raw feel and brain state predicates are in a suitably broad sense *theoretical* predicates, can we not conceive of a reduction of raw feel theory to brain state theory?

45. If the concept of reduction is construed on the model of the physics-chemistry case, then, as I see it, the answer is 'No.' For reduction in this sense is a *special case* of the identification of universals located with respect to two theoretical structures which are expected to merge. A second alternative must be taken into account. Roughly, instead of the primitive predicates of the reduced theory ending up as *defined* predicates in the unified theory —which is the chemistry-physics case—these primitive predicates could perfectly end up as *primitive* predicates once more in the unified theory. On this alternative, the *to-be-discovered* sense impression universals would be no more complex than the sense impression universals expressed by current sense impression predicates; they would have a different categorial framework, and be nomologically related to (but *not* analysable into) universals expressed by other primitive predicates in the to-be-achieved unified sense impression, brain state theory. While all these predicates—including sense impression predicates—would be physicalistic in a broad sense ($physical_1$) as belonging to a spatio-temporal—nomological framework of scientific explanation, they would not be physicalistic in that narrower sense ($physical_2$) in which to be physicalistic is to belong to a set of predicates just adequate to the theoretical description of non-living matter.[8] As I see it, then, in the never-never land of ideal brain state theory the logical space of sense-impressions would, so to speak, be transposed into a new key and located in a new context. It would not, however, have become internally more complex in the way in which the logical space of chemical properties became internally more complex by virtue of their identification with micro-physical properties. That is to say, there would be no increase in complexity with respect to what might be called the factual content of sense impression universals. Such increased complexity as occurred would be of a logical character. Roughly, the new sense impression universals would be exemplified not by single logical subjects (persons) but rather by a manifold of logical subjects which might be called—

borrowing a term without its philosophical commitments—sensa.[9]

46. But if sense impression or raw feel theory is to merge with brain state theory, the latter phrase must be used in its proper sense of "theory adequate to explain the properties of empirical brains as 'core persons' " and freed from any commitment to the idea that neurophysiological theory is of necessity a theory the scientific objects of which are nerves which are reducible, along with their properties, to systems of micro-*physical* entities in a sense which implies that all the predicates of an ideal neurophysiology would be definable in terms of micro-physical primitives none of which apply *exclusively* to micro-physical systems which are the theoretical counterparts of brains.[10]

47. Thus, if the objects of brain state theory are conceived to be reducible to micro-physical objects (however un-thingish) by an adequate micro-physical theory, the latter phrase must connote *not* 'micro-theory adequate to the explanation of *inanimate* physical objects' (as it often tends to do), but rather 'micro-theory adequate to the explanation of any physical object, animate or inanimate.'

48. It is my conviction that a theory which is to explain the properties of core persons will involve a family of families of predicates which would be a categorial transformation, but not substantive reduction, of raw feel predicates, and which would apply only to systems of scientific objects which are the theoretical counterparts at the most fundamental level of empirical brains. In other words I accept the identity theory only in its *weak* form according for as I see it, the logical space of raw feels will reappear transposed to which raw feels or sense impressions are states of core persons, for as I see it, the logical space of raw feels will reappear transposed *but unreduced* in a theoretical framework adequate to the job of explaining what core persons can do. In my opinion such a theory is not yet even on the horizon.

49. The plausibility of the more radical interpretation of the reducibility of neurophysiology to micro-physics rests on the fact that if one thinks of 'sense impressions' or 'raw feels' as theoretical constructs introduced for the purpose of explaining simple 'discriminative behavior' such as is found in white rats, then one

would indeed find no reason to suppose that the postulated states might not be conceived of as reducible along the lines described in 46. After all, we can conceive of—and even construct—machines which can perform these discriminations. It is therefore crucial to my thesis to emphasize that sense impressions or raw feels are common sense theoretical constructs introduced to explain the occurrence *not* of white rat type discriminative behavior, *but rather of perceptual propositional attitudes,* and are therefore bound up with the explanation of why human language contains families of predicates having the logical properties of words for perceptible qualities and relations. In this radically transformed sense, then, I would defend the thesis of concept empiricism according to which the content and structure of our *concepts* of the perceptible qualities of things is derived from the presence of this content and structure in the world—though I would reject the idea that the derivation involves an 'act of abstraction.'[11]

50. I shall conclude with a brief mention of other facets of the problem. Perhaps the most important of these is the fact that the logical space of the perceptible qualities and relations of physical things and processes on which that of the attributes and relations of raw feels is modeled, is, in an important sense, *closed.* Perceptible qualities and relations are, as the identity theory indirectly acknowledges, pure occurrent qualities and relations. They are neither dispositional nor mongrel states. To say of a physical object that it is red and triangular is not to ascribe a power or disposition to it, though it is, in a very strong sense, to imply that it has certain powers and dispositions. Now it is not the 'internal structure' of the families of occurrent perceptible qualities and relations which generate the demand for theoretical explanation, but rather the nomological structure of the changes and inter-actions of the physical things and processes to which these qualities and relations belong. Roughly, it is not such facts, expounded in a 'phenomenology' of sensible qualities and relations, as that to be orange is to be between red and yellow in color which demand scientific explanation, but rather such nomological facts as that black objects sink further into snow than white objects when the sun is shining.[12] And when physical theory explains the powers

and dispositions of perceptible things and processes by "identifying" perceptible things and processes with systems of micro-physical objects, the "identification" is not to be construed as involving a *reduction* of perceptible qualities and relations to the qualities and relations of scientific objects, but rather as a *correlation* of these two sets of qualities and relations by means of "bridge laws." On the other hand, a realistic interpretation of theoretical entities—such as has been presupposed throughout the argument of this paper—requires that physical objects and processes be reducible to, i.e. *identifiable with* (rather than merely correlated with) systems of scientific objects.

51. But how can it be maintained that perceptible things *but not the perceptible qualities and relations in terms of which all their empirical properties are defined* are, in the strict sense reducible? The answer, surely, must be that once again the chemistry-physics paradigm is inadequate.

52. Scientific Realism maintains the *in principle* replaceability of the framework of perceptible things by a framework of scientific objects which contains highly derived counterparts of the inductively established *causal properties* of the former. But while Scientific Realism grants that the framework of scientific objects also contains highly derived counterparts of the *occurrent perceptible qualities* of perceptible things, it need not and, if my argument is correct, must not hold that these qualities are reducible to, i.e. replaceable by, their counterparts in micro-physical theory —as in the chemistry-physics case. The intrinsic structure of their 'closed' logical space (§ 50) requires rather that they be *relocated.* This relocation involves a simultaneous move on the sense-impression front. For the qualities and relations which are irreducible to their counterparts in the micro-physics of the objects of perception, *are* reducible, i.e. identifiable with, the qualities and relations which, I have contended (§ § 45-8), must be postulated in an adequate theoretical explanation of the nature and function of sense impressions or 'raw feels.'[13]

Dept. of Philosophy
University of Pittsburgh
Pittsburgh, Pennsylvania

NOTES

[1] Herbert Feigl, "The 'Mental' and the 'Physical,'" *Minnesota Studies in the Philosophy of Science,* II (1957), 370-498.

[2] For the main outlines of the view I would defend, see "Abstract Entities," *Review of Metaphysics, 17,* 1963.

[3] For the difference between these two stages compare a newly minted theory which is still but a candidate for espousal, with an accepted theory which is in day-to-day use.

[4] The scholastics took the different, and ultimately unsatisfactory tack of holding that the characteristics are the same, but that the mode of exemplification is different. For an analysis of the Thomist-Aristotelian approach to this question see my "Being and Being Known," *Proceedings of the American Catholic Philosophical Association,* 1960, reprinted as Chapter II of *Science, Perception and Reality,* (London and New York, 1963).

[5] For an elaboration and defense of these distinctions, see Chapter II, sections 22 ff., and Chapter III, section VI of *Science, Perception and Reality;* also "Reply to Aune" in a forthcoming volume of essays edited by Hector Castaneda to be published by the Wayne State University Press.

[6] Compare the trivial move from 'shapes are properties of physical objects' to 'shapes are identical with certain properties of physical objects.'

[7] See my paper on "Theoretical Explanation" in B. Baumrin (ed.) *Philosophy of Science. The Delaware Seminar,* vol. 2 (New York, 1963); also "Scientific Realism or Irenic Instrumentalism" in the present volume, part 7.

[8] For the distinction here drawn between 'physical$_1$,' and 'physical$_2$' see "The Concept of Emergence" by Paul E. Meehl and Wilfrid Sellars in Vol. I of *Minnesota Studies in the Philosophy of Science.* According to the view I would defend, the to-be-discovered raw feel universals are physical$_1$ but not physical$_2$. Feigl, on the other hand, is strongly inclined to argue that they are physical$_2$ and to make this claim essential to any identity theory worth the name. We have had a running controversy on this point since the early days of the Minnesota Center for the Philosophy of Science.

[9] For a development of this point see *Science, Perception and Reality* pp. 100-5, 190-6.

[10] Compare the distinction drawn above (45) between physical$_1$ and physical$_2$ primitives.

[11] For an elaboration of this point see section III of "Philosophy and the Scientific Image of Man" in Robert Colodny (ed.) *Frontiers of Science and Philosophy,* (Pittsburgh, 1962), reprinted as Chapter I in *Science, Perception and Reality.*

[12] Cf. William Kneale, *Probability and Induction* (Oxford, 1949), pp. 78 ff.

[13] I have explored the conceptual problems involved in this 'relocation' of the perceptible qualities of commonsense things in a number of places, most recently in the essay referred to in footnote 11 above. See also the concluding sections of Chapter III ("Phenomenalism") in *Science, Perception and Reality,* as well as my second essay in the present volume.

COMMENTS

ISRAEL SCHEFFLER

I.

I will put my cards on the table right away: Professor Sellars thinks Feigl's "identity approach" is a "theory which is either very exciting but false, or true but uninteresting." I differ on two counts: First, the identity approach, is, in my view, not a theory at all but, at best, a poor prediction in *search* of a theory. Second, I think it is in *any* event uninteresting, in that we cannot decide, in advance, that it will or won't be realized. It follows that both Feigl and Sellars are (in my view) mistaken: Feigl in trying to claim that it *will* be borne out, and Sellars in arguing that it *won't* or *can't*. The issue between them, had best, I think, be abandoned in favor of detailed examination of specific proposals to deal with selected terms or theories about the mind and the body.

II.

That the controversy is pitched at too general a level is suggested by comparison with several other philosophical disputes over reduction. Is mathematics reducible to logic? Is biology reducible to physics? Is ethics reducible to science? Is the physical reducible to the phenomenal? These questions cannot be reasonably approached unless we are clear about the content of the reducing and the reduced branches in question. Even if we *get* clear about the *content*, reducibility is not an obvious matter, as e.g. *Principia Mathematica* shows: In general, ingenuity or discovery are required to set forth specific proposals on reduction which may then be evaluated in the light of available knowledge.

In the present case, neither the so-called "raw feel" content nor the so-called "brain-state" content is specified with anything even approaching the requisite precision. Nor, *a fortiori,* are we given any actual definition or principle of reduction to look at. In Feigl's eloquent account of the matter, "Any detailed account of the $\psi - \phi$ identities is a matter for the future progress of psychophysiological research. But in the light of the scanty knowledge available even today, it is plausible that only certain types of cerebral processes in some of their (probably configurational) aspects are identical with the experienced and acquaintancewise knowable raw feels."[1] As Professor Sellars sums it up, "The claim is, rather, that among the universals which would find expression in the predicates of a to be developed 'brain state' theory, some are identical with 'raw feel' universals." If there is any area in which the predictive confidence of science falters, it is in forecasting its own future vocabulary and theories, for if it could foretell them, it would have them. In any event, even supposing we could *somehow* reliably predict *certain reductive features* of future science, would not the main philosophical job be to formulate and articulate them? The present status of the identity claim thus seems to me to be that of an ungrounded, and, further, an utterly uninteresting prediction, which distracts us from our main task.

III.

Professor Sellars, as I read him, allows that "raw feel" terms may eventually be absorbed as primitives into some overarching micro-physical theory but holds that they will never be eliminable by definition within such a theory. I cannot persuade myself that I have grasped his arguments for this claim, however. He offers as his "fundamental point" that "raw feels are common sense theoretical constructs introduced to explain . . . perceptual propositional attitudes" rather than "white rat type discriminative behavior." The relevance of this point completely escapes me; perhaps he will clarify it in the discussion to follow.

In another place, he argues that "raw feel universals" . . . "cannot, of course, be identical with any universals expressed by empirical predicates *defined* in terms of the publicly observable features of the brain, for raw feels are pure episodes which are public only in the sense that others can infer that which is given to oneself." This argument I can understand but I cannot accept. For it rests on the principle that if a set of primitives is such that each is observable, then every predicate defined on the basis of this set must be equally observable, in the same respects. This principle seems to me false as stated, and no obvious, more restricted and superior version has occurred to me.

Finally, Mr. Sellars seems to rely on a special notion of synonymy underlying his anti-reductionist thesis. His *explanation* of synonymy in terms of identity of use is, in general, obscure, for the individuation of *uses* is no clearer than that of *meanings,* and his references to the reporting pattern and the consequence pattern do not help, for *patterns* are in the very same way obscure.

But it is the *appeal* to some sort of synonymy rather than its *explanation* which concerns me at the moment. The criticism of the identity approach that he thinks goes to the "heart of the matter" is: "How could a predicate defined in terms of neurophysiological primitives have the same use as (be synonymous with) a predicate which is not?" And the corresponding more sophisticated question he offers, in connection with his chemistry example, is, "How can the predicates of current chemical theory, which have no definitional tie to micro-physical primitives have the same use as any predicates of future micro-physical theory which will have such a tie?" These questions he considers to be relevant and indeed decisive in every case but that in which the identity approach appeals to identification of universals as yet to be discovered.

But surely, to require that synonymy be imposed as a *condition* of scientific definitions is to require too much. By the last argument cited we should have to reject as a clarification the Whitehead-Russell account of numbers, for example. And we should, in general, need to give up many alternative systems of definition for the very same concepts. In this vein, indeed, Goodman has argued that even extensional identity is too strong a requirement on definition, for it would, for example, force us to deny that points are definable as certain infinitely large classes of volumes, à la Whitehead, *and also,* alternatively, as certain pairs of lines. Some weaker form of correlation or isomorphism seems quite sufficient for definition, as well as for reduction. But the relevance of *definition* to *identity* must then correspondingly be weakened. In reducing non-phenomenal to phenomenal objects, according to Goodman, the phenomenalist is "proving that for the purposes at hand we need not assume that the less-than-observable region is anything other than the class of observable regions in question. This is quite compatible with similarly proving by means of alternative adequate systems that we could avoid the need for assuming that a less-than-observable region is anything but a complex of some other basic elements . . . the *reductive* force of a constructional system consists not in showing that a given entity *is* identical with a complex of other entities but in showing that *no* commitment to the contrary is necessary."[2]

If my argument is correct, then Mr. Sellars' general anti-reductionist thesis is misconceived. For even if reduction yields no synonymy, it is nonetheless no worse than reductions elsewhere in science, so that his pointing out this defect is itself "uninteresting." It is a separate question whether his case even for non-synonymy can be clarified and sustained.

To sum up, neither identity nor non-identity, *in the sense of* definitional *reducibility* or *non-reducibility,* is something we can, or ought to try to, decide for vague masses of future scientific content. The problem is, for the philosopher as for the fat man, to try to reduce now.

Graduate School of Education
Harvard University
Cambridge, Mass.

NOTES

[1] Herbert Feigl, "The 'Mental and the Physical'" in Feigl, Scriven and Maxwell, eds. *Minnesota Studies in the Philosophy of Science,* vol. II (Minneapolis, 1958), 457.

[2] Nelson Goodman, *The Structure of Appearance* (Cambridge, Mass., 1951), 26. (My Italics.)

4

SIR GEORGE THOMSON

SOME THOUGHTS ON THE SCIENTIFIC METHOD

Presented May 2, 1963

Ladies and gentlemen, I feel very odd and rather terrified because this is the first time I've ever deliberately lectured on a subject of which I am ignorant. I cannot claim any knowledge of philosophy. I've never been taught any, and although I've tried to read a few philosophical books, I don't believe very much of it has stuck. But I should like to say to begin with to the philosophers present that I do read some of the writings of members of their profession, that I'm over-awed by the subtlety and indeed the force of arguments which I am quite unable adequately to criticise and even less able to match. But I think in these problems there is something to be said for looking at them so to speak from the user's point of view. I believe that this is the technique in the salesman's world, and if I can be of any help in giving the user's point of view I shall not have entirely wasted your time.

The scientific method is a thing which practically everyone says is terrifically good. But we rarely get very much further in deciding what it is. And this is not surprising when you consider what it apparently is expected to do. How little really there is in common between the sciences. Some sciences are observational, such as, for example, astronomy, where apart from the moon, no one has yet seriously tried to set into action forces in the objects observed. On the other hand, you have the experimental science of physics, which might in some respects be regarded as rather similar, but in which it is possible to make laboratory experiments under carefully controlled conditions. Some sciences are mathematical like these two, others are only so to a slight degree. Such for example as botany. In some sciences, classification is highly important, and in other sciences it plays a quite secondary role. So it is a little surprising if one method can apply to all of them. And I question in fact whether it really does. What I shall be talking about primarily is the method of physics, but I think most of what I say will be applicable to some other sciences.

Now the first point I want to bring out is the question of what we are really trying to do. From what I've gathered from the philosophical accounts of science, the idea is that the object of science is to do something which I understand you must not call explanation, but for which that is the most obvious word to use. I doubt this. I think that the most interesting and exciting thing in science is the discovery of entirely new effects which no one has ever thought of before. Take for example, the science of electricity and magnetism—it is surely more important to produce a myriad of electrical phenomena from the sparks of the 18th century up to the transistors of the present day, than to understand however carefully and fundamentally just what happens when you rub a piece of amber and it draws up a bit of chaff. As a matter of fact, by what I regard as a very proper irony, remarkably little is known of this process. And I doubt if you will get a very much more illuminating explanation of it than could have been given you by Benjamin Franklin. I do not mean of course that the side of explanation is to be neglected; one wants to be able to understand what one is doing and there are varying degrees of explanation possible. I haven't time to go into this.

If you are interested, and I think we are, in making discoveries, there are two essential conditions; one is that there are things which can be discovered and the other is that there are things which can be discovered (I'm trying to be logical) and are also worth discovering. And this is not obvious. I'm inclined to think that the rise of science on two occasions—one, in the sixth century B.C. and the other in the Renaissance—were partly due to a sudden realisation that the world was larger and more complicated than had been supposed. And I believe in each case this was set off by geographical exploration, by the Greeks' penetration of the Euxine and perhaps also by further opening up of the western Mediterranean, and in the later case by the discoveries of the

Portuguese down the coast of Africa, leading to the discovery of the way to India and of course America.

Now I will come to what I think is perhaps the most important discovery; certainly one of the two or three most important scientific discoveries of the scientific method. This is that you mustn't be too proud intellectually. On the whole it is the *triviality* that is likely to lead you to the really exciting results. Take these examples, the attraction of chaff by amber, the attraction of iron by a mineral from Magnesia and the twitching of frog's legs. Now if with all your knowledge you'd been transmitted back to ancient Greece and in addition had been provided with an ability to speak colloquial classical Greek which probably no one now possesses, you might have been tempted to have had a talk with Plato. Suppose you had tried to persuade Plato that the best route to an understanding of what you considered to be the most fundamental things in the universe was to study these pieces of chaff and these little bits of iron and these legs of dead frogs. You can imagine what Plato would have thought of you. I doubt if he would have taken the trouble to be polite, but he might have been. Anyhow if you had been so rash as to try and argue the matter with him, he would have laughed you to scorn, and his arguments would have been absolutely conclusive to any sensible Athenian who happened to be listening to you. On his facts he would have been perfectly right. It is a great marvel that these trivialities can lead to such astounding results and make science possible. One consequence of this however is that the selection of things to do in science becomes an art. It becomes an art because there are far too many trivialities and only a few of them have promise. What is one to discard? It is a difficulty which any experimentalist is perpetually finding. You are always up against a dilemma that you find some very odd thing which you can't understand, shall you pursue this and really clear it up, or shall you go on with what you want to do? If you pursue it you'll find two or three days later, when you have discovered that it was something really trivial, that something else comes along and so on indefinitely. If, on the other hand, you say "no, I have made up my mind to do *this* experiment and *this* experiment is the one I'm going to do" you may then be in the position of the notorious gentlemen who was engaged in a research with a discharge tube and who noticed that photographic plates in the neighborhood were blackened. So with great common sense he moved the photographic plates further away and was able to continue his research without the disturbance of having discovered x-rays. There is no rule to tell you when to do which, at least none that I know. The art in fact, is to ask the right question and to go on asking the right question and when a righter question comes along to allow yourself to be diverted to it but not to do so unless the question is righter! And for this there is no rule. The same applies right at the beginning of a research in choosing the promising line. Not everybody is so fortunate as to be able to start a new line, but some people are very quick and intelligent in getting onto a discovery whose implication they realise fully. Often the greatest advance in an important line has been made not by the first discoverer but by the men who followed it up, with the greatest ability. Rutherford, for example, did not discover radioactivity, the Braggs did not discover the diffraction of x-rays.

I'm sure that it pays to look about quite a bit to select the right thing. In dealing with almost any problem it is essential to have an hypothesis. The hypothesis that is chosen is very important because if human beings, human minds being what they are, start off with the wrong idea it is extraordinarily difficult for them to get rid of it. Particularly if you've put a good deal of work into the subject—the more work you do the harder it becomes to change the idea. Thus the German physicists Hertz and Lenard worked on the subject of cathode rays and started with the conviction, quite a good guess, that these rays were a form of electro-magnetic disturbances in an aether. And although they made considerable experimental discoveries, they did not see that their discoveries were capable of at least two interpretations and that indeed the other interpretation that the rays were particles was on the face of it more plausible, so they went on convinced that their original idea was right. My father used to say, the great thing is to get hold of the right end of the stick.

But hypotheses always being liable to be false, one might argue that you should not have an hypothesis, that you should simply do experiments and so to speak add the results up and behave as Bacon advised the scientists of his age to behave: deduce consequences by some sort of automatic action on observed phenomena. Well, this in fact doesn't work. And one of the reasons why is that unless you have some idea as to what are the things that matter you cannot repeat your experiments. You see, no experiment can be repeated exactly. There will always be something different. If you repeat the experiment with the same apparatus as the first one, then that apparatus will be a little worn and even if it's only a monatonic layer that has been removed or added, this may make all the difference. If, on the other hand, you supply an entirely new apparatus, well this *is* a different apparatus and how do you know that it is entitled to behave in the same way? Or what about the motion of the planets? Most people in the Middle Ages who were working on the transmutation of the elements would have considered it essential to recall the position of the planets when they did their experiments. I have no reason to suppose that Rutherford did so, in fact I'm practically certain that he didn't! If he had he never would have got them twice in an identical position. He never would have repeated the experiment because the periods of the planets are incommensurable and they therefore never come back to the same position. There are other things, the importance of which may not be obvious. It is a well-recognised fact, you can only do delicate electricity experiments in London on Sunday and the reason is not any anti-sabbatarian behaviour of electricity, but merely because on Sunday the electrical disturbances of the underground are sufficiently slight and infrequent to make it possible to get really sensitive apparatus to work. These are things which you possibly might not think about. And therefore, it is clear that you cannot repeat an experiment in the absolute sense. But you may say that these aren't relevant (e.g. motion of planets). Then, of course, I ask what are relevant phenomena. And this is where you've really got to have a theory. What it really comes to when you say you repeat an experiment, is that you repeat all the features of an experiment which a theory tells you are relevant. In other words, you repeat the experiment as an example of the theory. And as long as you stick to the theory as to what is relevant then you are able to repeat the experiment. But you work only in terms of that theory.

If the experiment turns out different you will be tempted to say that your theory was wrong or inadequate and you will frequently put yourself straight, or hope to put yourself straight, by including some other circumstances as relevant which you had not previously considered. This obviously has some connection with induction, though I'm frightened of using the word. I'm perfectly aware that induction is not logically provable. I wonder whether it is in fact valid. I mean in this sense. Let's take a concrete example. You're testing an engine on the test bed (2,000 rev. per minute or something of that sort) and you've been going on for a number of hours and you've got a million revs, so the same thing has happened pretty accurately a million times. It is reasonable to suppose that unless you cut one of the wires or turn the gas off it will continue to do this. And the longer it goes on doing it the better chance you think it should have of going on a little longer. But people who know about engines know that in fact engines have a pretty definite test life and that something in fact usually goes wrong after say 100 or 500 hours on that particular kind. Your induction will not in fact give you the right answer in that case. There will come a time when the thing will stop. I don't know what one can make of it. I think common sense tells one that if the thing has in fact been running satisfactorily for 20 hours it's rather unlikely to stop in the next ten seconds. But if you were to ask if it were likely to run for another 20 hours I would have to admit that I haven't the least idea. I dare not venture a theory on this, but it does seem to me that in some queer way a kind of time scale comes in, and that what the observation shows is that the phenomenon is permanent for a time scale of the order of hours. You somehow or other guess or hunch that you will not be likely to have great changes in a time that's very small compared to this. But whether this is based on anything other than one's own experience of a comparatively limited number of things, whether this has any theoretical validity or not, I cannot say.

All existing theories, in physics at any rate, are demonstrably invalid unless there's been one published last week which the other theoreticians have agreed has got it! There is no existing theory which doesn't break down at some stage or another, in the sense that it either answers incorrectly or it refuses to answer, a question which it seems reasonable to ask it. In some cases the limitations in certain directions are well-known, are calculable. You can calculate the error involved in using the theory in a particular case. For example, one can calculate what the error is in applying the ray theory of light instead of Fresnel's wave theory. Here is a theory we had known since Fresnel to be wrong and we seemed to know its limitations exactly. But then came the discovery of the photoelectric effect where both theories fail. So there may be not only the limitations of the theory you know of but other limitations of the theory of which you do not know. I really think, and this does seem to me a point which is natural to an experimentalist, that certainty in even these philosophical things is a myth. And I believe the desire for it is a vice. People are certain, but usually when one knows that people are certain one is pretty certain that what they're certain about is wrong. And certainty in any philosophical sense seems to me to be almost obviously absurd. It's contrary to all one's common experience that you can ever hope to foresee everything that can affect the situation. And I do not believe that even if you use the most careful and refined mathematics you get much better. This of course is in line with what Gödel says, but it seems to me that even without Gödel's work a slight knowledge of history would enable one to come to the conclusion that because a thing is even as much proved as Euclid appeared to be for 2,000 years to highly intelligent people, it therefore will continue to appear proved for all time. And until the proofs which are now regarded as valid in mathematics have had their 2,000 years of testing and a good deal more to be on the safe side, I do not see that one is entitled to regard them as certain.

You will recognise that a great deal of what I have to say comes out of Karl Popper, but I have looked at it from a slightly different point of view.

I should now like to say a little about the limits within which one can speak of determinism. According to Heisenberg the behaviour of minute particles, or to put it more accurately, behaviour in which only a limited number of quanta are involved, is to some extent indeterminate. This at least is the orthodox view of physicists though you can find a few who disagree with it. Quite a number of people say that though this is very important in a way it is very much limited as it only applies to things whose size is small. In a sense this is true. I'm not suggesting for a minute that observations on the planets would show indeterminacy—but there are occasions in which quite large scale phenomena might be so affected. Modern technology is capable of amplifying an effect without limit, or rather the only limit is the power available at the final stage. And so you could have a device which depended say on the explosion of a radium atom ejecting an alpha particle within a given interval of time during which a Geiger counter is made sensitive, which would, shall we say, explode a hydrogen bomb in the middle of New York and create a war which would perhaps exterminate humanity. This could be made a thing which would depend on chance in the Heisenberg sense. Another class of cases has been studied by Max Born. Supposing, for example, that you have some gas in a vessel, there will be a large number of molecules all wheeling about. And suppose you introduce a marked molecule; it shall be marked by the fact that it's radioactive. Now ultimately it's going to blow up, and you can arrange to tell where it was when it blew up because you'll be able to see the track. This position is therefore observable. You know where the molecule starts from, you know the speed, and you wish to determine even roughly the place where it eventually reaches the walls of the vessel, where we can suppose it sticks till it blows up. It collides with the molecules of the gas and you must calculate how it bounces off them. All these molecules of course are moving also. Consider the first one it hits. If it hits it a bit to the left it will tend to go off one way, if it hits it a little bit to the right it will tend to go off the other side. But you would try to calculate so cleverly that you would know in fact it's hitting say just a little bit to the left. Well, as it goes on it hits another particle. This by the way is also moving. But you will have carefully observed its movement and you will know where it's got to. But if you progress thus through several billions of trillions of collisions, consider the accuracy which you require even assuming that these molecules behave like perfectly elastic billiard balls; even on Newtonian mechanics you will see that the possibility of making an exact prediction vanishes, or tends to zero. True you could, if Newtonian mechanics were correct, logically say "I know the collision of these things is determined. I have a theory which taught me so; this theory has never been contradicted; and I believe it as an act of faith." This would be a possible view. And we'd then say, the whole motion is determined—though we can't do the sums. But this is where the quantum theory comes in: since there is a limit below which Newtonian mechanics fails, this way out is not open to you. And you therefore get an indeterminacy of a slightly different kind from the normal Heisenberg indeterminacy—an indeterminacy which does indeed depend, ultimately and theoretically, on the quantum theory, but which does not depend on the magnitude h, Planck's constant. One can readily show that even if h were millions of times smaller than it actually is, prediction would still be impossible with a normal gas.

There are other cases in which an indeterminacy arises in comparatively large things from the quantum theory. For example, cases of instability. Supposing that you have a wire which is being stretched; the wire is being deformed with tensions uniformly distributed over it. At which point will the wire break? Well, usually of course it breaks at some weak point, and you must try to find that out and stop it. But let us suppose you've really taken trouble about it and made the wire properly uniform. Then the point at which the wire breaks will depend (since metals get weaker as their temperature rises) on some local "fluctuation" in the temperature distribution in the wire, in the sense in which this word is used in statistical mechanics. And this is by quantum theory, and indeed in practice without quantum theory, quite unpredictable; so here's another kind of unpredictability. Other instabilities of this kind are those that give rise to turbulence. As you know, air flowing over the wing of an aeroplane for instance flows over smoothly in a laminar flow till we get to a certain distance back, and then the motion becomes turbulent. We cannot say exactly where the turbulence will start. In practice I imagine it usually starts by some roughness in the wing. But it's highly probable that if you really wished to do so you could make wings so smooth that in fact the point at which turbulence started depended on some fluctuation in the velocity or temperature of the molecules of air or conceivably even of the wing itself. And such instabilities, if once started, tend to become permanent.

Consider the question "is the weather in long term predictable?". Undoubtedly in a good many meteorological conditions you're concerned with instabilities. A tornado springs up in one place, and I imagine that if a small tornado starts in a certain place it will grow and inhibit the growth of another one a little way away; it will absorb it. So that the exact position may depend on the start of quite a small turbulent effect. And this may well be small enough to come under the heading that I've already discussed. I believe therefore that a very much smaller amount of the world is in any useful sense of the word, and, I am inclined to think, even in a philosophical sense of the word, determinate than one is accustomed to believe. There are other cases, which are perhaps of interest from the point of view of attempts to decide whether living matter is determinate, in which the thing is so complicated that you have got to get so much information in order to have any basis to make your prediction—you need so many initial conditions, so to speak, no matter how good your theory is, in order to get your particular prediction—that the process of getting them would alter the situation they are meant to ascertain and make the prediction impossible in very much the same way as the Heisenberg theory asserts but for much larger bodies.

I now pass on to some of the questions about what you expect of a theory. And here I'd like to draw a distinction between what I call classical theories and what I call active theories. A classical theory is one in which the theory itself is definite and certain limitations at least are known and there may, if you're lucky, be no other limitations than those known. Such a theory is essentially used deductively. It's the kind of theory used in technology. Such a theory for example as the ordinary theory of elasticity. There's nothing mystical about it; you know it isn't right, you can give a pretty good idea where it is likely to go grossly wrong and so on. The other kind of theory is the active theory where you are trying to find out something beyond what you already know. For example, theories which are propounded about elementary particles. Now these are, of course, much more interesting theories from the point of view of a researcher. These are the theories which often introduce a model. This is probably necessary as a help to the human brain, but I say this with hesitation because the capacity of the brain of a good mathematician is something beyond my conception and I may be doing it an injustice. But for most people models are needed to help one to think. Now these models may be of many kinds, and have in the past been of many kinds. Among the first models in my sense of the term are the models used by the Greeks, Eudoxos and Ptolemy, to explain the motions of the planets. These were models based on geometrical considerations. They were based on the view that symmetry was a highly desirable thing, that the heavens were highly desirable and so presumably the heavens were symmetrical–they aren't, but the Greeks tried to make them so. They did it by explaining the actual motions as compounded of the motions of the most symmetrical things you can have, namely spheres. The results were very complicated. I think each of these theories required something of the order of 50 spheres.

The next stage historically, though I don't say mentally, was a mechanical type of theory which is illustrated by the theories of the aether which were constructed in the late 19th century, to explain the peculiarities of light, such things as the laws of Fresnel and so on. These are now quite out of fashion and are derided, but I think they are rather unjustly derided. I mean it's rather foolish to laugh at a bustle because ladies don't wear them now. They were very attractive to the young men at the time, as is clear from reading the novels of the age. And a mechanical theory although it is unattractive to the theoretical physicist now may not be so in another generation. So be careful not to deride things unnecessarily which have been successful in the past. We have now come to a mathematical theory of a rather peculiar kind. By a mathematical theory, I don't mean a theory which uses mathematics, they always do; but a theory based on mathematics as a model.

One kind of mathematics is chosen. Heisenberg for example chose matrix theory, some aspects of which have a similar pattern to some of the physical facts. Each element of a matrix, in the example, lies both on a certain row and in a certain column. Each line observed in a spectrum represents a transition of the atom from a certain initial state to a certain final state. An attempt is then made to find other parallels, features in the mathematics which will suggest physical occurrences. If these are known to happen the theory is somewhat strengthened; if one was not known to happen but is discovered after the theory has predicted it, the theory is very greatly strengthened. The process is tentative, and only deductive in the details. This kind of theory has a faint resemblance to the Greek view.

I listened not long ago to an inaugural lecture by Professor Paul Matthews at Imperial College in which he said early in the lecture that Kepler had recorded in his notebook his excitement and happiness because he thought he had found evidence linking the motions of the five planets then known with the five "regular" symmetrical solids. This was Kepler's idea of a perfect theory, almost too good to be true, as in fact it was! Later on in the lecture Matthews explained some of the theories now used to study the so-called "fundamental particles." Most of them are based on the symmetry properties of various curious "spaces." We seem to have come full-cycle, and why not?

I wonder though if we are not following the wrong Greeks and if Pythagoras has not more to teach us than Ptolemy or Kepler? He stressed the importance of integers; by the study of the sounds of vibrating strings he first solved, experimentally and imperfectly it is true, the first problem in what the mathematicians call "proper values" which are so fundamental in the quantum theory.

The history of the last 70 years of physics has been the history of the progress through physics of an idea the chemists had taken seriously a hundred years before, the idea of atoms and molecules. But the physicist has spread it much further, to electrons, to protons, to neutrons and to quanta. This basic idea, that in the last analysis one is dealing with integral, though often large, numbers of identical or closely similar objects not only underlies the whole of physics and chemistry, except for general relativity, but spreads in a modified form throughout the realm of biology. I need only point to the myriads of cells of each of many kinds, to the hairs of your dog, the leaves of a tree and indeed to the members of the human race itself. This "atomicity" as one may call it, is a fact of the world comparable in importance only with evolution, though some instances of it are so familiar that its strangeness is forgotten.

University of Cambridge
Cambridge, England

5

NORWOOD RUSSELL HANSON

NUMBER THEORY AND PHYSICAL THEORY: AN ANALOGY

*Presented October 29, 1963**

Philosophers of Science make much of the distinction between physics and mathematics, so much that it seems perplexing to read of Galileo, Descartes, Newton, Laplace, Maxwell and Eddington as apparently *not* comprehending the differences. Granted, consistent claims which have consistent negations differ sharply from claims which do not: but in some respects physics and mathematics *are* analogous. By stressing one of these **I hope** to suggest how the physical world could have seemed to some to be only an instantiation of solid geometry and the differential calculus. Are the differences between mathematicians and mechanicians now known to be so great that any attempt to delineate partial analogies must fail? Let's see.

Lagrange and Laplace are positioned within Newtonian Mechanics analogously to the way Hilbert is located within the development of Elementary Number Theory.

Consider three theories, θ_1, θ_2, and θ_3. Suppose someone sought to ground these in a single discipline—θ'. Success would consist in demonstrating that θ_1, θ_2 and θ_3, is each one completely generable from θ'. Allow that θ_1, θ_2 and θ_3 were "properly made" to begin with; they are "well-formed." From a few primitive or unanalyzable concepts, and Formation Rules, Axioms are fabricated. By substituting values for the variable terms in these Axioms one can derive (*via* Transformation Rules), expressions

* This essay was first published (without the technical appendix) in *Philosophy of Science 31* no. 4 (October 1964).

which are "theoremic", expressions which "unpack" from the Axioms.

Now, the *way* in which the values are determined distinguishes formal systems from scientific theories. Within the latter they are reached empirically: the "theorems" at the "bottom of a page" of scientific calculation are interpretable as "observation-statements."

θ_1, θ_2, and θ_3 are assumed, then, to be "well-formed." To render them as special cases of θ', one needs more general deductive techniques *via* which θ' can "envelop" θ_1, θ_2 and θ_3—themselves being now 'theoremic' with respect to the Axioms of θ'. Let it transpire that from the primitives, Construction Rules, Axioms and Transformations of θ' all parts of θ_1, θ_2 and θ_3 can be derived —including what were originally 'primitives' and 'axioms' within these latter. This is what "reducing to θ'" means.

Naturally, θ' must be "well-made" too. Therein lies the rub, the nub and the hub of my position. Independent criteria are necessary for evaluating θ'—criteria additional to the mere fact that θ_1, θ_2 and θ_3 can be deduced from θ'. If θ' were inconsistent, the other three theories would follow by strict deduction. But so would their negations! Consequently, a first criterion might be that θ' cannot be accepted if it is demonstrably self-contradictory; for then θ_1, θ_2, and θ_3 would "follow from" it only in a deductively degenerate sense. This is not the sense invoked when one speaks of theories (e.g., Huygens' Optics and Faraday's Electromagnetics) being "founded in" or "enveloped by" or "deducible from" some more basic theory (e.g., Maxwell's General Electrodynamics).

Granted, a self-contradictory θ' may stand on the same deductive ground as a consistent θ' so far as entailing θ_1, θ_2, and θ_3 goes. But one would like to show also that θ' is itself consistent, complete, non-redundant . . . etc., in short "well-made." *These* are the algorithmic criteria we wish any "more fundamental" theory to exhibit.

From this abstract account consider now the actual development of mathematics. It is known that 150 years ago the mathematical disciplines were several and unconnected. Some geometries had been "reduced" to general algebraic form. But the branches of higher Analysis, wherein $\sqrt{2}$, $\sqrt{-1}$, and π were deployed—these were not seen to have any direct logical connection with the principles of Elementary Number Theory. An "Arithmetization of Mathematics" was achieved by Gauss, Weierstrass, Dedekind and Peano. They showed the intricate concepts within Higher Analysis to be "reducible" to general notions of arithmetic—the ideas of *number, zero,* and *successor,* plus some set-theory [as when real numbers were identified (by Dedekind and Frege) with sequences of rationals]. Peano went further; he formulated 'axiomatically' those propositions ("the Postulates") from which all of Elementary Number Theory, and hence all of higher mathematics, was generable.

But number-theorists perceived that in Peano's θ' the notions of *number, zero,* and *successor* were treated as primitive—hence, not further analyzable. Frege and Russell sought a more fundamental θ'' ('super prime') from which the θ's ('primes') of Peano, Dedekind and Weierstrass could themselves be unpacked. The primitives in *Principia Mathematica* are just those of *negation* and *disjunction* (plus quantifiers of arbitrarily high order). Sheffer went still further; in *his* θ''' ('super-super prime') the *negation* and *disjunction* of PM themselves unpacked from the even more primitive "neither . . . nor . . .".

Studies of fundamental arithmetic θ's loosely grouped to form a discipline: Elementary Number Theory (henceforth ENT). One of the great contributors to ENT was David Hilbert, whose 'program' is representable in terms of the foregoing. If all specific mathematical theories, θ_1, θ_2, and θ_3 . . . are themselves to be "arithmetized" (à la the θ's of Weierstrass, Dedekind and Peano) then some further things must be demonstrated about the ultimate θ' itself.

'Why?' Perhaps the inability to establish θ''s consistency and completeness, only reflects shortcomings in θ_1, θ_2 or θ_3? Were any of these undetectably incomplete then no theory θ' (which encompasses them) could be complete either. Is it not enough that θ''s theorems and the joint set of those of θ_1, θ_2 and θ_3 are the same?

Not quite. It still remains that if ENT were self-contradictory, then all of mathematics would *necessarily* follow. But such an ENT would not then be an advance comparable to the "Arithmetization of Mathematics" in the 19th century—which is not to be characterized as the finding of ever more general contradictions from which mathematics might be unpacked! What Hilbert sought to show was that besides generating diverse disciplines (θ_1, θ_2 and θ_3), ENT (i.e. his θ') could independently be revealed as a consistent, a complete, and a well-made theory.

Classically, when seeking to establish an algorithm as consistent, it is undertaken first to demonstrate that it is *decidable.* A theory is decidable when any expression made in accordance with its Construction Rules is such that either it is, or it is not, a theorem [that is, these Rules, determine what will be acceptable as 'well-formed formulae'; the classical objective is then to demonstrate that there is no well-formed formula f of which it cannot be shown either that f is, or is not, a theorem of the theory θ'.] Thus, every expression "well-made" in θ' must be decidable within θ'; i.e., everything meaningful for θ' must be demonstrably true, or demonstrably false in θ'. That is the classical 'decidability' objective.

Only this can "close" an algorithm: ideally this should be established before seeking the algorithm's consistency. Because, if some expressions acceptable to θ''s Construction Rules were nonetheless *not* decidable—it could not be established that nothing inconsistent lurks within θ'. Perhaps some well-made but undecidable, expression of θ' has (p. $\sim$ p) as a consequence, despite having been unpacked from well-formed prior expressions. One can ensure against this only by "closing" θ' to begin with—by providing a procedure through which everything well-made *via* θ''s Construction Rules is decidable in θ'.

An early part of Hilbert's objective, then—since he aspired for a general consistency proof—involved showing that his θ' (ENT) was *decidable throughout;* that everything well-made according to ENT is provably *part* of ENT. Transformation Rules alone would then reveal any acceptably-formed expression either as true within θ', or as definitely not so.

This sets the stage for Kurt Gödel's work of 1931. His theorem can be encapsulated thus; *no θ' can, if it is consistent, demonstrate its own consistency. Its non-completeness is thus a consequence.* That is, if any system which includes ENT is assumed to embody an effective decision procedure [i.e., a 'mechanical' technique for deciding of any well-made expression either that it is, or is not, a theorem]– then such a system can always be shown to generate inconsistencies. *So* no logical system which includes ENT embodies an effective decision procedure. *Ergo* no logical system which includes ENT can consistently demonstrate its own consistency. Any such system is *essentially* incomplete.

It is not my objective to explore the Hilbert program or the Gödel Theorem. I've sought only to suggest that when one undertakes to "axiomatize" apparently-independent disciplines within one fundamental discipline, θ', further criteria concerning the suitability of θ' must be met. Hilbert sought mathematics' conceptual foundations. Gödel's Theorem shows that not all of what Hilbert seeks can be found, at least not in the unqualified manner typical of early programmatic statements of Formalism. For just assume that it can be found–that the decision-procedure required in a general consistency proof is itself a constituent of ENT: if ENT is then assumed to be complete it is inconsistent, and if it is assumed consistent it is incomplete.

Thus, simply enveloping θ_1, θ_2 and θ_3 within a more "fundamental" θ' may not be enough. It ought to be demonstrable that further things about θ' obtain–the things Hilbert sought to demonstrate of ENT, and which Gödel demonstrated could not be demonstrated.

A loosely analogous development structures the history of physics. This is not strange–given the intimate connections between pure mathematics and fundamental physics. Before the 17th century, physical science was a cluster of apparently independent disciplines. Benedetti, Galileo, Beeckmann and Tartaglia had created *Mechanics* and *Ballistics*. Copernicus, Rheticus, Brahe and Kepler had given us *Celestial Mechanics*. Snell, Descartes and Huygens had formulated much of *Optics*. Crude theories of the *Tides* were available; some *Hydrodynamics* was known. Gilbert's work on *Magnetism* was familiar; and philosophical speculations on *Motion* abounded. But, as with 19th century formal science, these disciplines worked with independent concepts, Principles and inferential frameworks. By the mid-17th century, then, there were many independent scientific theories like our θ_1, θ_2 and θ_3.

But there was no more fundamental discipline, θ', from the postulation of whose primitives and Axioms θ_1, θ_2 and θ_3 could be shown to follow as consequences. There was no Elementary Physical Theory (EPT), θ'' from which e.g., Mass-point Mechanics, Ballistics and Celestial Mechanics could be generated.

Between 1665 and 1687 Isaac Newton created such an EPT. His θ' was presented as the *Principia Mathematica Philosophiae Naturalis.* Here (ideally) was a fundamental "Force-and-Gravitation-Theory" from which most of the previously independent theories of Mechanics, Ballistics, and Celestial Mechanics followed —as observational consequences.

This constitutes Newton's greatness. He was certainly no Planetary Theorist in the traditional sense; he was not a "coordinate-fixer" or an "orbit-shaper" as had been Ptolemy, Copernicus and Kepler. He was not, like Stevin, Toricelli, or Viviani, an experimentalist. He was not, like Brahe and (often) Boyle, a "data-recorder." He was primarily a systematizer; an architect, not a bricklayer. Newton discerned the axiomatic foundations of many of those apparently independent, physical disciplines extant when he was born. His θ'—"Force-and-Gravitation-Theory"—fused several θ_1s, θ_2s and θ_3s . . . so as to form but theorem-subsets within his great EPT built on the Laws of Motion, of Central Forces, and of Universal Gravitation. As to his linking of mechanics and gravitation—some may question the legitimacy of this. Thus Truesdell notes that "90% of today's specialists in this area are as skeptical of Newton's theory of gravitation as Euler was. A working summary, yes, but a physical theory, no." This is a telling observation—but compatible with the *historical* recognition of Newton's theory as *the* great EPT, the θ' of all time. Insert the astronomical values for the variable terms, and Newton's fundamental formulae transformed into general propositions from which could be unpacked myriad predictions of the states of celestial bodies—as well as retrodictions of their past states! Similarly within Mass-point Mechanics and Ballistics.

The initial test of Newton's synthesis was to check whether the already-verified observation statements within "derivative" physical disciplines (θ_1, θ_2, θ_3 . . .) did really follow from calculations in Force-and-Gravitation-Theory, his θ'. As students of Lagrange, Clairaut, Euler, Laplace and Gauss know, the verification of θ' generated exciting moments within Analytical Mechanics. In crisis after crisis, Newton's theory ultimately triumphed, sometimes at the moment it was to be abandoned. True, delicate readjustments had continually to be made (and are still being made). Techniques were subtly refined; the observed values were 'plugged into' the theoretical variables in increasingly sophisticated ways. But Newton's θ'—as developed by Lagrange and Laplace and especially by Euler and Gauss—proved itself, just as did the Weierstrass-Dedekind-Peano "arithmetization of mathematics." Newton attempted a corresponding "mechanization of physics." His *Principia* fused a diversity of physical theories into one well-made unitary discipline. From its assumptions, axioms, principles, rules and inferential techniques, the descriptions and predictions of most apparently independent physical phenomena were generable. Newton had "reduced" gobs of physics to a single θ'—Force-and-Gravitation-Theory. Here was a 'Hilbertian Ideal' in part realized within the theoretical physics of the 18th and especially the 19th century.

Is there anything corresponding to a "Gödelian critique" within later developments of Newtonian mechanics? Not really, perhaps. But there are additional criteria used in assessing the suitability of Newton's θ'—criteria loosely comparable to the desired proofs of consistency and completeness within Hilbert's θ'. Obviously, these additional criteria for EPT cannot be identical with any of those for ENT. Proofs of consistency and completeness define the Hilbert ideal. Such demonstrations are impossible where *any* physical theory is concerned—although many philosophers and physicists continue to speak of orthodox quantum theory as being 'incomplete'—as if it could be something else. A physical theory can never be 'closed' and rebuilt as a self-contained symbolic game, McKinsey and Suppes to the contrary notwithstanding. Further observations must always 'be to the point,' they must be relevant. All observation-statements unpackable from a physical theory must be synthetic—i.e., their negations must be consistent. Their acceptability thus cannot ever be a matter solely of their form. Since a physical theory conveys factual *information,* there *must* be other possible states of affairs which it claims do not obtain. Hence our certitude *vis-à-vis* anything 'at the bottom of the page' in EPT must be different from our corresponding certitude with respect to anything 'at the bottom of the page' in ENT. It is always possible that we made mistakes in our initial conditions, or in recording our observations. The suitability of our values for the variables within EPT is forever time-dependent: better techniques, sharper eyes, finer mathematics—might tomorrow require us to alter the numbers in our computations about e.g., lunar motions; (we now know of 'libration' and define π differently from the ancients). In short, EPT must remain empirically vulnerable always. The suitability of our choices of values for variables within ENT, however, is not time-dependent in any important way. So, whatever may be the criteria with which we assess the further suitability of EPT, they cannot be identical with those Hilbert once envisaged for ENT—abstract proofs of consistency and completeness.

Our inquiry has reached this point: ENT and EPT, since they are like any θ' in that they encapsulate less 'general' disciplines, $\theta_1 \rightarrow \theta_n$, should meet additional conditions *independently* of their encompassing $\theta_1 \rightarrow \theta_n$. What are these? For ENT they would have consisted (ideally) in providing a general decision procedure, a self-administering consistency proof, and (derivatively) a completeness proof; this was the "Hilbert ideal." [Gödel, although initially in search of this same Hilbertian grail, ultimately exposed it as a mirage.] EPT, however, being logically different from ENT, cannot be subject to these criteria—a consistency proof and a completeness proof. What are *its* additional criteria, given that it can deductively generate θ_1 (e.g. Terrestrial Mechanics) θ_2 (Ballistics) and θ_3 (Celestial Mechanics)?

The additional condition to be met by Force-and-Gravitation-Theory is the provision of a *Stability Proof.* What does this mean?

Consider any five 'boring' laps encountered during the Indianapolis Speedway "Big 500" Race. The laps between the 45th and 50th last May 30th were thus describable: once the big cars settled down they ran their circuits at even spaces with engines delivering less than full power. These are the "breathing" laps, the "coasters"!

Suppose an ingenious mathematician designed a calculus faithful to the kinematics and dynamics of such laps. A "Keplerian" geometry of the moving cars could be generated; their angular velocities in the turns, their accelerations on the straightaways, their apparently retrograde motions as viewed from overtaking racing cars—all this might be captured within a rigorous symbolism by our theoretician. His calculus might even permit retrodictive and predictive remarks about the "middle" laps in Indianapolis races ten years ago, ten years hence!

How could physicists be sure, from the *Principia* alone, that Newton's θ' was not merely a calculus like this? How could they be sure that the planets will not simply "shut down their engines" moving into some celestial garage after these few hundred present circuits are completed? What is it about the *system* of objects to which Newton directed attention which distinguishes it from that encountered during the "500"? An earth tremor could halt the latter. Would a "sky tremor" do the same to our solar system? Suppose a comet pierced our planetary array on its way from Polaris to the Magellanic Clouds—would not this disrupt the ordered gyrations of the system, much as the "500" might be stopped by an enormous electromagnet at one of the turns?—one which killed magnetos, unbalanced camshafts, and even deflected the cars?

To this hypothetical taunt the history of astronomers would have answered:

> no—our planetary system is *stable.* Disturbances impinging from without result in perturbations which, owing to the inherent stability of our "quasi-gyroscope," will be dampened and absorbed into the dynamic rhythm of the planetary machine.

[It should be noted that since it might just have made sense to suppose (1) the solar system is a local instability in a stable universe, or (2) the solar system is locally stable in a generally unstable universe—the answer given above skirts a very large-scale commitment. Newton would have accepted that commitment; for him if the planetary system decomposed this would be a *feature* of the degradation of the universe as a whole. Historically, assumptions about the system's stability were wholly general—just as our local assumptions about the roles of the Second Law of Thermodynamics have vast astrophysical consequences.]

But the 'possibly unstable' taunt was haunting!, just as the 'possibly inconsistent' taunt haunted Hilbert. To *presuppose* that our system is stable—a presupposition as common to Aristotle, Ptolemy, and Sacrobosco, as to Copernicus, Brahe, Kepler, and Galileo—this is less satisfying than a proof; just as a *presupposition* of arithmetic consistency is less satisfying than what David Hilbert aspired for, although no mathematician has ever doubted that arithmetic is consistent. We know what unstable systems are like; a spinning top has an equilibrium unlike that of a stationary top balanced on its point. So the 'stability' supposition is not primitive to Newton's theory in that alternatives were inconceivable. Our "Indianapolis 500" example might be generalized; what we observe in the heavens may be but a fleeting configuration of whirling bodies in but a temporary vortical phase of their interminable travels through space.

Newton perceived the problem. He felt obliged to meet those "additional requirements" for his θ', to which he had reduced the independent physical disciplines interlocked within the *Principia.* We considered the objections to reducing all of mathematics to an *inconsistent* θ'. Analogous objections would be lodged against EPT if apparently-distinct disciplines were reduced to a θ' which was not itself demonstrably 'stable'; a theory which descriptively encapsulated but a passing configuration of nearby celestial objects, and which was powerless to prove otherwise would be a sorry θ'. No; Newton feels that his Force-and-Gravitation-Theory is fundamental to the "nature of things." This had to be demonstrated. His θ' had to be more than an ephemeral descriptive calculus; indeed it was soon felt (e.g. by Euler and Gauss) to provide a mechanical *understanding* of what kinds of objects planets, satellites, stars and comets really are! But it was not enough just to *say* this. One ought to prove it! The celestial mechanician should show that, because of the other dynamic properties of the system, it is *in its nature* to be stable; it will persist *ad indefinitum* despite myriad disturbances and shocks from without.

Demonstrating this is the "problem of the proof of planetary stability." It is loosely analogous with "the problem of the proof of arithmetic consistency"—since both proofs constitute "additional conditions" to be met by θ's which systematize less general, and apparently distinct θs, $_1$, $_2$, $_3$,. To stress this has been the objective of this paper. From here on, although analogies abound, divergencies mark developments of ENT and EPT. For although EPT has had its Hilberts, it has never yet had a Gödel—even though the ambitions of early Newtonians for EPT have required more complex, intricate and elusive techniques than the original bold conceptions would have suggested.

Newton could not piece together the required proof. He explicitly leaves it to God's infinite wisdom and omnipotence to keep planetary matters running aright. The Master Mechanician is expected to dampen disturbances so that they never enjoy an oscillatory growth sufficient to disrupt the dynamics of our solar system. Oscillations, remember, can begin with tiny deflections and expand to explosive proportions: slightly unbalanced camshafts have slowly decomposed massive engines. The makers of aircraft instruments devote most of their research to refining "dampening techniques." But in 1687 the theories of small oscillations, and of dampening techniques, had not yet affected the astronomer's quiescence. [Besides, with the Mechanical Philosophy harassing Theology from all sides, what could have been more fitting for a devout physicist than to leave the stable maintenance of our planetary system to the Divine Engineer?]

This was like leaving it to God to ensure that arithmetic—since so much was based on it—was provably consistent and complete. Apparently some learned divines and despondent dramatists, were as appalled by Gödel's discovery as John Donne had been with the Mechanical Philosophy. But, given the historical line from Weierstrass—through Dedekind—to Peano, formal theorists could not let the "arithmetization of mathematics" terminate in appeals to God's algorithmic stewardship. If ENT was to be the universal foundation-stone, then *we* had to know that it was "well-made" for the task. Faith in God's consistency and completeness could never replace a proof of ENT's completeness and consistency. Similarly with the historical sequence from the Parisian Impetus Theorists–through Galileo–to Newton; if the growing constellation of independent disciplines was really to be founded on Newton's θ', this also had to be demonstrably "well-made" for the task. Why 'reduce' θ_1, θ_2, and θ_3 to a fundamental θ' if one had then to assume again what was *already* assumed in θ_1, θ_2, and θ_3. Copernicus, Brahe, Kepler and Galileo all construed the planetary system as stable and conservative, in this reflecting their attitude towards the entire universe. Newton the 'Axiomatizer' initiated a program which, ideally, ought not to have had to assume the same things earlier theorists had to assume. So just as the 'arithmetization of mathematics' was thought by many to be unfinished without consistency and completeness proofs for ENT (*by* ENT), so similarly, the 'analytical mechanization of physics' was regarded as incomplete without a stability proof for EPT and its paradigm subject-matter–our planetary system–the universe *in simulacrum.* Laplace's ideal is expressed thus:

> Astronomy is a great problem of mechanics in which the elements of the motions are the arbitrary constant quantities. The solution depends upon the accuracy of the observations, and upon the perfection of the analysis. It is very important to reject every empirical process, and to complete the analysis, so that it shall not be necessary to derive from observations any but indispensable data.

It is within this framework that Laplace's efforts to write a Stability Proof must be located.

The neglected demonstration occurs in *Mécanique Céleste.* There Laplace argues (in effect) that *if* the planetary system were multiply-periodic–if it were quasi-gyroscopic, then the orbits would tend towards circularity, despite Kepler's discovery that they are all ellipses. Most of the *de facto* orbits are but small deviations from perfect circularity–deviations which could, however, constitute just those small oscillations which grow progressively to disrupt entire systematic configurations. What needs to be shown is that our system is such that its constituent orbits grow more circular and, hence, the system gyroscopically more stable—rather than progressively more elliptical and hence disruptive; in 10,000 A.D. the orbit of Halley's comet (if it still be with us) should be less eccentric than now *if* our solar system is really stable / is really multiply-periodic / is really a Eudoxian-gyroscopic / is really such that all its apparent secular or short-range disturbances are ultimately periodic or repetitive over great spans of time). This is *the* insight of Laplace.

Thus even Kepler noted something wrong with the "coupled motions" of Jupiter and Saturn. Halley suggested an acceleration of the former and a retardation of the latter: in one millennium Jupiter should have been displaced forward by 57′, Saturn aft by 2°19′. Here is the kind of secular-progressive displacement which might threaten the stability of the entire planetary system. Within years such a gyrational imbalance would begin to decompose it. [Imagine a helicopter's rotor-blades, one of which slowly moved further away from another: the result will be a disastrous instability.] Could such progressive displacements be accommodated within Newton's theory? Euler tried in 1748 and in 1752; at one time he even suggested the mechanical feasibility of secular perturbations—those which always 'progress' in the same direction, without apparent resolution or periodicity. In 1763 Lagrange addressed the Three-Body Problem—as first enunciated by Newton. He sought a *general* analytical technique for computing the displacements of any three objects due to gravitational interattractions between them. The usual method was to calculate this perturbation initially for two of the bodies [*via* $F = G\,(m_1\,m_2)/r^2$], and then to add "correction terms" *ad indefinitum* because of the displacement suffered by both due to the attraction of the third body (but ignored in the initial computation.) This general and analytic solution has been as much a Sangreal within EPT as has been a general consistency proof within ENT. Lagrange generated novel methods and applied them to Jupiter and Saturn. He *did* find a secular term for both. But it was too small.

Enter Laplace. He re-examined all the tiny higher-order terms neglected in earlier computations of the planets' mean motions. They all cancelled out—like the printed border on the edge of a dinner plate, one which does not quite 'join up' the first time around, but which would fit perfectly were the pattern drawn around the edge (say) another five times. The *continuous* dislocation of these planets thus had to be *nil*—just as the continuous "dislocation" of the plate-pattern is *nil* on the fifth circuit. Then Lagrange extended all this by demonstrating that the mutual attractions of all the planets *in toto* could not produce any secular progressive changes in their mean *distances* from the sun and the periods of revolution; these latter could be subject to periodic variations only. This was a monumental achievement. The road towards a general stability proof seemed now well paved. For, Lagrange's division between secular and periodic motions had become fundamental to analytical mechanics. Sharply distinguished were those perturbations which were progressive (hence increasingly disruptive of a system's "quasi-gyroscopic" and multiply-periodic attunement)—and those different perturbations which were ultimately repetitive and thus only complex additions to orthodox descriptions of planetary orbits. To reveal all apparently rhymeless disturbances in our planetary movements to be perturbations of the latter kind (aspects of higher-order dynamical patterns), to expose all secular aberrations as being ultimately periodic—this was the essential first step in any general, analytic stability proof. Lagrange made that step.

A successful Stability Proof would thus assume all individual planetary motions to be singly periodic; the motions of the system as a whole would then be demonstrated as multiply periodic, even though a complete cycle might take centuries. Secular disturbances, then, would be only shortrange descriptions, the complete account being so multiparametrically-periodic that the observed motions seem more like arbitrary cosmic whims. Imagine e.g. 6000 spinning roulette wheels with a phosphorescent point on each—all affixed to a lopsided Ferris Wheel, which is itself set up on a railway-turntable. On a dark night the ultimate rhythmicity of such an intricate and complex set of motions may well be disguished by sheer complexity. Who could predict how many times the turntable would spin before all the roulette wheels, and the Ferris Wheel, were simultaneously returned for an instant to their original starting positions? Yet if all these motions were *systematically* connected in some way, such prediction would have to be possible in principle.

With respect to Jupiter and Saturn, Laplace approached their observed accelerations and retardations as if they were just such phenomena—small segments of a long, intricate, periodic cycle. By 1784 he perceived that any five revolutions of Jupiter 'equaled' any two of Saturn; the planets resumed their coordinates with respect to each other, and with respect to the sun-and-fixed-stars, after each two revolutions of the latter or (what takes the same length of time) any five revolutions of the former. After each 59-year period the two planets will always meet again at almost the same place in the ecliptic. Two tiny terms of the third order (usually neglected in ordinary calculations) return in the positional descriptions for all computations involving 59-year repetitions, *and* these terms recur in precisely the same way. If analytically unperceived the continued cancellation of these terms grows into discrepancies between the (unadjusted) predictions, and the observations of planetary latitudes. The "conjunction position" of the planets shifts to other longitudes, as noticed by Kepler, and Halley and Newton earlier. This (apparently) progressive effect reverses after each 450 years! Thus the perturbations, originally thought to be "secular," constitute a protracted rhythmic oscillation having a period of 900 years. With this important calculation all ancient and modern observations of Jupiter and Saturn were suddenly well represented. As Laplace said:

> The irregularities of the two planets appeared formerly to be inexplicable by the law of universal gravitation—they now form one of its most striking proofs.
>
> [*Mécanique Céleste,* p. 324]

In effect, a *general* stability proof for celestial mechanics would demonstrate that every secular perturbation, even those initiated *de novo* by some energy-source from outer space, will ultimately be "dampened" within the repetitive rhythm of some periodic perturbation pattern—some of these requiring millennia for their full cycle.

Naturally, were a supernova to pierce our galaxy near the sun, the planetary system would decompose implosively. The compensatory mechanism Laplace envisages would be insufficient to dampen a disturbance of that magnitude. But all Laplace seeks is what every planetary theorist wanted—a demonstration that 'sufficiently' small disturbances to our planetary configuration will not surge in an oscillatory growth, so as to disrupt the entire system's dynamic geometry. Similarly, a well-designed aircraft instrument will incorporate a "dampening mechanism" such that external shocks of *sufficiently small* magnitude will be absorbed within the device's energy-train without perceptible effects on the accuracy of the readings. The most perfect chronometer, or directional gyro, will cease to function under a steamroller. But the sudden dislocations such instruments encounter during 'normal' use will not magnify during transfer through the internal mechanism of these devices. This is all celestial mechanics, and Laplace, desired to demonstrate of our solar system—that it was a shock-proof machine, not also a catastrophe-proof creation.

This, then, is a major component within the "Laplacean program." Its comparison with a major component of the "Hilbertian program"—the consistency proof—has been delineated. Laplace's stability proof was originally thought to have been wholly successful—just as Hilbert never doubted that his objectives *vis-a-vis* a consistency proof for ENT would be achieved. Gödel formulated the absolute critique of the later. There is no corresponding critique of the former—no sudden death of an ideal! But a protracted history of unsuspected complexities has modified the Newtonian ideal of stability—and the Lagrangian-Laplacian proof of it. This, although in principle an analytical Stability Proof is possible. Thus the ideal of analytical security for EPT never suffered the unexpected blow Gödel dealt to ENT. But it is no longer what it must have seemed to Newton.

The Hilbertian program cannot, as a matter of principle, ever be fully realized; my understanding of Gentzen indicates no rebuttal to this. He so broadens the *idea* of the Hilbertian program that it is far from clear that Hilbert would identify it with his own early pronouncements. The Laplacean program can, in principle, be fulfilled, but then again only in the manner which invites saying that Democritus' ideals have also been fulfilled in the complex lists of Elementary Particles (at least 36 in number) we find today.

The Democritean ideal, the Hilbertian ideal and the Laplacean ideal, if they survive at all now, do so in profoundly altered forms.

Consider what Laplace does in his "proof." I will use the expressions "multiply-periodic" and "quasi-gyroscopic" indiscrimately [even though not every multiply-periodic system is also quasi-gyroscopic]. These are the visions of stability Laplace invokes. A physical system ϕ which is quasi-gyroscopic, or multiply-periodic, will be such that any external shock or disturbance, if less than some given magnitude ε [the value of ε is a fundamental property of ϕ]—this shock will be absorbed, or dissipated, within the periodicities and gyrations of the system.

A *general* stability proof for our solar system as conceived in the 18th century would consist in showing that it is multiply-periodic, or quasi-gyroscopic, in precisely this sense. How did Laplace proceed to set out such a proof?.

Laplace simply *deduces* the consequences of the assumption:

> *The planetary system is multiply-periodic; it is a quasi-gyroscope in the Eudoxian sense.*

It follows from this that small oscillations *will* be dampened, never to grow beyond the capacity of our planetary mechanism to contain them. Growing oscillations would never be observed, granting this assumption; all perturbations would slowly be absorbed into complex periodicities.

We do in fact observe all this. Aberrations originally thought to be initiating a limitless disruption—e.g., the Jupiter-Saturn problem—turn out to be resolvable in quasi-gyroscopic terms. Secularities are *observed* to be but short-range aspects of periodicities.

Do these observational facts confirm the original assumption? In some sense *yes*. This is the "hypothetico-deductive" sense in which *any* assumption is (insofar) confirmed when its consequences are observationally verified.

The ideal case here is where *further* consequences of the assumption (perhaps initially unsuspected) also turn out to square with the facts. The assumption thus gains *independent* and *diversified* support. Thus the logical oddity of Laplace's undertaking: there is nothing unsuspected about the observed periodicities in planetary motions. These regularities are what first led to the conviction in all of Laplace's predecessors that the planetary system was stable! Hypothetico-deductively, then, Laplace's Stability Proof seems degenerate. Observations of those periodicities which led originally to the capital assumption that the system *is* stable—these themselves follow as the signal consequences of Laplace's assumption. In short, the assumed premise and the derived observational conclusion are so intimately connected (semantically) that the proof seems sterile.

Slightly more formally: to consider the solution of $\ddot{y} = -\sin y$, assume y is small. Then $\ddot{y} = -y$, $y = \alpha \cos t + \beta \sin t$

Hence:

$$y\,(t) \leqq [y(o)]^2 + [\dot{y}(o)]^2,$$

so that if y(o) and $\dot{y}$(o) are small, so is y(t).

But every mathematician will remark this as wholly unrigorous. *This 'proof' assumes y to be small at all times in order to prove that in fact it remains small if it starts out small (and if $\dot{y}$ starts out small too).*

Thus to say "the planetary system is stable" just *is* to say "disturbances from without will not in fact disrupt the system; periodicity is the pattern." Observing examples of this latter *will* support the assumption, just as observing men die will support that "all men are mortal." But is the general assumption *proved* thereby? The question is particularly barbed when the observational examples unpacked from the assumption are the very ones which led to the assumption initially. Within the context so familiar to Lagrange and Laplace could anything so "merely empirical" ever conclude with *quod erat demonstrandum?*

There are other consequences of the stability assumption—but they were not readily verifiable in strictly astronomical terms. Hence, in the absence of *independent* support for the stability assumption (i.e., that y remains small at all times, and $\dot{y}$ too), Laplace's argument must be adjudged weak, deductively and hypothetico-deductively. Since his argument is to some extent inductive it cannot be a strict proof; continually confirming the observational consequences one unpacks from an assumption does not strictly prove that assumption. And since these same observational facts are both genetically responsible for, and also inferentially generable from the stability assumption, i.e., that y remains small, this cannot constitute a very confident example of empirical confirmation either.

This mildly negative appraisal of Laplace will be no surprise to anyone who knows something of 18th century mechanics. But the argument schematized above is accepted as almost a dogma of religion by most physicists. (Cf. e.g. Jeffreys' *Methods of Mathematical Physics.*) But it seems suspect in every way.

What would a less contestable form of Laplace's proof have been like? It would have shown analytically that our planetary system, with all its normal motions and unexpected perturbations, is representable not only *now* via the Force-and-Gravitation-Theory embodied in the Newton-Laplace θ'—the system must be proved always to be so representable. Not in the sense that unless it is θ' will no longer apply to it, but rather in the sense that the system is *in fact* such that θ' constitutes its most fundamental description and explanation. What is required is an analytical proof—i.e., an explicit demonstration expressed as analytic functions—that the solar system must be multiply-periodic, that it must be representable as a type of Eudoxian quasi-gyroscope. It must be demonstrated *as a consequence of the* Principia, or else as some needed addition, that nothing short of a celestial catastrophe could possibly disrupt the smoothly functioning mechanism analytically encapsulated within θ'. The bodies composing our planetary group must be shown to constitute a unified mechanism of which it can be proved that all disturbances to the system are 'absorbable' in Force-and-Gravitation-terms.

Consider now what this last request is a request for:

It seeks nothing short of a *general analytical solution to the multi-body problem.* What else could one be seeking here? What else other than a general proof that given groupings of bodies are not only describable *here and now* in Force-and-Gravitation terms, but that nothing *could* occur to make such terms inapplicable? Despite the multi-parametric influence each such body has on all the others (and *vice versa*) we should still be able analytically to generate all future state descriptions for the entire constellation:

that is Laplace's quest. It is one thing to calculate the perturbational interactions between earth and moon in two-body terms, and then to figure in the sun's attraction on both—and then Mars' attraction on all three [each time correcting all earlier computations by redescribing the initial states]. It is another thing to have a general technique through which all future states of planetary objects *en bloc* are describable so as exactly to underwrite an ideal stability proof.

Weierstrass and Poincaré gave their energies to this part of Laplace's program. They succeeded in refining the technique of "successive approximations." In finding values for the variable terms in [$F = \gamma(Mm)/r^2$] one will first determine *M* and *m* as if these were masses of otherwise unperturbed bodies. But weighing *M* and *m* is accomplished only by assuming some value for F. As intimated, both *M* and *m* will be perturbed by gravitational influences of other objects, e.g., their own satellites, asteroids, planets, the sun, comets . . . etc. As one computes the influences of these other objects on our original *M* and *m,* assessing the interactions between these latter will have to be adjusted and readjusted *ad indefinitum.* But so will further adjustments follow when we consider the corresponding descriptive determinations involving the satellites, asteroids, planets, and comets, whose perturbational propensities we *assumed* in order to introduce corrections for *M* and *m.* There is no theoretical terminus to this. One can, however, get successively closer to the "true" perturbational situations within any multi-body complex *via* this technique.

This method is inductive. It cannot generate the Q.E.D. lusted after by Laplace. Although our solar system's being multiply-periodic may become plausible through this technique, such a conclusion can never be "deduced" from a series of successive approximations.

Laplace's "Proof of Stability in the Solar System" cannot be a proof within the orthodox mathematical tradition; not if he assumes on faith just what his algebraically less-sophisticated predecessors had also assumed on faith—viz., *that the solar system was stable.* To assume as he did that a system is multiply-periodic just *is* to assume its dynamical stability. That we can then unpack a cluster of observation-statements corresponding to the facts—this does not decide whether our observations of planetary motions are of an ephemeral conspiracy amongst 'celestial wanderers,' or of the workings of the stable machine our planetary system actually may be. Since our astronomical observations are compatible with *either* hypothesis, they cannot *prove* one of them, as against the other—especially when one's "proof" begins with the very assumption one seeks to establish, viz., y(t) will remain small.

What then, since Laplace? Newton had demanded that the planetary system be "multiply-periodic, or quasi-gyroscopic." As I've argued, he *assumed* this for θ' in order to prove it. Distinctly unconvincing. In Celestial Mechanics today this would be felt to be unrealistic. Just because Laplace used trigonometric functions in his long calculations, this does not mean all his successors had to as well. They did not.

There are scores of Stability-concepts, somewhat weaker than what Laplace wanted, whose proofs (for our solar system) are quite amenable to sound mathematics. With respect to the algebra set out earlier, mathematicians have long since found rigorous proofs which require no explicit solution, and apply to any system of the form:

$$\dot{y} = f\,(y,u)$$
$$\dot{u} = g\,(y,u)$$

in which f and g are analytic functions. In general, these, proofs are such that *their solutions depend continuously on the initial values.* Indeed, a most elegant alternative to the technique of successive approximations is given in § 147 "The Question of New Integrals" on p. 268 of Moulton's *An Introduction to Celestial Mechanics* (2nd ed., New York 1914)—and there have been other attempts at analytical solutions to the n-body problem, both before and after Moulton.

It has sometimes been urged that there cannot be a strictly analytical solution to the n-body problem. There is extant, so far as I know, no demonstration of the *non*-existence of such a solution; work such as Moulton's drives one towards the opposite decision—even though the last word has not yet been said. As for the notorious Poincaré-Painlevé theorem—this refers to solutions analytic in the *masses.* The theorem reads: There is no strict solution to the n-body problem which is set out as analytic in the masses involved. But Siegel has remarked that this conclusion has no relevance for the existence of solutions for particular masses (like planets). Moreover, the Poincaré-Painlevé computations are purely formal—as is the more recent work of Cherry. Painlevé's effort is indeed no more than a programmatic abstract. There is no sound proof known for this much-publicized theorem.

So, Laplace's ideas *vis-à-vis* stability in EPT have succeeded, or at least they have been constructively transformed—in a manner distinguishable from what transpired in ENT. Nonetheless, success or failure aside, the *occasion* for a Stability Proof in the former case and a Consistency Proof in the latter arose in that EPT and ENT functioned as θ' for other 'derivative' disciplines. Thus the loose analogy which structures this paper.

APPENDIX

Demonstrations of the stability of dynamical systems have been fraught with perplexities, despite all the elegance. If one has a set of simultaneous linear differential equations for n variables, and then eliminates these equations in succession so as to favor one variable, the process can terminate in a differential equation for that one. Substitute $e^{\delta t}$ for it: the result is then an equation for δ; if there is a repeated root then there will be a second solution for $te^{\delta t}$. Should such a thing happen in the theory of small oscillations it would transpire that a repeated value of δ will lead to terms of the form t cos κt, t sin κt ($\kappa = i\delta$) and, save for specially designed initial conditions, *a small oscillation will grow indefinitely*. This is another way of putting Newton's and Laplace's problem: the planetary system is always representable as such a set of simultaneous linear differential equations. What is to prevent a small external disturbance from provoking a small internal oscillation of this sort (as when a rock in the road is responsible for an oscillatory vibration in a car's speedometer-needle) such that the latter grows to disruptive proportions? This has never been found to happen in our system. But *could* it? Newton felt that the Deity would not allow it. But, if it did come about it would contradict the fundamental principle that if the planetary system's potential energy is a minimum in the position of equilibrium (the Lagrange-Laplace assumption), and the initial displacements and velocities were sufficiently small (but not zero!), then there will of necessity be a limit that no displacement can ever exceed. Otherwise engineers could, by clever constructions, 'feed' small oscillations into a suitably designed mechanism generating thereby *any* amount of oscillatory energy at the terminus of the power train—a conclusion which conflicts with every Conservation Principle known. Whatever it is that prevents the construction of a *perpetuum mobile* (of the First Type) also prevent extra-systematic disturbances from generating intra-systematic dislocations any greater than the original disturbance. But what then of the theory of small oscillations which allows for progressive enlargements of oscillatory amplitudes?

Laplace was deeply troubled by this. Routh ultimately found a resolution in 1877 [*Stability in a Given State of Motion;* and compare Heaviside, *Electrical Papers* I, p. 529]. If the system in question is not dissipative and the roots are unequal, the zeros of the minor of any element in the leading diagonal separate those of the original determinant [Cf. Weierstrass, *Monatsber. d. K. Akad. d. Wiss.,* Berlin (1858): *Werke* I, 233-46. Cf. also Lamb, *Higher Mechanics* (1920) 222-26 and Bromwich, *Quadratic Forms* (1906)]. If, then, the determinant has a factor $(p^2+\alpha^2)^k$, every first minor contains the factor $(p^2 + \alpha^2)^{k-1}$. When we evaluate the contribution from the initial conditions to the operational solution

$$[\text{i.e. } X_m = \frac{E_{rm}}{\Delta} a_{rs} (p^2 u_s + p v_s)]$$

a factor $(p^2 + \alpha^2)^{k-1}$ will cancel out. We will be left only with a single factor $(p^2 + \alpha^2)$ in the denominator. This will happen with *every* repeated root; the interpretation will contain *only* terms of the forms $\cos \alpha t$ and $\sin \alpha t$. Thus varying u_s and the v_s will alter ratios of the coefficients of these trigonometric factors for different coordinates; but they will never introduce terms like $t \cos \alpha t$ or $t \sin \alpha t$—the trouble makers.

Consider now not a *general* initial disturbance, but one confined to the period equation. Suppose the equations of motion are these:

$$\ddot{x}_1 + f\dot{x}_1 + c_1x_1 - b\dot{x}_2 = 0$$

$$\ddot{x}_2 + f\dot{x}_2 + c_2x_2 - b\dot{x}_1 = 0$$

[where $c_1<0$, $c_2<0$, $f\geqq 0$, and b is large enough to ensure stability when $f=0$]

Assume solutions proportional to $e^{\delta t}$, must then satisfy

$$\delta^4+2f\delta^3+\delta^2\ (c_1+c_2+b^2+f^2)+f\delta\ (c_1+c_2)\ +c_1c_2=0,$$

from which it follows at once that

$$\Sigma\delta=-2f\leqq 0$$

and that

$$\Sigma 1/\delta=-f\,(1/c_1+1/c_2)\geqq 0.$$

The roots for $f=0$ may now be chosen as $\pm\ in_1$, $\pm\ in_2$ [where $n_1>n_2$]. For $f>0$ (but very small) let the roots be $\pm\ in_1-\alpha_1$, $\pm\ in_2-\alpha_2$, to order f.

Then:

$$\alpha_1+\alpha_2=f>0$$

and

$$\frac{1}{in_1-\alpha_1}+\frac{1}{-in_1-\alpha_1}+\frac{1}{in_2-\alpha_2}+\frac{1}{-in_2-\alpha_2}\doteqdot-\frac{2\alpha_1}{n_1^2}-\frac{2\alpha_2}{n_2^2}>0$$

Such inequalities are consistent only if α_1 and α_2 have opposite signs: in fact, since $n_1>n_2$, $\alpha_1>0$ and $\alpha_2<0$.

Hence any system which is kept stable by gyroscopic action *only* is such that small friction will always produce instability. The fastest free vibrations will be damped, but the slower ones will increase in amplitude with time (to achieve which the system's 'gyroscopic energy' must be rechanneled into the growing slow oscillations, upsetting thereby the system's 'gyroscopic stability').

If, in the case of our planetary system, we postulate a near-gyroscopic configuration, and neglect friction altogether, then all the roots are purely imaginary: hence the system is stable! If $c_{rs}x_rx_s \geqq 0$, in the presence of very little friction $(a_{rs}\dot{x}_r\dot{s}_s + c_{rs}x_rx_s)$ will decrease and all oscillations will be gradually damped down! Such a system will then be called *secularly stable*: as with our planetary system, imbalance-oscillations will become periodic over long periods, and their amplitudes will decrease.

But if the quadratic form is not ≥ 0 and the system is kept stable only by the gyroscopic terms, the slower oscillations must increase in amplitude, leading ultimately to profound changes in the character of the overall motion. This kind of system will then be *ordinarily stable,* but *secularly unstable.*

Newton's worry consisted in having no argument against this latter contingency in the case of our solar system. Laplace's triumph consisted in his recognition of the form of the problem. To prove the planetary system stable, then, he had to assume $c_{rs}x_rx_s \geq 0$. And he had to postulate but very little friction in the system. If either of these suppositions are challenged, he must then infer that the stability of the solar system is not the result of gyroscopic terms merely—that forces other than gyroscopic determine the stable cohesion of the system.

This kind of argument is methodologically fascinating. In a way Laplace, to the degree that he reasoned thus, has not really proved anything about our planetary system as it exists *de facto*. Rather, he has deductively unpacked the semantical content of the suppositions: "Grant that the planetary system is quasi-gyroscopic, in balanced 'vortical motion' around its own freely-moving (but physically unoccupied) center of mass; and grant that very little friction is operative in the configuration; and grant that in its trigonometric description $c_{rs}x_rx_s \geq 0$." From this springboard one easily leaps to the conclusion that all oscillations will gradually be damped down, even those initiated by the sudden introduction into the system of quite violent and dislocating energy spasms from 'outside.' Any oscillation that does not thus dampen down was too large to begin with!

The reasoning is magnificent: *if* our system were as supposed, *then* it would be secularly stable!

But for what reasons ought we to grant such suppositions about our planetary system? If we simply argue that by doing so a proof of the system's stability follows, the reasoning is not only circular, but viciously so. This would give us no reason for being convinced of systematic stability other than a merely hypothetical kind—and no reason for making our 'gyroscopic' assumptions other than that the consequences are reassuring. Newton did as much by leaving it all to God!

No, we need some independent argument either in favor of the system's quasi-gyroscopic nature, or else in favor of the system's stability—from either of which we can then infer to the other! *But we cannot at once assume both what is to be proved and also that from which the proof is to proceed.*

What other reasons are there, therefore, for supposing the initial suppositions about our planetary system to be fulfilled? The answer is clear: we must find independent *theoretical* data from which it will appear that our planetary system's gross motion is comparable to something quasi-gyroscopic, e.g., something like our 6000 epicyclically-mounted roulette wheels on a Ferris wheel, itself mounted on a railway turntable, which in its turn . . . etc. But to show the latter really is gyroscopically stable (if it is) one must undertake an immensely complex computation, determining the force components of each roulette wheel, and the nature of the many dynamical systems it forms with other roulette wheels to which it is mechanically linked, and the relationships obtaining between the mass-centers of every such roulette-couple, and then the compositional effects of all these on the gross gyroscopic balance of the Ferris wheel, which in its turn must be . . . etc. Clearly this is the multi-body problem in gyroscopic-engineering terms. But this is precisely the kind of thing for which Laplace must find a general analytical solution in gravitational-dynamical terms if he is ever to proceed with his further deductions.

To show the planetary system stable in such terms, then, one must first address the multi-body problem. Since a general solution to this problem is not yet indubitably secured, the 'gyroscopic' springboard assumed in any practical demonstration must be pieced together by successive approximation technique—the ultimate result of which can never be a *quod erat demonstrandum* of the Euclidean type.

Ergo, the proof without which Newton's synthesis must be adjudged seriously incomplete, may not yet have been finally given. One's convictions as to the stability of our solar system indeed, may not actually *be* founded on any formal demonstration. Ultimately it may rest on an immense accumulation of inductive considerations. These may be quite sufficient for planetary theory. But we must remark to what degree the logic of the situation, and our expectations withal, have changed.

Dept. of Philosophy
Yale University
New Haven, Connecticut

ARMAND SIEGEL

ON CHRONICALLY UNSOLVED BASIC PROBLEMS IN PHYSICAL THEORIES

Comments on Hanson's paper of Oct. 29, 1963

I would like here to cite several parallels in physical theory to the failure to provide a solution to the stability problem of Newtonian Force-and-Gravitation Theory. There are, I think, enough of them to suggest that this phenomenon may be typical in physics. If this is so, it would lend support to the "loose analogy" Prof. Hanson has tried to establish in the preceding paper. I wish, however, to retain the emphasis on the word "loose," and I think this is in keeping with the spirit of Prof. Hanson's presentation.

Since there is nothing more unpalatable intellectually than a parallel that is pushed too hard, I shall restrict myself to a rather simple line of thought in the selection and interpretation of these parallels. Most striking about the unsolved planetary stability problem is that it was, and is, a "scandal": After all, the stability of the solar system has been, throughout the civilized history of the human race, nothing less than a household fact; major religions, dead and living, have been based on its seemingly miraculous nature. Yet this phenomenon, so plain to see without the use of any instrumentation at all, and which *should* follow from Force-and-Gravitation Theory, has eluded demonstration by the greatest minds of mathematics and physics for nearly three centuries!

An interesting but secondary aspect of similarity which motivates my selection of parallels is a qualitative one. Not only are they "scandals," not only do they pose difficult challenges to the theories in which they occur, but they all also have to some extent the specific aspect of *stability* problems.

My first parallel is the failure of quantum mechanics to provide a rigorous mathematical derivation for the formation and stability of crystals. In this case the problem is indeed literally a matter of stability. Consider that the great initial triumph of quantum mechanics was to explain the stability of *atoms* and *molecules* consistently with the use of only the (attractive) Coulomb force. (Let us not cavil here at the lack of an exact theory for anything but a two-particle system; the development of approximation methods for larger molecules is so considerable, both theoretically and practically, as to remove this question from the category of a scandal.[1]) Quantum mechanics is held to contain, implicitly, the explanation of the larger problem of the stability of solids, including the crystal, which is the simplest type of solid; and this is even held to be one of its merits. Yet there is no rigorous proof for this. Work on the problem of crystal structure is based on the adiabatic approximation. This is an assumption applied so strongly in practice that it amounts to assuming crystal stability in advance; the true achievements of the quantum theory of solids start only at this point, and deal mostly with the details of solids on the *assumption* of a stable lattice.

Yet, like the stability of the solar system, the stability of crystals is a household fact, requiring no apparatus at all to demonstrate. I will agree that quantum mechanics provides a qualitative interpretation: There is the Pauli Principle, which explains the mutual repulsion of electrons without the ugly introduction of a force tailored for the purpose. This, combined with the Coulomb attraction, may be said to make the existence of a crystal trivially evident from a commonsense point of view. But the same kind of argument can be given for stability in celestial mechanics. We are speaking of matters of rigorous deduction, such as are claimed to be possible in the "exact science" of physics.

My second parallel is from the field of statistical mechanics, and also concerns a household fact. It is the failure of statistical mechanics to provide a theory of phase changes (melting of solids, vaporization of liquids, etc.). In this case there is a mitigating factor: At least models have been constructed in two cases (the Einstein Condensation and the Ising Model, the latter solved by Onsager) which describe phase changes in a mathematically exact and consequent way from fundamental hypotheses. These models are, of course, not realistic systems, but they at least show the compatibility of *some* phase changes with the fundamental theory in principle. As in quantum mechanics, the great progress (theories of superfluidity and superconductivity, and of imperfect gases) is made on a semiempirical basis. In this example the rôle of stability is less direct; but phase changes have to do with the transition between stable phases, and implicit in a theory of phase changes would be a proof that the phase which results from a change of external conditions is indeed a stable one.

The third parallel, also from statistical mechanics, is the ergodic problem. It is remote from any household facts, but on the other hand occupies an even more fundamental place in its theory than the previous examples. More progress has been made on this than on any of the other problems I have mentioned. Yet regardless of the significant progress that has been made in solving it, the circumstances of this progress and the time it has consumed entitle it to rank quantitatively with the unsolved problems. It is clear from reviews of the subject (Khinchin[2], Truesdell[3]) that the process of solution has consisted of a progressive recasting of the formulation of the problem, which indicates its extraordinary subtlety. Some of the names of those who have been associated with it will also indicate its magnitude: Boltzmann, Clerk Maxwell, Gibbs and Poincaré, among others.

The ergodic problem may be described in terms used by Professor Hanson: It arises in the attempt to derive a θ_1 from a more fundamental θ_a; in this case, to derive statistical mechanics from classical mechanics. Classical mechanics is, of course, not quite sufficient for the purpose; some assumption about randomness is needed. But the ergodic hypothesis, in whatever form, is an attempt to reduce the degree of randomness to a minimum, namely to the assumption of the *a priori* homogeneous distribution of events in time.

The ergodic problem is very intimately tied in with a stability problem, viz. the problem of statistical equilibrium. In fact, a sufficient characterization of the equilibrium distribution of physical systems emerges from the ergodic property which the ergodic hypothesis aims to establish.

It is not too strong to call all of these "consistency problems"—they are attempts to prove consistency of the theories with external phenomena, or even, just like the consistency problem of mathematics, with other theories (the latter is the case with the ergodic problem). In physics, the inability to construct consistency proofs may be, as nearly every physicist believes, merely a matter of computational difficulty and not a matter of principle. But this difficulty is so immense that its implications for physics are akin to what they would be if they were difficulties in principle.

For the net effect is that, from a rigorous standpoint, the correctness of these physical theories is not established, and this state of affairs is a chronic one. To be sure, there is the "agreement with the experimental facts" which is claimed for all the currently accepted theories of physics. I do not mean to minimize the accomplishments of physics in this respect, but merely to indicate that this agreement with semiempirical methods establishes only that the theory (which usually provides a model for these methods) is a gratifyingly useful tool, not something valid *in toto*. Suppose it were to turn out that all the theories I have cited were actually inconsistent. It would then be cause for surprise that physicists were able to agree so well on the structure of these semiempirical theories.

And that is the point: Since proof of the consistency theorems is lacking, there must be something wrong with the picture of the major physical theories as a repository of present knowledge and a determinant of future knowledge, as if all the significant results of future research (apart from those which will arise from theories as yet unpropounded) are already present in embryo in them.

Although I am not at all ready to resign myself to a breakdown of the deductive method in physical science, I am prompted by this state of affairs to discuss seriously other approaches to the role of theories. The motive of the deductive approach is, I think, to have a completely automatic procedure in which the "human element," including matters of taste, personality, opinion, etc., can be essentially eliminated. On the other hand, if consistency proofs are not available, one must admit the importance of these fortuitous elements. Although they are not conspicuous in the quantitative sciences, they are present nonetheless. The reason they are not conspicuous is that there is always in these fields a "short-range" logic; that is, consistency in a restricted range of subject matter is entirely possible. If the various specialties are sufficiently self-contained, with little overlap, there may be no visible inconsistency. And if they share common points of view with one another as well, the illusion of perfect consistency can well exist. The major theories in physics do seem to confer this benefit of providing a common point of view.

As a tentative summary then: The rôle of theories is to some extent to act as a common storehouse of methods for the solution of problems; as tested languages, well adapted to the interpretation of the phenomena we find interesting; as collections of images and ideas which furnish a common point of view through which scientists are able to understand each other and work together without demonstrable *in*consistency, and consequently without wasteful, paralyzing controversy.

These ideas point to a characterization of science along operational lines in a social-dynamic context. Part of such a characterization might be: Science is a method for the understanding and manipulation of phenomena which uses theories and deductive methods, but subordinates them to the overriding aim of achieving agreement on methods by the practitioners of science; these methods in turn are to produce results which are agreed to be interesting. The effectiveness of a theory is measured by its concordance with the data, again as agreed by practitioners; and by some degree of support by enlightened outsiders as to the importance of the results.

The above definition tries to avoid two dangers the deductive method is also meant to prevent: The demoralizing effects of too much scientific controversy (by emphasizing the necessity of devices for the achievement of agreement); and the possibility of the emergence of a pseudo-scientific priesthood, dedicated to perpetuating itself through possession of a mystery and a cult devoid of substantial content (by—somewhat gingerly—requiring the "support of enlightened outsiders").

It surely must contain other inadequacies I am not aware of at this preliminary stage. But I would argue that the relinquishment of belief in "automatic" theories could be fruitful in the deeper study of science. It could stimulate the study of science itself as a phenomenon in the social sciences. It would furnish a basis for ridding science of the invidious exclusion of the social sciences by the physical sciences, and of the excessively sharp distinction between the quantitative and non-quantitative sciences. I realize that those psychologists, psychoanalysts, economists and others who have been working fruitfully without reducing everything to a quantitative basis may feel no need whatever for the prestige of being accepted by the physical scientists. But it might be good for the physical scientists if they did so anyway.

Dept. of Physics
Boston University
Boston, Massachusetts

NOTES

1 But against even this easy-going attitude we can quote the authoritative John C. Slater, who points out in the Preface to *Quantum Theory of Molecules and Solids* (New York, 1963), that the real accomplishments of atomic and molecular structure theory have nearly all been based on semiempirical methods. "The goal [of the non-empirical methods of quantum mechanics] is still far off . . ."

2 A. I. Khinchin, *Mathematical Foundations of Statistical Mechanics* (translation), (New York, 1949).

3 C. Truesdell, *Ergodic Theory in Classical Statistical Mechanics*, in *Proceedings of the International School of Physics "Enrico Fermi," Course XIV*, edited by P. Caldirola, New York, 1961).

6

Everett Mendelsohn

PHYSICAL MODELS AND PHYSIOLOGICAL CONCEPTS: EXPLANATION IN NINETEENTH CENTURY BIOLOGY

Presented November 19, 1963

I

Theodor Schwann closed his volume of *Microscopical Researches* (1839) with the proposal of a new "mode of explanation" of living things which he believed to be in harmony with many already known organic phenomena. That he was very proud of his new theory is made abundantly clear; it was set off, in a special part of the volume (section three); and attention is called to it in the prefatory remarks: "At the conclusion of the Treatise," Schwann reminded his readers, "I have attempted a theory of organisms."[1] In order that no confusion exist (and perhaps to safeguard the proper recognition of his positive scientific achievements) Schwann purposefully excluded all theoretical and hypothetical statements from the body of the work itself.

> "The theory has at least this advantage, that by its aid any one may form a precise idea for himself of the organic processes, which may lead [conduct] to new researches; such a theory may therefore be of use, even if assumed to be decidedly false. It contains the principles of the organic phenomena, both of the healthy and diseased organism."[2]

What emerges after these remarks strikes the reader by its sobriety. In measured terms Schwann advanced a modest proposal by way of a "physical" explanation of the formation of the cell. There was no ringing denunciation of the old vitalism or *Naturphilosophie* as was to come later from DuBois-Reymond and Rudolf Virchow. Instead there was a step by step build up of a new "mode of explanation" which if adopted, however, left the reader with no alternative but to seek the explanation of organic phenomena through a reduction to physico-chemical causes. Schwann was explicit, the teleological view of nature was "admissible [only] when the physical can be shown to be impossible."[3]

As I looked over these statements by Schwann and analysed some related ones in a recent paper,[4] I began to ask whether it was possible to recognize any trends toward physico-chemical reductionism in the development of physiological theories in the nineteenth century. Here, after all, was Schwann setting the stage for a dramatic change in the conception of what the living organism was and he was doing it within the framework of a study of cells. Could anything be learned about the "structure of scientific change" by examining the fine structure of this adoption of a reductionist explanatory model in physiology?

The study, which is now underway, has moved in two directions simultaneously. The first has involved a careful examination of the published writings of Theodor Schwann and his immediate colleagues, and where possible an analysis of the extensive collection of manuscript materials relating to Theodor Schwann.[5]

The other part of this study is at once simpler and more difficult; it consists in asking the question—were there various ways in which physics and chemistry became part of the explanatory models of nineteenth century physiology, what were they like, and why the difference? Owsei Temkin, in a brief but provocative paper indicated a marked difference in the approach to bodily functions as they were examined by French and German physiologists in the middle decades of the century.[6] We might almost speak of a national style in physiological explanation since the attitude adopted and type of experiment undertaken was on the whole quite distinct. We would, however, have to point out that France had its Germanic types (men like Dutrochet) and Germany had its Franks (in the person of Rudolf Virchow and others). The point is fairly simple, the major tendency in German physiological theories was reductionist, while the French, characterized by Claude Bernard, were well aware of the importance of physicochemical techniques and laws for understanding biological phenomena, but insisted none the less that there was need for special biological laws.

The history of physiology can hardly be charted in terms of a single advancing front, from Harvey to Helmholtz or Helmholtz to the most recent Huxley (Andrew) to gain scientific fame. One can more nearly observe wave-like fluctuations in which two repetitive trends can be recognized. There were the periods of strong belief in the continuity of nature and laws of nature from the inorganic through the organic realms. These were matched by periods during which it was assumed that nature was neatly divided into discrete realms each of which was governed by separate laws with no necessity for similarity in the content or structure of these laws. These two trends traditionally have been reduced to the terms *mechanism* and *vitalism* and the adherents of one or the other called *mechanists* and *vitalists* as a short-hand way of characterizing the theories of nature proposed. It became clear to me, however, that these terms had lost their usefulness so far as providing an adequate classificatory system for nineteenth century physiologists. Owsei Temkin had already encountered the difficulty and found himself referring to "vitalistic materialism" in an effort to encompass a point of view found among the early nineteenth century French experimenters.[7] The difficulties become compounded when we try to deal with the organic chemists, like Liebig, and their excursions into physiological experimentation. The reason that this short-hand breaks down in the nineteenth century is the very same reason that the physiology of this period is so challenging to the historian: by the middle of the century, physics and chemistry had become indispensable for the study of organic functions and the construction of physiological concepts.[8] Much more important than the characterization as either vitalist or mechanist is the role played by physico-chemical theories, on the one hand, and vital forces or special biological forces on the other hand, in the construction of the explanatory models offered for physiological phenomena. What this means for the historian is that he must look not only at the professions of "faith" made by the scientists, but also directly at the theories constructed and the relative importance of their various components and come away not using the simple classifications of mechanism and vitalism, but instead presenting an analysis of the explanatory models, suggesting which were the critical and which were the expendable components and where possible indicating the way in which the concept was understood by contemporary investigators.

Before proceeding with a comparison of the French and German theories let us return to Schwann and examine both the background of his new "mode of explanation" and to the extent possible, its intent. Both the French and German physiologists of the mid-century reflected a move from earlier vitalistic and/or teleologic explanations of nature, explanations which were, however, quite different. Although we can recognize an almost continuous tradition of teleological and typological thought in biology, the late decades of the eighteenth century witnessed the emergence of a very strong metaphysical trend in biological explanation. The German *Naturphilosophie,* an extension into science of the Romantic philosophy of Schelling, Herder, Fichte, et al., left its mark on almost every German biologist of the day.[9] The writings of these biologists indicate, in almost every case, a strong defense or attack on the philosophical approach in biology. Rudolf Virchow characterizes the reaction of the opponents when he pointed out that many ideas were discredited because they had roots in *Naturphilosophie.* He bitterly recalled "the spell which philosophy, *Naturphilosophie* in particular has for a long period cast over science." In the response which followed he said scientists "fought against *a priori* speculation . . . rejected systems" and came finally "to rely only on experience."[10]

Schwann himself was only one step removed from *Naturphilosophie.* And it is this step which seems of critical importance if we are to understand the emergence of the physicalist approach in physiology. The pivotal figure is Johannes Müller, the physiologist and anatomist who was the teacher not only of Theodor Schwann, but also of Emil Du Bois-Reymond, Ernst Brücke and Hermann von Helmholtz, the men who brought the reductionist philosophy to its fullest expression.

One factor which really should not be overlooked is sociological in nature and concerns the emergence in Germany, at just this period, of the teaching-research laboratory. This innovation in science education and practice arose in physiology at exactly the same time as it did in organic chemistry. The results are no less marked, for there grew up around Johannes Müller (as also about Justus Liebig) a group or "school" whose members took up the problems common to the laboratory. Müller emerges as a towering figure in physiology, not so much for his own scientific discoveries, as for the training he provided to a whole generation of men who went on themselves to carry out new physiological experimentation and even more importantly, to occupy the professorships in physiology at many of the German, Austrian and Swiss universities. Like Liebig, and almost contemporaneous with him, Müller also developed an important outlet for the publication of research results and new theories. The *Archiv für Anatomie und Physiologie* and the two volume *Handbuch der Physiologie des Menschen*[11] are filled with contributions made by men while still students or assistants in Müller's laboratory.

An examination of Müller's life shows that he commenced his scientific career as an advocate of *Naturphilosophie,* an approach which he did not reject until 1827.[12] By that time, he had already completed a series of studies of "the mind" which he treated from a physiological point of view. These provided the basis of the new theory which became widely known as the law of specific nerve energies. Müller recalled that his own teacher Carl Rudolphi had been quite critical of his acceptance of *Naturphilosophie* and of his studies on "the more abstract subject of the physiology of the mind, rather than with the investigation of the anatomy of the organs of the senses, such as that of the eyes of insects and spiders."[13]

A sampling of the spirit of *Naturphilosophie* is important for later comparison of the violence of its rejection by the Germans, to the much milder swing away from the vitalistic biology of Frenchmen like Bichat and Barthez. This difference in reaction to the past probably provides one clue to the difference in attitude towards physico-chemical theories in the mid-century. *Naturphilosophie* is best characterized, or should we say caricatured, in the works of Lorenz Oken, a man of many sided genius and great influence in the German scientific community. [Founder of Deutsche Naturforscher Versammlung and editor of *Isis*.] Adopting what his admirer Richard Owen called the *a priori* method in biology,[14] Oken poured forth a series of ideas many of which were repeated over and over again in the early years of the century and definitely influenced what biologists did and the way in which they explained nature and its activities. "The sphere," claimed Oken, "is . . . the most perfect form;" [a not unfamiliar claim] "for it is the primary, the divine form. Angular forms are imperfect. The more spherical a thing is in form, by so much more perfect and divine is it." Then he adds his own note. "The Inorganic is angular, the Organic spherical."[15] This idea reappeared many times in the succeeding decades, with Johannes Müller himself expanding upon it.[16] Incidentally, this conception of organic sphericity probably reinforced some of the poor microscopic technique of the period and helps explain the readiness of microscopists to see all of nature as being made up of minute globules.

Speculation, some of it uncanny in retrospect, was a hallmark of the approach to nature of the *Naturphilosophen*. But it was due to this guided speculation—not wild speculation—that these philosophical biologists achieved some of the oft-mentioned happy insights. Richard Owen joins in claiming for Oken a prior discovery (or invention) of the cell theory.[17] The basis for these claims are a series of statements, not unlike the following, which can be found in several of Oken's writings. "Organisms," he said in the simplest terms,

> "are a synthesis of infusoria. Their generation is none other than an accumulation of infinitely numerous mucous points, infusoria. In these the organisms have not forsooth been at once wholly and perfectly depicted as on the smallest scale, nor contained in a state of preformation; but they are only infusorial vesicles, that by different combinations assume different forms and grow up into higher organisms."[18]

There is no doubt that this is speculative biology. The question remains, however, was it suggestive as well?

Many of those who had briefly been adherents of *Naturphilosophie* came away almost as though rubbing their eyes in wonderment at what they had previously believed. Their characterizations, as they were passed on to their students, were in the sharpest language. Liebig referred to "the degenerate philosophical methods of investigations, as they had been embodied in Oken, and still worse in Wilbrand," which he felt, "had a most pernicious influence on the branches of natural science, for it had led in lecture and study to a want of appreciation of experiment and of unprejudiced observation of nature, which was ruinous to many talented young men. From the professorial chair," complained Liebig, "the pupil received an abundance of ingenious contemplations; but, bodiless as they were, nothing could be made of them."[19]

Johannes Müller, like Liebig, came to reject *Naturphilosophie,* and like Liebig he nevertheless maintained a conscious commitment to the necessity of including the concept of a vital force in the explanatory models he constructed for physiology. Organic function and structure could never be explained by physical laws alone, special organic laws would always be necessary. How the organic or vital forces originated or how matter first became endowed with these special forces is, said Müller, "beyond the compass of our experience and knowledge to determine." This is a problem, he claimed, which "is not a subject of experimental physiology, but of philosophy. Conviction in philosophy and in natural science has entirely different bases; the first suggestion here, therefore, is not to be led away from the field of rational experiment. We must be content," he concluded, "to know that the forces which give life to organic bodies are peculiar, and then examine more closely their properties."[20] In his researches, Müller adopted an almost positivistic approach to natural phenomena, experimenting and recording in much the same way his French counterpart Magendie worked. He seemed even more aware, however, of the physical and chemical tools which had become available; and unlike Bichat who had denied any role for physics and chemistry in biological explanation, Müller was receptive to, and encouraged physicochemical researches on organic functions. However, he warned, don't expect the laws of inorganic chemistry to explain the chemistry of the organism—of that task they are incapable.

In practice, Müller's commitment to a force not reducible to those of inorganic nature raised some serious problems. He had adopted a theory, current at the time, which claimed a fundamental difference between organic chemical molecules and inorganic molecules, the former conforming to special laws. When Wöhler's synthesis of urea became widely known (post-1828), Müller clearly had to respond to the new experimental results. (We will not push the question of whether experimental results change a man's views as to what constitutes an acceptable type of explanation). Wöhler had synthesized urea, a waste product of the organism and a molecule generally regarded as being organic, from ostensibly non-organic sources. Müller's response in this instance was characteristic. "I know Wöhler well", he seemed to be saying; his "experiments afford the only trustworthy instances of the artificial formation of these [organic] substances." [21] After reviewing the whole process, which was still quite new, and admitting its accuracy, Müller turned to assess its implication. "Urea, however, can be scarcely considered as organic matter, being rather an excretion than a component of the animal body. It has not perhaps the characteristic properties of organic products." Rather than give up his concept of special chemical forces for organic molecules, Müller was willing to create a whole new class of compounds. These existed at the borderline between organic and inorganic substances and formed something like a gray area into which questionable compounds—that is seemingly organic ones which had been successfully synthesized—could be placed. Special organic forces remained essential to the basic explanatory models developed for physiology by Müller. "Light, heat, and electricity" do, he claimed, "influence the compositions and decompositions going on in organic bodies, as they do those in inorganic bodies; but," and here came the important reservation, "nothing justifies us in regarding without further inquiry any one of the imponderables—namely heat, light, and electricity,—as the final cause of vital actions."[22]

II

Müller was at the height of his creative work in physiology when Theodor Schwann came to his Bonn laboratory as a young student in 1830. Three years later, after Müller's move to Berlin, Schwann rejoined him, first to complete his medical degree and later as an assistant. At the time Müller was clearly a vitalist and the first volume of his *Handbuch,* which gave detailed expression to this point of view, was on the presses. Yet, he led his students into exploration of the very problems through which they went on to develop a radically new idea of what constituted an acceptable physiological explanation. Schwann took up the problems of muscle contraction, digestion and cellular structure, function and development; Du Bois-Reymond, another student, was led directly to his investigations of the electrical phenomena of the nervous system; and Helmholtz was encouraged to explore the nature of heat production in muscle and nerve. Each of these men examined physiological processes for which Müller had invoked special non-physical forces in his theories and hypotheses. These men emerged from their *initial* studies believing that the vital functions under exploration could be successfully explained with recourse only to physico-chemical laws.

Theodor Schwann is best known to us as the co-inventor of the cell theory. However, I would contend that during his own lifetime he was equally influential as the initiator of the new physico-chemical approach which marked German physiology during the fourth and fifth decades of the nineteenth century; what I am saying is that the influence of his positive science was matched by that of his philosophy of nature and further that some of the most interesting elements of his experimental work reflect his attempt to establish new criteria for physiological explanation.

Early in 1835, some time after he had joined Müller as an assistant, Schwann entered in his notebooks his intention of making physical measurement of physiological phenomena.[23] He mentions glandular secretion and muscle contraction, both, incidentally, indicated as problems by Müller. In October 1835 he outlined a series of experiments on muscles and in November he carried them out. The results were first reported in a paper read to the Deutscher Naturforscher Versammlung in September 1836. A summary appeared shortly thereafter in Oken's *Isis*[24] and a lengthy description of the experiments was included by Müller in volume two of his *Handbuch* which appeared in 1837.[25] Neither the *Isis* report nor

Müller's recapitulation carry any philosophical overtones. The experiments themselves are really not very startling. Working with muscle from a frog Schwann attached one end to a lever system whose other arm was attached to a recording device on one side and a facility for increasing the weight on the other. A nerve remained attached to the muscle and was used for stimulation. A series of stops was included which prevented the muscle from being overstretched and which also could be finely adjusted, so that the active length of the muscle was known. The object of the experiment was to correlate the length of muscle necessary to balance the force created by the addition of weights pulling against the normal contraction. What he found was that a uniform increase in length of muscle was attended by a uniform increase in the force of contraction. The "law" he derived from his observations was that "the power of the muscles was found to become greater in the same ratio as they were less contracted, or to diminish as the contraction of the muscle was greater."[26] This was the same law which was held to regulate the force of elastic bodies, a point which led Schwann in his oral presentation to suggest that muscle could be considered an elastic body with a changeable force of retraction.[27]

My guess is that this experiment would long since have been forgotten if its author had not believed that he was involved in doing something new, and if the philosophy which motivated this work had not been spelled out in detail as part three of the book on cell theory. But Schwann was engaged in a campaign and this simple experiment on contractile force of muscle was seen as one (perhaps the first) important step. In retrospect Schwann claimed that this experiment was part of a larger argument against Johannes Müller and his use of vital or organic forces in biology.

In a letter (22 December, 1858) in response to an inquiry from Du Bois-Reymond, Schwann contended that even as a student of Müller's in Bonn in 1830, the direction of his thoughts had become markedly different from that of his master.[28] Although there is some reason to doubt that Schwann's own philosophy of nature was clarified by 1830, it is quite clear that Schwann thought of his muscle experimentation as a direct challenge to the concept of a vital force. He refers in his letter to his pleasure at seeing Müller substituting the experimental method in his attempts at characterizing the energies and forces of the body. Müller was, after all, introducing a correct approach in place of *Naturphilosophie,* but, complained Schwann,

> "I was still not satisfied with his vitalistic ideas. The physical approach which *I introduced* into physiology and essentially the search for a real [*wirkliche*] explanation of the phenomena (something impossible in the other type of interpretation) was unknown to Johannes Müller."[29]

In an attempt to remove any ambiguity as to his retrospective appraisal of his actions, Schwann continued, "The first thing that I published along this line was the research on the relationship between the force of a muscle and its degree of contraction and further I gave a report on it at the Naturforscherversammlung at Jena in 1836." Schwann then adds a phrase which is very hard for us to understand: ". . . To my knowledge this was the first time that a clearly vital process had been mathematically treated and included in a numerically expressed law." That he was not the first to render a physiological function in mathematical form was surely apparent to Schwann. Just why he chose not to seek strength for his approach by calling upon authority is not wholly clear; except that he may have wanted to avoid any association with the gross physicalism of the iatro-physicists and iatro-mathematicians, which had been rather thoroughly discredited by the late eighteenth century.

There is very good evidence to indicate that as early as 1836 Schwann thought he was engaged in a movement to drive vitalism from biology. And of equal importance with Schwann's own assessment, his contemporaries credited him with just this motive. Emil Du Bois-Reymond, the most outspoken protagonist of the reductionist philosophy in biology, had come to Müller's laboratory at about the point that Schwann himself was leaving to accept a professorship at Louvain. In his obituary of Müller, whom he ultimately succeeded as Professor of Physiology at Berlin, Du Bois-Reymond credited Schwann with just the leadership that Schwann himself had claimed.[30] Du Bois-Reymond went even further and alluded to a new school of physiology, with Schwann *"an der*

Spitze," which brought to a close the long period of Müller's influence and replaced the vitalist approach in physiology with the molecular mechanical outlook.

But the most telling evidence of Schwann's role in changing the nature of explanation in physiology comes from letters of the young Du Bois-Reymond, written while he was still a student in Berlin. One such letter, composed in 1841, declares unashamedly: *"Schwann ist für mich eine Art von Halbgott."* Then he continued, referring directly to those elements of which Schwann himself was most proud: "His way of carrying out experiments is unique. The theoretical part of his book which I reread recently is marvelous beyond measure. What a style! This sort of calm and clarity does not come to be a second time."[13]

Whether deserved or not Schwann had very definitely acquired the reputation of having been the first spokesman for the physicochemical approach in physiology; when the Berlin Academy of Sciences had a celebration in his honor in 1878 they too referred to his pioneer efforts, commencing with his first research on muscle, in introducing the physical sciences into biology. They concluded that these attempts placed him, in order of chronology, the true head of modern school of physiology.[32] Schwann himself was partially responsible for this claim, which he re-emphasized in retrospective comments which are probably stronger than the early evidence can sustain. But then the new outlook had gained success, so why not remove all question from the early efforts. In his 1858 letter to Du Bois-Reymond, Schwann claimed:

> "Already in Berlin I had drawn up an application of the method of physical explanation of the phenomena of animal life in which I directly combatted the views of Johannes Müller on 'vital energy' of tissues and nerves."[33]

The certainty of viewpoint represented here (letter of 1858) is in some contrast to a letter which Schwann wrote to his brother Peter, a student of philosophy and theology, in 1835, which shows Schwann still searching and examining the status of vital forces.[34] In a manuscript which I believe to be the draft of another letter to his brother, although probably of slightly later date, Schwann continued his philosophical examination of the role of vital forces in the organism.[35] He talked of all the possible types of forces which can be identified, reduced them to two, but rejected the second. The first he pointed to was described as the kind which "pushes another body in front of it and continues to do so in the same direction, unless interfered with by another external force." This, he said in a very pointed manner, "is the type of all natural forces, without exception." He placed in the second category forces which suddenly and without external cause change their direction. On the basis of the distinction which he drew between the two types of forces he formulated an axiom from which, he said, "derive all my efforts to reject vital energy." Schwann then proceeded to chastise his correspondent and all the others who had become used to assuming the existence of self-purposive forces which they believed necessary for the explanation of the functions of the organism. "However," claimed Schwann, "my work has just the purpose to show that this is by no means necessary." Simple forces would suffice to explain organic functions. The real difficulty, Schwann reminded his correspondent, was that, "in our entire outlook we differ as much as it is possible. The reason for this" difference he found to lie "in our method of arriving at conclusions. I readily believe," Schwann continued, "that all philosophers will look at the matter in the same way as you have, however, I am therefore by no means persuaded that I am wrong. The philosophical and the scientific [*naturwissenschaftliche*] mode of knowledge here stand against each other. Which," Schwann asks, which "one of the two is right?"

Fairly early in his scientific career (just how early is not yet certain) Schwann began asking, and finding answers to a series of questions which were designed to explore—to expose—the concept of a vital force. The commitment he developed was aimed at banishing vital forces from biology; and his new theory of the organism which became the third and final part of his first book, the *Mikroskopische Untersuchungen,* took issue with many of the essential parts of Müller's concept of vital or organic forces. Basic to Schwann's cell theory was his painstaking demonstration of the similarity, on the cellular level, of all organisms, that is both plants and animals; and his further contention, backed up by a fair amount of evidence, that the processes of cell formation and growth are similar in both animals and plants. He takes one further step of great interest, this is to remove to the level of the cell, rather than the total organism, the fundamental vital processes, nutrition and growth. This move has two consequences, the first is to directly challenge the then prevailing view of the organism as the bearer of powers greater than found in the sum of its parts. Müller had cited with approval the contention of Kant: " 'The cause of the particular mode of existence of each part of a living body resides in the whole, while in dead masses each part contains this cause within itself,' " and added that it is the concept of the unity of the whole which distinguishes the organism from other seemingly organized bodies.[36] On the contrary, argued Schwann, "every elementary part possesses a power of its own, an independent life."[37] He is then able to bring the argument to the point at which he can introduce his new mode of explanation. "The question, then," Schwann noted, "as to the fundamental power of organized bodies reduces itself into that of the fundamental cells."[38] This of course is just where he wants the discussion, away from the ambiguities and complexities of the total organism and at the level of what he calls the "plastic" and "metabolic" phenomena of cells. (This latter term incidentally is his invention: "implying that which is liable to occasion or suffer change.")

Having argued that the fundamental organic processes occur at the level of the cell and not that of the whole organism, Schwann moved on to propose his hypotheses of the mechanisms involved. I have referred to these hypotheses elsewhere as "Schwann's Mistake"[39] but suggested that they served an explanatory function beyond the bounds of the cell itself. Schwann had proposed to explain cell formation and growth as a process of crystallization, that is, to adapt the explanation governing inorganic crystals to the development of cells. It must have been with hidden pleasure that Schwann used this wholly physical explanation, for his teacher Johannes Müller had explicitly said that "organic combinations . . . never occur in a state of crystallization in organized bodies during life."[40] True, the nature of crystallization, as Schwann explained it, had special features, but his intent was clear; he was hoping to establish the view "that the fundamental powers of organized bodies agree essentially with those of inorganic nature, that they work altogether blindly according to laws of necessity and irrespective of any purpose."[41] Schwann insisted at least upon a strict determinism. He believed that the adoption of physical explanations was much more conducive to the objects of science but he protected his outlook against being routed by new discoveries by pointing out "that, when speaking of a physical explanation of organic phenomena, it is not necessary to understand an explanation by known physical powers . . . but an explanation by means of powers which operate like physical powers, in accordance with strict laws of blind necessity, whether they are also to be found in organic nature or not."[42] In his conclusion that "organisms are nothing but the form under which substances capable of imbibition crystallize"[43] Schwann had succeeded in his efforts to propose an explanatory model for physiology which matched his desire to banish vital forces from the organism and reduce vital functions to recognizable physical causes. He realized that his explanation "involves very much that is uncertain and paradoxical"[44] but it was the type of explanation which he believed to be acceptable, rejecting teleological or special organic causes so long as physical ones were possible.

Not all German physiology followed Schwann's lead in insisting on physical explanations and only physical explanations for biological phenomena, but a large enough group adopted this approach in their experimentation and theory building so that mid-nineteenth century German physiology gave the impression of being reductionist. During this very same period, French physiology emerged with an alternate series of explanatory models which not only left room for, but insisted upon, the role of special biological causes, laws, or forces. I have not yet reached any conclusions about the sources of this difference in approach, but as has been pointed out by both Temkin and Shryock it spills over into many related areas. For example, French biologists tended to ignore the microscope and other instruments, which had great use in biology, while their German colleagues successfully exploited them as well as developing important new ones [witness Helmholtz and the ophthalmoscope].[54]

III

Surely one explanation of the different use to which physics and chemistry was put in French physiology can be found in the vitalist traditions which the new generation of French physiologists had to overcome or build upon. There is no sign of the utter contempt with which *Naturphilosophie* was rejected when German physiologists moved in new directions. François Xavier Bichat had left a legacy of special vital forces to French physiology and had sowed seeds of distrust of the physical sciences. He had pointed to the remarkable difference between "laws of physics" and "powers of life" claiming that the latter incessantly vary in intensity, energy and development and throughout their course assume a thousand modifications. "On the contrary," he noted, "physical laws are invariable. . . . Attraction is a physical power; it is always in proportion to the mass of brute matter in which it is observed; sensibility" on the other hand, Bichat pointed out, "is a vital power, but in the same mass of matter, in the same organic part its quantity is perpetually changing." [46] The invarying way in which physical laws presided over matter, Bichat indicated, enabled the construction of formulas of calculation in this realm. But, he continued,

> "applied to the actions of the living body, mathematics can never give us formula. The return of a comet, the resistance of a fluid in traversing an inert canal, the rapidity of a projectile may be calculated; but to calculate with Borelli the force of a muscle, with Keil the velocity of the blood, with Jurine and Lavoisier the quantity of air which enters into the lungs, is to build upon a quicksand an edifice solid of itself, but necessarily decreed to fall for want of a foundation."[47]

Bichat knew his physiology well and cited example after example of the variability of vital phenomena and could justly conclude that "the instability of the vital powers, is the quicksand on which have sunk the calculations of all the Physicians of the last hundred years." His contributions to positive science were great enough so that even later generations who disobeyed his strictures against using the physical sciences never totally rejected his work and hardly ever used critical language when referring to him. In his notes to Bichat's *Physiological Researches on Life and Death,* François Magendie gently reminds the reader that the physical sciences were not wholly without benefit to the study of living phenomena:

> "Bichat often complains in his works of the injury that has been done to the physiological sciences, by the attempts that are made to facilitate the study of them by means of physics. He was not competent to decide the question, not having sufficient data in the sciences, the use of which he reprobated; the most that he should have said, was that a bad application had been made of them. Even this reproach was too general to be just. No doubt, mankind have been led into errors by attempting to support on slight foundations a science which was still in its infancy; but even in the time of Bichat it could not be denied that it was to the progress of these same sciences, that was owing the explanation of many important phenomena; that by it was ascertained what takes place in respiration, and by what means a living body always supports itself between certain limits of temperature, &c."[48]

There is no use of terms like "pernicious influence" or other pejorative comment in the references that Magendie or even Claude Bernard make to the works of Bichat. Bernard made a separation between philosophy and physiology as practiced by Bichat. He commented that "in a word Bichat erred, like the vitalists his predecessors, on the theory of life; but he did not err in his physiological method."[49] So important were Bichat's ideas that, in spite of his vitalism, Bernard could claim of them that "they produced a profound and a universal revolution in both physiology and medicine."[50] Perhaps the influence of Bichat's vitalist philosophy of nature was offset in the minds of French physiologists by just the type of careful physico-chemical experimentation referred to by Magendie. The experiments by Lavoisier and Laplace and the latter ones by Lavoisier and Séguin represent some of the earliest fruitful attempts to explain physiological functions, in this case respiration, animal heat and transpiration, using the theories of contemporary physical sciences. This alternate tradition might well have provided the new generation of physiologists with just the balance necessary to avoid having to totally reject the old in order to replace it with something new.[51]

One other tradition to which Temkin calls attention is that of the French *idéologues* whose ideas, he suggests, were widely current among French scientists of the period. Temkin notes that Magendie's assertion "that physics and chemistry had their legitimate place in physiological research" was not incompatible with the contentions of the idéologues.[52] In the Preface to his early *Précis Elémentaire de Physiologie,* 1816, Magendie indicated the attitude which was to govern most of his later physiological experimentation:

> "I have not, however, neglected the possible and useful application of physical, mechanical, and chemical principles to the phenomena of life. Perhaps the application may be found somewhat different from what has been heretofore proposed, for I have taken every care to arrive at the greatest possible degree of exactness."[53]

Hardly a wild embrace of the physical approach to biology; instead it is a cautious pledge not to ignore the benefits of physical and chemical analysis, as had Bichat, but neither would physics and chemistry provide the models upon which physiological explanations were to be fashioned. From the beginning to the end of his scientific career, Magendie would have it both ways, using physical analogies for biological activities and again warning his readers not to "fall into the . . . excess of seeking to explain everything by the laws which hold sway over inorganic matter. . . . Nature," he was wont to remind his students, "has not up to the present permitted man to raise the veil which hides from him the understanding of vital phenomena."[54] But much more important than Magendie, the philosopher of nature, was Magendie the experimentalist, the observer of nature, the continuous searcher for new facts. So strong did this seeking of new results become that it seemed to be done to the exclusion of theory building, leaving his biographer with the impression that his later work was a "mere conglomeration of experimental results."[55] Magendie stepped away from the argument of whether or not the organism could be explained in physico-chemical terms and emerged in the tradition of the medical positivism of the nineteenth century.[56] The type of experimentation which brought Magendie his greatest recognition can be classified as vivisection, experiments on the living organism. Knowledge of the processes of life, he continually asserted, can be gained only by experiments on living animals.[57] This attitude and mode of operation is what characterized French physiology in the middle decades of the nineteenth century and made it distinct from the general practice in Germany.

Seldom in the history of science do we encounter a figure who was as conscious of the philosophical implications of his activities and had more regard for the methodology employed in his work than Claude Bernard. A student of Magendie, Bernard came to embody in his own experimentation the two powerful traditions present in physiology. He made full use of the tools of physics and chemistry, but also depended upon vivisection and experimentation on live organisms. In fact he tried very consciously to establish the proper relationship between the physico-chemical and the organismic approach to physiology, and never tired of reminding his students that "life cannot be characterized exclusively by either a vitalistic or materialistic conception."[58] He directly opposed the trend he noticed in the science of his German contemporaries and denied that all organic processes could be reduced to physico-chemical laws. "The vital properties of tissues," he said, "are manifested through the physico-chemical properties of the milieu but are not created by them."[59] Instead he pleaded for the recognition of special biological or organic laws, commensurate with the complexity of biological phenomena. These laws would have the same property of determinism, which governed all science for Bernard, and definitely would not rely on the active intervention of a vital principle.[60]

Bernard recognized that the only way in which experimentation could proceed in physiology was through examination of the physico-chemical conditions of the vital processes and through determination of the structure of vital organisms carried out through vivisection, through histological studies and through chemical analysis. This meant for Bernard that the explanatory models constructed for physiology would not contain only physico-chemical laws, although he was quick to deny that there existed any more than one physics and one chemistry for both living and non-living substances.[61] As an example of organic activities which for Bernard cannot be reduced to physico-chemical causes are the organotrophic phenomena, the processes by which living forms originate and evolve. Clearly implied here is the notion of direction, which can be affected by physical and chemical means and which can be studied with physical and chemical techniques, but which cannot be explained by physical and chemical causes alone. The organotrophic phenomena bring Bernard face-to-face with a first cause which as a scientist he believes he must refuse to treat. What he will deal with are the manifestations of the first cause as found in biological activities. His statement of the difference is quite direct:

> "The first cause of life provides the evolution or the *creation of the machine;* but the machine, once created, functions by virtue of the properties of its constituent elements and under the influence of the physico-chemical conditions which act upon them."[62]

Bernard's attitude emerges as that of an anti-reductionist but not a vitalist in the sense of a Bichat or even a Müller. "I admit," said Bernard, "that manifestations of life cannot be wholly elucidated by the physico-chemical phenomena known in inorganic nature. . . . Biology," he added, "must borrow the experimental method of physico-chemical sciences, but keep its special phenomena and its own laws."[63]

When he turned to the study of cells, referred to often as *éléments organiques,* Bernard encountered what for him was a truly physiological unit. "Cells," he claimed, "are the primary elements of living matter,"[64] while life itself was a result of the contact of the organism or cell with its environment.[65] Bernard was aware that the same set of chemical elements was to be found in living substances and non-living material. This of course was true of cells, "they are composed of chemical elements, but" they enjoyed a special type of organization such that "when we analyse them chemically we destroy their vital properties."[66] The cell membrane then marked the boundary beyond which the physiologist could not fruitfully penetrate and still claim to be studying physiological phenomena. "The cell alone lives,"[67] proclaimed Bernard, all other bodily functions exists to serve the cell.

If the cell was the typical physiological unit as seen by Bernard, the relation of the cell to its environment was a typical biological relationship. "The inner environment," emerged in Bernard's studies as "the true physiological environment."[68] It was the level at which the phenomena of life could be most efficiently altered. The environment of the cell would yield to the physico-chemical analysis which the living unit itself could not sustain. Physiology, then, as Bernard came to define it, "embraced two distinct types of studies; study of the organisation and study of the environments within which the organized or living bodies manifested their properties."[69] Explanatory models to successfully encompass the varied phenomena of life had to cover both the organism and its environment. The latter might be successfully reduced to physico-chemical causes, the former required special biological or organic laws which could no more be successfully reduced than a physical law like gravity could be further reduced. Bernard surely would have agreed with the stricture of Rudolf Virchow who claimed that it was long past time to give up "the scientific prudery of insisting that all organic processes be reduced to the resultant of inherent molecular forces."[70]

Bernard in France, Schwann in Germany, took their theories of the organism in different directions. Schwann reacting against the past utilized physics and chemistry to *repudiate* the earlier German biological traditions. From his efforts a whole generation of physiologists emerged embracing physicalism as the acceptable guide for concept formation in physiology. In France, Bernard *absorbed* the past of biology and welded to it the new theories and techniques of physiological chemistry. The physiology which he spawned fashioned explanatory models capable of encompassing both physical and special biological causes.

Committee on the History of Science
Harvard University
Cambridge, Mass.

NOTES

* An earlier version of this paper was presented to the Boston Colloquium for the Philosophy of Science, Cambridge, Mass., November 19, 1963. Support for this research has been generously given by the National Science Foundation.

1 Theodor Schwann, *Mikroskopische Untersuchungen über die Uebereinstimmung in der Struktur und dem Wachsthum der Thiere und Pflanzen,* Berlin, 1839, p. xvi. The English quotations have been derived from the translation made by Henry Smith for the Sydenham Society, *Microscopical Researches into the Accordance in the Structure and Growth of Animals and Plants,* London, 1847, p. xviii. In the following, page numbers from the translation will be followed by those of the German edition in brackets.

2 *Microscopical Researches,* p. xviii (xvi).

3 *Microscopical Researches,* p. 190 (225-226).

4 E. Mendelsohn, "Cell Theory and the Development of General Physiology," *Archives Int. d'Hist. Sci.,* no. 65, 1964.

5 Marcel Florkin has already called attention to much of the manuscript material, but his purpose has been primarily to round out the picture of Theodor Schwann's life and achievements. My own interests are more limited and center about the task of finding clues to the sources for the new approach developed by Schwann. See M. Florkin, *Naissance et Déviation de la Théorie Cellulaire dans l'Oeuvre de Théodore Schwann,* Paris, 1960.

6 O. Temkin, "Materialism in French and German Physiology of the Early Nineteenth Century," *Bulletin History of Medicine,* vol. 20, pp. 322-327, 1946.

7 Temkin, "Materialism," p. 323.

8 This point has been detailed in my recent study, E. Mendelsohn, *Heat and Life: The Development of the Theory of Animal Heat,* Cambridge, Mass., 1964.

9 The necessity for a fuller study of this influence is evident. See E. Mendelsohn, "The Biological Sciences in the Nineteenth Century: Some Problems and Sources," *History of Science.* vol. 3, 1964.

10 Rudolf Virchow, "Standpoints in Scientific Medicine (1877)," translated in *Disease, Life, and Man, Selected Essays of Rudolf Virchow,* translated with an Introduction by Lelland J. Rather, Stanford, 1958, p. 142.

11 Johannes Müller, *Handbuch der Physiologie des Menschen,* Coblenz, 1835, 1837, 2 vols.

12 The most interesting biographical study is that prepared by Emil Du Bois-Reymond, "Gedächtnissrede auf Johannes Müller (1858)," *Reden,* vol. 2, pp. 143-334, Leipzig, 1887.

[13] Johannes Müller, "On the Life and Writings of the Late Professor Rudolphi," *Edinburgh New Philosophical Journal,* vol. 25, p. 227, 1838.

[14] Richard Owen, "Lorenz Oken," *Encyclopedia Britannica.*

[15] Lorenz Oken, *Elements of Physiophilosophy,* trans. by Alfred Tulk, London: Ray Society, 1847, p. 29.

[16] Müller, *Handbuch,* vol. 1, p. 22.

[17] Richard Owen, "Lorenz Oken."

[18] Oken, *Physiophilosophy,* p. 190.

[19] Justus von Liebig, "An Autobiographical Sketch," trans. by J. Campbell Brown, *Smithsonian Report,* 1891, p. 260.

[20] Johannes Müller, *Handbuch.* The English quotations are based on the translation by William Baly, *Elements of Physiology,* 2 vols., London, 1838, 1842. Citations will give the volume and pages from the translation followed by those of the German edition in brackets. Müller, *Physiology,* I, 18 (I, 18).

[21] Müller, *Physiology,* I, 3 (I, 3).

[22] Müller, *Physiology,* I, 4 (I, 4).

[23] See Florkin, "Schwann chez Müller, Ses travaux de l'année 1835 sur la physiologie des muscles," *Rev. Médicale Liège,* vol. XI, pp. 640-646, 1956.

[24] Theodor Schwann, "Untersuchungen über die Gesetze der Muskelkraft," *Isis,* 1837, p. 524.

[25] Müller, *Physiology,* II, 903-906 (II, 59-62).

[26] Müller, *Physiology,* II, 906 (II, 62).

[27] Schwann, "Muskelkraft," p. 524. Later experimentation, it should be pointed out, has found Schwann's formulation to be wrong.

[28] Cited from Marcel Florkin (ed.), *Letters de Théodore Schwann (1810-1882),* *Mémories de la Société Royale des Sciences de Liège,* ser. 5, t. II, fasc. 3, p. 138.

[29] *Lettres Schwann,* p. 142. (Italics added.)

[30] Du Bois-Reymond, "Gedächtnissrede Müller," p. 206, 219.

[31] Estelle Du Bois-Reymond, *Jugendbriefe von Emil Du Bois-Reymond an Eduard Hallmann,* Berlin, 1918, p. 88.

[32] Cited in *Lettres Schwann,* p. 216; and also reprinted in Du Bois-Reymond, *Reden,* vol. II, p. 551.

[33] *Lettres Schwann,* p. 143.

[34] *Lettres Schwann,* "An Peter Schwann, 7 März 1835," pp. 31-2.

[35] A copy of this manuscript generously was made available to me by Dr. P. Böhringer, director of L. Schwann Verlag, Düsseldorf, who has a sizeable collection of materials relating to Theodor Schwann.

[36] Müller, *Physiology,* I, 18-19 (I, 19).

[37] Schwann, *Microscopical Researches,* p. 192 (228).

[38] Schwann, *Microscopical Researches,* p. 193 (229).

[39] E. Mendelsohn, "Schwann's Mistake," *Ithaca: Actes Xe Congrès International d'Histoire des Sciences, 1962,* Paris, 1964.

[40] Müller, *Physiology,* I, 20, 22 (I, 20).

[41] Schwann, *Microscopical Researches,* p. 187 (221).

[42] Schwann, *Microscopical Researches,* p. 190 (225-226).

[43] Schwann, *Microscopical Researches,* p. 215 (257).

[44] Schwann, *Microscopical Researches,* p. 215 (257).

[45] Richard H. Shryock, *The Development of Modern Medicine,* New York: Knopf, 1947, pp. 169-170.

[46] Marie François Xavier Bichat, *Physiological Researches on Life and Death,* with notes by François Magendie, trans. by F. Gold and G. Hayward, Boston, 1827, pp. 77-78.

[47] Bichat, *Physiological Researches,* p. 78.

[48] Bichat, *Physiological Researches,* p. 80.

[49] Claude Bernard, "Définition de la Vie," *La Science Expérimentale,* Paris, 1878, p. 162.

[50] C. Bernard, "Définition de la Vie," p. 162.

[51] Compare Bernard's comments as found in "Définition de la Vie," p. 163 and *Leçons sur les Phénomènes de la Vie Communs aux Animaux et aux Végétaux,* Paris, 1885, t. I, pp. 8-9.

[52] Owsei Temkin, "The Philosophical Background of Magendie's Physiology," *Bulletin History of Medicine,* vol. XX, p. 28, 1946.

[53] François Magendie, *Précis Elémentaire de Physiologie,* Paris, 1816-17, 2 vols. I have used the English translation of John Revere.

[54] Cited by J. M. D. Olmsted, *François Magendie, Pioneer in Experimental Physiology and Scientific Medicine in XIX Century France,* New York, Schuman, 1944, pp. 203-204.

[55] Cf. Temkin, "Magendie's Physiology," p. 35.

[56] Temkin, "Magendie's Physiology," p. 35.

[57] Olmsted, *Magendie,* p. 143.

[58] Bernard, *Phénomènes de la Vie,* p. 38-39.

[59] Claude Bernard, *Principes de Médecine Expérimentale,* ed. Leon Delhoume, Paris, 1947, p. xvii.

[60] Bernard, *Phénomènes de la Vie,* p. 51.

[61] Claude Bernard, *De la Physiologie Générale,* Paris, 1872, p. 191.

[62] Claude Bernard, "Du Progrès dans les Sciences Physiologiques," *La Science Expérimentale,* Paris, 1878,, pp. 53-54.

[63] Claude Bernard, *An Introduction to the Study of Experimental Medicine,* trans. by H. C. Green, New York: Dover, 1957, p. 69.

[64] Claude Bernard, *Leçons sur les Propriétés des Tissus Vivants,* Paris, 1866, p. 22.

[65] Bernard, *Introduction,* p. 75.

[66] Bernard, *Tissus Vivants,* p. 23.

[67] Claude Bernard, *Pensées, Notes Détachées,* ed. Leon Delhoume. Paris, 1937, p. 50.

[68] Bernard, *Introduction,* p. 76.

[69] Bernard, *Tissus Vivants,* pp. 6-7.

[70] Rudolf Virchow, "Cellular Pathology (1855)," *Disease, Life, and Man,* p. 87-8.

COMMENTS

ERNST MAYR

Dr. Mendelsohn has analyzed with sensitivity and great insight the emergence of a new system of explanation in 19th century biology. He presented a well balanced selection of quotations from the writings of Johannes Müller, of Theodor Schwann, and of their associates, and traced the origin of the new ideas. Since he knows the literature of that period so much better than I, it will be best if I limit my comments to the interpretation of the currents and counter-currents of that period. As in any other historical event there are, of course, several interpretations possible concerning the reasons for the shift in Schwann's thinking. For the sake of the argument, I shall attempt an interpretation which differs somewhat, at least in emphasis, from Dr. Mendelsohn's analysis. I would not want to claim that it is a superior explanation, but merely that it is a legitimate alternative. In view of our uncertainty about the intellectual currents in the first half of the 19th century, it would seem worthwhile in future researches to make a comparative exploration of both alternatives.

It is extraordinary how many new movements, indeed perhaps all of them, start as a rebellion against that which preceded them. Since there is no doubt that a new interpretation of physiological phenomena emerged in the minds of Müller's students, Schwann, Du Bois-Reymond and Helmholtz (with Schwann recognized as their leader) we must ask: "What did Müller's students rebel against?" Clearly it was something in Müller's world of ideas, but what? Was it really his attachment to *Naturphilosophie,* as it might appear on the surface? Personally, I rather doubt it. Müller, himself, had rejected this influence at an early age, and his inaugural lecture on October 19, 1824, was already distinctly anti-Schelling and hence anti-*Naturphilosophie.* To be sure, Müller was still pro-Goethe and hence pro-idealistic morphology, but this is something quite different from *Naturphilosophie*. Before taking up what the new generation rebelled against, let me add an aside on *Naturphilosophie*.

When we look at the scene contemporary with Müller, we do not seem to find a great following for *Naturphilosophie* among scientists. Indeed it is my impression that, except for Oken and Carus, there was not a single respectable anatomist, physiologist, zoologist or botanist whom one could truly label a disciple of Schelling or follower of Oken. I admit that I have not made a special study of their writings, but from what little I know of their work, all the leaders of contemporary science, not only Rudolphi, but also Retzius, Rathke, Meckel, Tiedemann, Ehrenberg, and Purkinje were rather sober, descriptive and uncommitted scientists.

Romanticism had a greater impact on poetry, literature, painting, and music in Germany than in any other country in the world. *Naturphilosophie* was the expression of Romanticism in philosophy, but its application to science was singularly unsuccessful and, I add, unimportant. Many of the ideas often ascribed to *Naturphilosophie* originated in the 18th century (Bonnet) or earlier, even though they were later incorporated into *Naturphilosophie*. Still others, such as vitalism, have really nothing to do with *Naturphilosophie*. We shall come back to this later.

What, then, in Müller's thinking was Schwann opposed to? I think there are two points, both of them appropriately stressed by Dr. Mendelsohn. The first is any and all mixing of philosophy and science. This is clear from Schwann's correspondence in 1835 with his brother Peter, and from Virchow's later statement. Idealistic morphology and other "interpretations" borrowed from philosophy were for Schwann as objectionable as *Naturphilosophie*.

More specifically, the rebels objected to the recognition of any kind of vital force, later often referred to as "vitalism." Vitalistic ideas existed long before the days of *Naturphilosophie* and have persisted long after the demise of *Naturphilosophie*. Indeed, a belief in vital forces has been part of the philosophy of many a biologist who had not a trace of connection with *Naturphilosophie*. It is extraordinarily difficult to discuss vitalism in an objective, detached manner. The closer we come to the recent period, the less successful have been the attempts to revive a vitalistic interpretation of phenomena of life. Vitalism as a possible theory of biology has been dead for some 40 or 50 years, as has been the entire argument of mechanism *versus* vitalism. The situation was totally different in the days of Johannes Müller and Claude Bernard. In those days, a belief in vital forces was, curiously enough, in many cases a sophisticated reaction to crude theories of mechanism. It was not a special philosophy and it had no separate manifesto. Usually it consisted merely in the statement of a working biologist that he was willing to study the phenomena of life in a detached, objective manner, and if he found an inexplicable residue, an unresolved *X*, so to speak, exclusively connected with living organisms, he would call it a vital force, or whatever terminology he used. This was actually a rather sober attitude, and it characterized Claude Bernard and Müller in his mature days, as well as many other reputable biologists in the middle of the last century. This approach was quite different from the later ones of a Bergson or Driesch, who used vitalism as a broad platform on which to erect the large edifice of a whole philosophy of organisms. The vitalism of the working physiologists in the first half of the 19th century was a philosophy of resignation and scepticism. It was based on an admission that there was a residue that could not be resolved.

I believe it was this resignation in Müller, the consequence of his disappointment with *Naturphilosophie,* against which his young students rebelled. It may sound paradoxical, but the writings of Schwann and his associates give one the impression that they were actually romantic in their hearts. To be sure, they were not romantics of mysticism and obscurantism, as Schelling and Oken, but reductionist romantics, romantics who dreamed of reduction of all biological phenomena to physical and chemical phenomena and laws, who were not willing to compromise, who were not willing to set aside an unknown or an unknowable. However, it was not for very long that the reductionists held the field. Dr. Mendelsohn has quoted quite perceptively Virchow's statement of 1855 in which he castigates "the scientific prudery of insisting that all organic processes be reduced to the resultant of inherent molecular forces." Du Bois-Reymond's later "Ignoramus, ignorabimus" is another illustration of the ultimate resignation that caught up also with this rebellious generation.

Dr. Mendelsohn asked the question why there was such a difference between, on one hand, the reserved approach of the French physiologists, of the mature Müller, and of Virchow, and on the other hand, the radical, uncompromising approach of the young German physiologists. No doubt there are a number of reasons for this difference, but the most important one probably is the dominant philosophy of the period preceding the stated scientists. If there was a radical movement first, the followers would be moderates. Where moderates had been dominant, radicals followed them. In the case of the French physiologists, their moderation was the reaction to the extreme mechanism (reductionism) of Lamettrie and Baron Holbach. In the case of Müller, it was the reaction to *Naturphilosophie;* in the case of the young German physiologists, it was the reaction to the compromising moderation of the middle-aged Müller, now thoroughly cured of his Naturphilosophical phantasies. Perhaps the most fascinating aspect of the developments at that period is the rapidity of reactions and counter-reactions. Within a period of less than 10 years, we have Müller's reaction to *Naturphilosophie* and Schwann's reaction to Müller's resignation and scepticism. Not much later follows Virchow's reaction to the youthful reductionists.

One aspect of the reductionist approach fascinates me particularly. Now, in the 1960's we see clearly that there is no vital force and that it is possible to interpret even teleonomic phenomena in terms of physics and chemistry as the product of genetic information coded in the DNA of the germ cells. As a consequence, we are *a priori* biased against any author admitting vital forces. Yet, if we look back at explanations in the first half of the 19th century, we find that the reductionists in their crude materialism ("muscle action like that of any elastic substance") are actually much farther from the truth than their vitalist opponents. All one has to do is to replace the X (= vital force) in the formula of the vitalist by certain newer findings and one has essentially an interpretation that can still stand. There is an obvious parallel here with the situation in evolutionary biology during the first two decades of the 20th century.

Their differences in philosophy affected the scientific methodologies of the physiologists. The reductionists did not care whether they worked on a muscle inside an organism or in an isolated "dead" muscle. The vitalists, in order to remove the factor *X* from their variables, took great pains to do their experiments only on parts of living organisms. By eliminating the complicating factor of death, they were much better able to study physiological phenomena *per se*. How wise Claude Bernard and Müller were in their insistence on this method is indicated by the fact that in principle this is still the prevailing method in physiology, except at the strictly molecular level.

You may feel that I have gone too far in turning the tables and labeling the reductionists as romantics and the vitalists as the sober, sceptical scientists. Perhaps I have, but this different emphasis is necessary in order to bring home an important point, and in order to help to establish a genuine balance in our interpretation. It sometimes takes a real effort to put oneself back into thinking of a long past period, and yet one has to do this in order to make just value judgments. To judge an 1830's scientist in the hindsight of 1960 may add much to an analysis, but cannot give complete balance.

Let me add one final point. When I stress the deficiencies of Schwann's reductionist approach, I am not trying to devalue either its uncompromising honesty nor its heuristic power. I think it can be shown again and again in the history of biology that the oversimplified, often crude, often atomistic approach almost invariably produces more results and results more rapidly than the more sophisticated, more holistic approach, which is perhaps most useful after the reductionist approach has exhausted the pathways that are accessible to it. The immense success of the reductionist approach of Schwann, Du Bois-Reymond and Brücke, is a matter of history.

Museum of Comparative Zoology
Harvard University
Cambridge, Mass.

7

J. J. C. SMART

CONFLICTING VIEWS ABOUT EXPLANATION

Presented December 12, 1963

I have considerable sympathy with some ideas on explanation which have been put forward by P. K. Feyerabend and by Wilfrid Sellars. These ideas are not easy to understand or to evaluate, and part of my purpose in this paper is to state the views of these philosophers in a perhaps rather oversimplified form and to compare them with one another and with the more orthodox view which we find in Ernest Nagel's book *The Structure of Science,* whose account I shall probably oversimplify too. I ask these three writers their indulgence for any such simplifications of their ideas, and I would also ask those who read this paper to go back to the original works and check what I have to say against the *ipsissima verba.* In this way I hope indirectly to promote understanding and criticism of these three philosophical positions: oversimplification can be a step towards a full understanding, since a complex and subtle view can easily get misunderstood while the reader is still lost among the complexities and subtleties. A simplified presentation can therefore help to bring such a reader back towards understanding, provided he heeds the warning to go back to the originals for himself.

Nagel is a notable exponent of the position which Feyerabend and Sellars are concerned to attack. He maintains a rather orthodox theory of scientific explanation and its relation to common sense. A leading feature of this view is that it holds *correspondence rules* to be needed in order to relate the concepts of scientific theory to those of common sense. Correspondence rules are also, on this view, often needed to relate one scientific theory to another:

e.g. classical thermodynamics to statistical mechanics. The picture is that of a number of relatively autonomous but interrelated theories, with common sense at the bottom, then theories such as classical thermodynamics, and then above these more remote theories, such as kinetic theory, and then above these even more remote theories, such as quantum mechanics, and so on. Theories can be connected to one another by means of the correspondence rules. Another important case of the hierarchy is that of common sense–classical mechanics–quantum mechanics. The "correspondence principle" relates quantum mechanics to classical physics, which gives the theory of the measuring instruments, and this in turn is anchored to observation reports in common sense language.

Feyerabend rejects this picture of science. According to him a theoretical language can be its own observation language. When I look at the book which is before me I can describe what I see as a parallepipedal swarm of atoms. When I carry a heavy suitcase, and if I know the general theory of relativity, I can say that I feel the curvature of space-time. According to Feyerabend, observation statements are not distinguishable from theoretical ones by the fact that they contain special observation terms: they are distinguished pragmatically, in that they are statements to which we may assent quickly, relatively reliably and without calculation or inference. They must also presumably refer to particulars or be existentially quantified. (This pragmatic characterisation of observation statements is to be found in Carnap's *Testability and Meaning,* but perhaps tacitly abandoned by him in later writings. See Feyerabend (2) pp. 36-42.) A man observes that the voltage across a circuit is 15 volts, not that a gadget marked "voltmeter" has its pointer opposite a mark beside which is inscribed the numeral "15." The latter observation would be of no scientific interest, and of course a gadget marked "voltmeter," and indeed which *was* a voltmeter, could easily, in certain circumstances, have its pointer opposite the "15" mark and yet without the circuit across which it was placed having a potential difference of 15 volts. Again, a scientifically educated person actually can observe sugar on his plate *dissolving,* and not just "disappearing from view" or "melting" as the nonscientific person frequently and erroneously will say.

One reason why Feyerabend wants to say that a theoretical language can be its own observation language is that he thinks of our ordinary unreflective common sense as a no longer tenable theoretical structure. (Consistently with his own view he sees common sense as itself a theory which is its own observation language.) The relation between common sense and (say) the atomic theory is like that between classical thermodynamics and statistical mechanics. The first of each of these pairs does not have to be linked to the second by means of correspondence rules: rather, the first drops out and is simply to be *replaced* by the second. For example Feyerabend says boldly, and I think correctly, that classical thermodynamics is simply false. So why not also common sense? That classical thermodynamics is false can be seen if we reflect that statistical mechanics implies that given a long enough time span things will not *always* happen as predicted by the classical theory. Moreover even within ordinary time spans, we see classical thermodynamics to be refuted by the phenomenon of the Brownian motion. (See Feyerabend (2) pp. 65-66.) It is of course not *directly* refuted: the Brownian motion can not even be understood if we are limited to classical thermodynamics. But the Brownian motion can be explained and its details predicted by the statistical theory, and this is one reason (among others) for preferring the statistical theory. On the basis of the statistical theory we can deduce that the motion of the Brownian particles violates the second law of thermodynamics, as stated in the classical theory. It follows that unless we are to fall into logical inconsistency we cannot retain both classical and statistical thermodynamics. *A fortiori* it is necessary to deny that classical thermodynamics is deducible from statistical mechanics. If classical and statistical thermodynamics are inconsistent with one another, then for the good health of the statistical theory, the classical one had better not be deducible from it.

It is often difficult in philosophy to know whether or how much a dispute is a substantive one, and not just one of emphasis or of what is thought interesting or important. But there does *seem* to be a substantive issue between Feyerabend and Nagel here, in that Nagel holds that classical thermodynamics should be deducible from statistical thermodynamics, partly by correspondence rules, and not just thrown away and *replaced* by the newer theory. Nagel makes the growth of science look like a growth by accretion, rather like that of an Australian city, where outer suburbs are continually added on and (until very recently at any rate) the central region does not change much. (There may of course be an *occasional* pulling down. Consider the rejection of the phlogiston theory. But by and large the picture is that of science building on what is already there.) Feyerabend makes the growth of science look more like that of New York: old structures, such as the old Park Avenue and Pennsylvania Station, are pulled down to make room for new ones. Even on Feyerabend's view, of course, this does not mean that the old and rejected theories are unimportant: often in fact we could not have got to the new theories without having invented the old ones first, just as the new Park Avenue depends for its construction on a financial and social structure which is in part caused by the old street. Nagel would argue, however, that the need to retain the older theories is not just a practical one. He might question, for example, whether we could even in principle do celestial mechanics straight away within general relativity, instead of using general relativity to correct the still indispensable theory of Newtonian gravitation. What would celestial mechanics look like if done purely relativistically?

At this point someone might retort that the issue between Nagel and Feyerabend is *not* a substantive one. New York can and does get on without the old Park Avenue and it could and will get on without the old Pennsylvania Station, but in science we just can not in fact get on without the old theories. For example it would be impossible in practice to use general relativity in celestial mechanics, though it may be needed to explain isolated anomalies (as with the perihelion of Mercury). It would seem to follow, then, that we do need both the old and the new theories and that we do need rules of correspondence between the theories. Feyerabend must be right in saying that these rules can not be such as to permit strict deduction of one theory from another (since the theories will in general be strictly inconsistent with one another). Nevertheless surely there must be some approximate rules of correspondence between the two theories, for example identifying phenomenological temperature of a gas with average kinetic energy of molecules. Though the old theory, being inconsistent with the new theory, can not be strictly deduced from it, surely the approximate correctness of the old theory must be deducible from the new one. And perhaps this was all that Nagel really meant to claim when he made deducibility a criterion for the reduction of one theory to another.

If this is in fact all that Nagel claims Feyerabend should be able to agree with him. He himself states that the new theory must explain why the old theory predicts the facts to the accuracy with which it does predict them. (Here Feyerabend agrees with Sellars. Feyerabend (1).) For example, general relativity should be expected to make the same predictions as Newtonian theory in those cases where Newtonian theory has been found to work. The theorist can tell, on theoretical grounds and from his knowledge of the old theory, just when it is safe for *practical and predictive* purposes to use the old theory. There is therefore no reason why he should be expected to use general relativity to compute the nautical almanac, and moreover the observational tests of the old theory in circumstances in which the two theories can be expected to agree become *ipso facto* tests of the new theory too. To Nagel the distinction might seem rather a fine one between: (1) rejecting *in toto* an old theory but arguing that in most cases it will give the same observational results as the new theory, and (2) keeping the old theory and joining it to the new one by means of bridge laws. For Feyerabend and Sellars the distinction might seem a much more important one. Moreover I have already suggested that Nagel could easily weaken his requirement of strict deducibility, and stipulate merely that the approximate truth of the old theory be deducible from the new one. Perhaps Feyerabend would say that truth can not be approximate but only predictive value can be: that the old theory is useful he will not deny, but he could still claim that this is more a matter for engineers than for natural philosophers.

As I have said, Feyerabend's attitude to common sense is similar to his attitude to such theories as classical thermodynamics which are to be replaced by more adequate theories. That is, he thinks of our common sense conceptual scheme as a false theory. Of course common sense notions are useful, and for this reason they have survived after thousands of years of linguistic and cultural evolution. Nevertheless the fact that they are useful does not in any way imply that they are true. Within a restricted domain of application a false theory may be more useful than a true theory precisely because it is false. Consider Newtonian mechanics, as applied to most of the planets, though not to Mercury. Being not quite true, it is enabled to be considerably simpler. The true theory, general relativity, is not so simple or easy to work with, for it has to take care of subtleties which are only rarely of practical importance. In the same way, common sense might be metaphysically false and yet have been of more practical usefulness than might have been a more correct theory. According to Feyerabend, a theoretical language can be its own observation language: the observational facts could be, and ideally should be, expressed in the language of the new theory itself. Indeed the observation sentence "this liquid has a temperature of 50°C" as stated by the theorist of classical thermodynamics will have a different meaning from that of the same sentence as stated by the theorist of statistical mechanics. Neither, of course, will have the same meaning as the common sense sentence, "the greyish looking liquid in that tube reaches a scratch with the numeral '50' written beside it."

But if a sentence which verifies the one theory has a different meaning from the apparently identical sentence which refutes the other theory, how can it be a crucial test between the two? Feyerabend's answer is that the usual notion of the crucial experiment is misconceived—a theory is a good theory if it explains all the facts in its own terms, and provided that no other theory explains all these facts and renders yet other facts testable too. (A theory can fail by concealing facts: Feyerabend points out that experimental techniques are not fine enough to enable us to test phenomenological thermodynamics by means of the phenomenon of Brownian motion, which leads to experimental test only within the conceptual scheme of statistical thermodynamics.) Thus the new theory must be at least as wide in its scope as is the old one.

As against this view of the matter, many people feel the following uneasiness. Surely, they will think, in a crucial experiment there *is* a neutral fact which belongs exclusively neither to one theory nor to the other theory and which decides between them.

Thus in the crucial experiment (the observations of the advance of the perihelion of Mercury) which helps to decide between Newtonian theory and general relativity, there are various facts of the form: Mercury is now *there* and yesterday it was *there*. Instead of Mercury being at *A* (where it ought to be on Newtonian theory) it is found to be at *B*. Now surely, it will be said, the question "Is Mercury at *A* or is it at *B*?" can be asked independently of commitment to Newtonian or Einsteinian theory. As against this, Feyerabend would reply that the concept of spatial position at a time has a different meaning within relativity theory from that which it has in Newtonian theory, in that in the former but not in the latter case it is something relative to a frame of reference. In this sort of way he would argue that the question "Is Mercury at *A* or is it at *B*?" means something different to the Newtonian from what it does to the Einsteinian, and so it is not a neutral fact which is independent of the two theories. The above answer will of course not quite do as it stands if the choice is conceived to be not that between general relativity and Newtonian theory but that between general relativity and Newtonian theory as modified by special relativity. But here I am concerned with the sort of answer that Feyerabend might be expected to give, and I am confident that in this last case too he could easily make a plausible case for meaning change. (I do not know whether modern linguistics would justify us in calling the above phenomenon "a change of meaning", but if not I doubt whether it would matter for Feyerabend, since he is not necessarily using the expression "change of meaning" in the sense in which the linguist uses it.)

But still one may feel a resistance. Surely it is a neutral fact that in order to see Mercury we have to point the telescope over the top of that tree say, and not, as predicted by Newtonian theory, over the top of that chimney pot. And surely one can talk of trees, chimney pots, and telescopes in a way which is independent of the choice between Newtonian and Einsteinian theory. However Feyerabend could well concede that we use Euclidean geometry and non-relativistic optics for the theory of our telescope. He would say that this is not the real truth about our telescope, the tree, the chimney pot and the planet, but nevertheless it is legitimate to think in this way in order to discuss the observational test of general relativity, since we know on theoretical grounds that our prediction will be unaffected (up to the limits of observational error) if we avail ourselves of this computational convenience.[1]

But, I have heard it objected,[2] what about some biological hypothesis? Suppose, to take a rather unlikely example, that some genetical or biochemical theory were tested by observing whether tigers have stripes. Surely we can observe whether or not the tigers have stripes without in any way at all committing ourselves to any of the rival genetical or biochemical hypotheses. Feyerabend's reply is, I think, that this is indeed so with theories or hypotheses of limited scope but this is not so with theories of universal scope, which apply to all objects in the world. (Such as Newtonian mechanics, general relativity, or quantum theory.) Since theories which apply to everything must apply to our measuring instruments and observational devices, it follows that such a theory can be, and indeed ought to be, its own observation language.

There is clearly something appealing about this. We feel that there would be something wrong with relativity if it could not in principle give the theory of light rays and the pointing of telescopes. Quantum mechanics would be deficient (perhaps indeed it is deficient in this respect) if it could not give an account in its own terms of the theory of measuring instruments. Nevertheless when we try with the aid of very concrete examples to see how a theory could be its own observation language we feel qualms again. Suppose one observes a track in a Wilson cloud chamber. Here "Wilson cloud chamber" is an expression in the common sense language of the permanent in change, not of the tenseless language of space-time. How in laboratory contexts could we talk the relativistic tenseless space-time language? We should have to say something like: "Time-slice of me (4) at t_2 observes time-slice at t_2 of photographic plate (4) which is like time-slice at t_1 of cloud chamber (4)." (The (4) in brackets after a word is to indicate that the word is being used to denote a four-dimensional space-time object. The expression "time-slice" is meant to be used as in Woodger's *Theory Construction* (1939).) Could such a language exist independently of common sense language? Could a baby grow up using a tenseless four-dimensional space-time language? Perhaps this is a psychological question only. Well, could we, as adults, for observational purposes use such a four-dimensional tenseless language? I incline to think that we could, but it is hard to be sure since the question has not sufficiently been investigated with the help of concrete examples.

There are other features of scientific discourse which make it hard to see just how observational talk could be done within the theoretical language itself. Just what would be the meaning of the word "galvanometer" as used in this sort of way? Does "galvanometer" mean something like "galvanometer-shaped swarm of electrons, protons, etc."? Feyerabend would have to say that in modern scientific discourse "galvanometer" does mean something like this. Moreover he could make out something of a good case. If you asked a physicist "What is that thing on the bench?" and he replied "It's a galvanometer", and if you then asked "Do you mean that it is an object consisting ultimately of electrons, protons, etc.?" it is very likely that he would reply "Yes". (I myself, as a philosopher, would wish to say "Yes" here.) Nevertheless, even supposing that the complete theory of our measuring instrument can be expressed, say, within quantum mechanics, can we seriously envisage talk about particular measuring instruments in the scientists' own particular laboratories *purely* in quantum mechanical terms? It certainly seems hard to envisage in concrete terms a theoretical language functioning as its own observation language.

This brings us to what I conceive to be the position of Wilfrid Sellars. He concerns himself with two different and mutually irreconcilable world pictures. The first he calls "the manifest image" and the second he calls "the scientific image". The manifest image is articulated in terms of our ordinary everyday conceptual scheme, the one so lovingly analysed by G. E. Moore and P. F. Strawson. Sellars agrees that it is important to analyse the manifest image in roughly Strawsonian fashion. (For convenience I shall therefore sometimes call ordinary language, perhaps philosophically tidied up a bit, a "Strawsonian" language.) The reason why Sellars thinks it important to analyse the manifest image is not, however, because he thinks that it is a true picture of the world. If anything gives the truth about the world it is the scientific image (either present science or some future, improved science, I suppose). Included in the manifest image is what we learn from those sciences, such as taxonomic zoology or stimulus-response psychology, which do not depart from the categories of a Strawsonian language, as does, for example, general relativity. The manifest image will also have concepts of common sense unobservables, such as minds. What gives the manifest image its philosophical importance is that it is the conceptual structure in which we naturally think, and in which we can not help thinking so long as we do not deliberately and consciously determine to think in terms of scientific theories. As soon as we relax our scientific theoretical guard, back comes the manifest image. According to Sellars, and contrary to Feyerabend, the manifest image is quite indispensable. It is only through the manifest image that we can get to theoretical science at all: the scientist in his laboratory must think in terms of the manifest image, for he must describe his apparatus partly in Strawsonian terms. Not only must he think of himself as putting copper sulphate crystals in water: he must also inevitably think of himself as putting a blue powder into a liquid such as is found in lakes. What would be the full quantum mechanical description corresponding to the above? So within his laboratory the scientist needs the language of the manifest image. Here Sellars agrees with Nagel, but where, in effect, he seems to differ from Nagel is that according to his view the sentences of the manifest image are not deducible from the theoretical ones (together of course with statements of initial or boundary conditions). They had better not be, on Sellars' view, because the manifest image is not the truth about the world. What must be deducible are particular truths about the world stated in the theoretical language itself, though there are also certain rules which enable one to connect the language of the theory to that of the manifest image in a rather looser way than that of deducibility within a language.

Sellars therefore occupies a position interestingly intermediate between those of Nagel and Feyerabend. According to Nagel common sense (or much of it) is both true and indispensable. According to Feyerabend it is neither true nor indispensable. According to Sellars it is (methodologically) indispensable but false.

I have some tendency to prefer one of either Feyerabend's or Sellars' positions to that of Nagel. For it does seem to me that common sense language does to some extent present a false theory about the world. To take one important feature for an example: common sense makes a partition between space and time which Minkowski has shown to be untenable. However, I would be inclined to defend the scientific truth of common sense more than Feyerabend or Sellars would. (For example I think that our common sense notion of colours can be analysed in terms of the discriminatory reactions of human beings, so that it can be fitted harmoniously into the scientific picture of the world.) I agree with Sellars that the manifest image is incompatible with the scientific image, even though I do not agree that it is quite as incompatible as he thinks it is. I believe that the choice between Sellars' and Feyerabend's positions will have to be made on the basis of empirical knowledge not yet available to us. It might turn out to be the case that a roughly Strawsonian conceptual scheme is innate to us. It might well be that babies are born with the neuronal interconnections in their brains to be such that they can be taught a Strawsonian language and no other. So that if we are to go beyond this common sense stage we must do so with brains hopelessly vitiated by the metaphysics of the manifest image. So we must go through life as Sellars argues we must, that is with two different and inconsistent images of the world. While as rational men we may prefer the scientific image we can not help continually relapsing into the manifest image.

Feyerabend, on the contrary, seems to adopt the alternative view that we could go straight to the scientific image without basing ourselves first on the manifest image. See his footnote 92 on p. 76 of his (2). As against the objection that the language of common sense is needed for introducing the scientific language, he asks why the scientific language could not be learned from scratch. He says: "Is it really asserted that what is impossible for a small child will be impossible for a philosopher . . .?" On the hypothesis of a certain innate brain structure in the child, it might well be impossible. I may of course be exaggerating the differences between Sellars and Feyerabend. For on pp. 67-9 of his (2), Feyerabend puts forward the view that the meaning of a theory is deepened when we contrast it with its alternatives, and if so, then surely the manifest image is needed in order to provide a contrast with the scientific image, and Sellars stresses the importance of such a contrast between the manifest and scientific images.

I believe that the hypothesis of an innate bent towards a Strawsonian conceptual scheme may well be true. It is somewhat analogous to a suggestion which Noam Chomsky has made. This is that a tendency to conjecture the hypotheses corresponding to the rules of syntax of natural languages is built in to us, and that on purely inductive grounds a child could never learn the rules of a language in sufficient time.

On the other hand if the hypothesis of an innate conceptual structure is false, then it may be that one day our children will grow up as little Feyerabendians and not as little Strawsonians.

University of Adelaide, Australia

NOTES

[1] Nagel would defend the truth of geometrical optics in this sense: it is a method of relating object to image after light has passed through lenses and been reflected from mirrors, and it is the truth, Feyerabend notwithstanding, because it does correctly give these relations.

[2] I think by David Stove, in an illuminating conversation which I had with him and with G. C. Nerlich and W. A. Suchting, at the University of Sydney, when we discussed Feyerabend's ideas.

REFERENCES

Carnap, R., "Testability and Meaning," *Philosophy of Science 3* (1936) 419-471 and *4* (1937) 1-40.

Chomsky, N., Review of Skinner's *Verbal Behavior, Language 35* (1959) 26-58.

Feyerabend, P. K., (1) Comment on Sellars' paper in H. Feigl and G. Maxwell (eds.) *Current Issues in the Philosophy of Science,* (New York, 1961).

(2) "Explanation, Reduction and Empiricism," in H. Feigl and G. Maxwell (eds.), *Minnesota Studies in the Philosophy of Science,* Vol. III, (Minneapolis, 1962), 28-97.

Nagel, E., *The Structure of Science,* (New York, 1961).

Sellars, W., *Science, Perception and Reality,* (London and New York, 1963). See especially Chapter 1, "Philosophy and the Scientific Image of Man," and Chapter 4, "The Language of Theories." (Chapter 4 originally appeared in the same volume as Feyerabend (1).)

Strawson, P. F., *Individuals,* (London, 1959).

Woodger, J. H., *The Technique of Theory Construction,* (International Encyclopedia of Unified Science, Vol. 2, No. 5, Chicago, 1939).

COMMENTS

WILFRID SELLARS

SCIENTIFIC REALISM OR IRENIC INSTRUMENTALISM

Comments on J. J. C. Smart

1. Professor J. J. C. Smart [1] correctly locates my views on the topics he discusses, taken as a whole, somewhere in the middle between those of Ernest Nagel and Paul Feyerabend. Any such one dimensional ordering of course does violence to the complexity of the issues and the many combinations of agreement and disagreement this complexity permits. But with respect to the large scale contrast he is drawing, I think that Smart is right in the following sense: I tend to agree with Nagel or Feyerabend on each of the major points, but to agree with neither on all of them. Roughly, I tend to agree with Nagel on methodological topics, with Feyerabend on ontology, and with each of them in turn on key epistemological issues.

2. Since, of the two, Nagel has given the far more comprehensive and systematic account of theoretical explanation,[2] one which not only continues the classical tradition on this subject but is itself already a classic, it is clearly good strategy to begin by concentrating on those features of Nagel's account which bear on the questions raised by Smart, and then proceed to examine the somewhat more impetuous, but challenging and illuminating, views which Feyerabend has expressed on these issues.

3. The theories which raise in paradigm form the philosophical puzzles with which Smart is concerned are of the type which postulate 'unobservable' micro-entities to explain the behavior of the 'middle size' objects of perceptual experience. Not all theories, even as scientists use the term, are of this type, and many subtle philo-

1 In "Conflicting Views about Explanation."
2 In his *The Structure of Science* (New York, 1961).

sophical problems are raised by theories of other types. Some of these additional problems are also raised by micro-theories—e.g. the problem of the relation of mathematically idealized concepts to coarser grained concepts admitting of perceptual discrimination. The following discussion, however, will focus attention by and large on problems *peculiar* to micro-theories.

4. Before I turn to Nagel's views on micro-theoretical explanation, it will be useful to make a preliminary comment on Feyerabend's notion [3] that 'common sense' is a 'theory' which, like all theories, should be abandoned by scientists, except for practical purposes, when a better theory comes along. The term 'theory' is one of those accordion words which, by their expansion and contraction, generate so much philosophical music. And while I agree with Feyerabend that the conceptual framework of 'common sense' is, in the last analysis, false, I find it confusing rather than clarifying, for reasons which will emerge in the course of the argument, to speak of this framework as a false *theory*.

5. Now if one uses the phrase 'the framework of common sense' in such a way that it includes common sense *beliefs* in its reference, then this framework would, indeed, include not only empirical generalizations, but theoretical fall-out from earlier science and even proto-theories of its own. But not all subject-matter dependent universal propositions to which common sense is committed are properly characterized as *beliefs*. There are many principles about physical objects and the perception of them ('categorial principles' they might be called) which are constitutive of the very concepts in terms of which we experience the world. And while I agree with Feyerabend that this does not exempt them from criticism and possible replacement, it is, as I see it, incorrect to compare this possible replacement to the replacement of a corpuscular by a wave theory of light, or dephlogistication by oxidation.

6. What would the conceptual framework of common sense be a theory *of?* Granted that common sense beliefs *within* this framework include proto-theories about specific subject matters capable, in principle, of being characterized without the use of the vocabulary of these proto-theories, what is the framework itself a theory

[3] Cf. Smart, pp. 159, 162.

of? A *trivial* answer is, of course, available namely, "Common sense objects and events." But this would be like answering the question "What is atomic theory a theory of?" by saying "Atoms." A *false* answer is also available, namely, "Sense impressions."

7. If we distinguish, in the spirit of the classical account, between the 'internal' and 'external' subject matters of micro-theories, so that in the case of the kinetic theory of gases, for example, molecules and their behavior would be the 'internal' subject matter of the theory, and gases as empirical constructs defined without reference to molecules its 'external' subject-matter, then, as I see it the conceptual framework of common sense has no *external* subject-matter and is not, therefore, in the relevant sense a theory *of* anything.

8. Feyerabend, of course, emphatically rejects phenomenalistic accounts of experience according to which the primary objects of knowledge are sense impressions. Thus he rejects the move according to which physical object statements stand to sense impression statements as, according to the classical account of theories, statements about populations of molecules stands to observation framework statements about gases. Thus he rejects, and rightly so, the idea that the framework of physical objects is *in the classical sense* a theory of which the external subject matter is sense impressions. Nevertheless, there is a sense in which for Feyerabend the framework of physical objects is a theory *with respect to* sense impressions, for he emphasizes that it is in terms of the framework of physical objects that, in perception, we conceptually respond to the impressions of sense.[4] Since, as he sees it, the language in which we *respond* to sense impressions is different from (indeed, incommensurable with) the language in which these sense impressions themselves are described,[5] there is no reason, he feels, why the scientist should not train himself to respond to his sense impressions by still

[4] "Explanation, Reduction and Empiricism," *Minnesota Studies in the Philosophy of Science*, Vol. III, p. 35.

[5] In developing his 'pragmatic theory of observation' which will be examined below, he emphasizes that the meaning of an observation predicate is not determined by the sensory state to which it is a response. This point is spelled out in great detail on pages 38-42 and 94 of the essay referred to in the previous note. He goes so far as to say on p. 39 that "the fact that a statements belongs to the observational domain has no bearing on its meaning."

another incommensurable language—the language of micro-physical theory. In *this* sense, one theory *with respect* to sense impressions would be replaced by another.

9. From this point of view, the correspondence rules which, according to the classical analysis, correlate micro-theoretical constructs with observation framework counterparts are excess baggage, as is the very contrast between internal and external subject matters. The important thing is to get the language of the theory connected *not* with another language, but with the sense impressions of the scientist. And because the concept of language-language correspondence rules and the related distinction between internal and external subject matters play an unimportant role in Feyerabend's account of micro-theoretical explanation, the only *important* sense in which, for him, a theory is a theory *with respect to* something is a matter of its being a behavior orienting perceptual response to that something. In *this* sense, though not in the ordinary sense, the framework of physical objects is a theory with respect to sense impressions. It is this conception as I see it, which underlies Feyerabend's willingness to speak of common sense as a theory.

10. But before these remarks can be put to use in a general assessment of Feyerabend's views, we must return to our general strategy and to Nagel.

II

11. According to Nagel's reconstruction,[6] the following elements must be distinguished. They are more or less explicitly present in the actual practice of micro-physical explanation and would be explicitly present in a micro-theoretical counterpart of Plato's Ideal State:

(a) The deductive system T_D

(b) The *partial* interpretation of T_D by correspondence rules which correlate expressions in T_D with empirical predicates in the observation framework. By doing so, these rules correlate theorems in the

[6] *The Structure of Science*, pp. 90ff.

theory with general propositions formulated in terms of these empirical predicate. Some of these propositions are inductively confirmed (empirical laws); none of them are disconfirmed; some are hypotheses whose possible confirmation is placed by the theory on the scientific agenda.

(c) The model: a *complete* interpretation of T_D

Of these items, (a) (b) are substantive parts of the theory. Whether, and if so in what sense, this is also true of (c) will be explored subsequently.

12. Let me go into these features in somewhat greater detail. In the first place, Nagel thinks of the non-logical vocabulary of the deductive system, T_D as consisting of predicate variables.[7] This, I believe, is a serious mistake which is rooted in an inadequate conception of what makes a non-logical predicate meaningful. It stands, as we shall see, in the way of a genuinely realistic interpretation of micro-theoretical entities.

13. Certainly the form of words

> If two Ps are on an L, then there is another P which stands to them in R

could be a statement function containing predicate variables 'P' 'L' and 'R'. It *would* be so if it was intended to substitute for them or submit them to quantification. But this *need* not be the intent and, in the case of pure geometry and micro-theories, it is *not* the intent. Clearly, it is only if, and to the extent that, we have predicate constants with the ideal syntactical articulation of mathematical geometry that we can formulate true statements which satisfy mathematical statement functions in the sense that they have constants where the statement functions have variables. But the whole point of the empiricist explanation of ideal geometrical predicates as elements in a deductive system is to account for the existence of such ideally articulated conceptual structures. If, however, the predicate expressions of a deductive system are *essentially* variables, the statement

7 Ibid., p. 91. See also p. 141 ff.

functions of the system *could* have no statement counterparts with predicate *constants* for which they were true—surely an incoherent notion.[8] The moral is that the logical space of pure geometry must contain predicate constants as well as predicate variables, and this means that strictly speaking implicit definition pertains to predicate constants.

14. It must, indeed, be granted that a non-logical predicate constant which isn't connected with extra-linguistic objects is not, in the full sense, meaningful. But not being extra-linguistically meaningful must not be identified with being a variable.

15. Thus, an adequate philosophical treatment of the predicate expressions of a pure geometry must specify a primary use in which they constitute a special class of predicate constants. How is this special class to be characterized? The answer lies in remembering that a pure geometry becomes a physical geometry by virtue of being connected by correspondence rules with empirical predicates pertaining to perceptually discriminable features of physical things.[9] Since the implicit definition of predicates by the postulates of a deductive system is not sufficient to give these predicates extra-linguistic meaningfulness, we can say that *qua* implicitly defined they are *candidate* constants. Their role is not that of being substituted for or quantified over but that of being available for connection with extra-linguistic fact.

16. As I see it, then, just as the predicate expressions of a pure geometry construed as a system of implicit definitions are not variables but candidate predicate constants, so the predicate expressions in a micro-theory *qua* deductive system are not to be

[8] To make the same point in a simpler context, the individual variable 'x' in 'Fx' would be a sham if, although the predicate 'F' were conceptually consistent, there *could* be no individual constant for which 'Fx' was true.

[9] These remarks apply to the idealized geometrical space of macro-physical objects. The point they enable me to make about the status of the predicates of micro-theories remains valid even though, as a final step in our philosophy of physical geometry, we bypass Space as a postulated quasi-physical *Unding* consisting of a continuum of geometrical objects (points) which satisfy non-logical predicate constants, and use the correspondence rules to connect operational geometrical measures and locations directly with appropriate structures in the domain of real numbers. The possibility of this latter move clarifies the sense in which Space, as contrasted with actual and possible physical relationships (overlappings and coincidences) among its 'contents,' is an *Unding*. Additional problems are raised by the physical geometry of micro-physical entities.

construed as variables, but as *candidate predicate constants* which, *to put the matter in first approximation,* get their extra-linguistic meaningfulness from the correspondence rules which connect them with constructs in what Bergmann has called the 'empirical hierarchy.'

17. Now Nagel sees that the idea that micro-theoretical predicates are variables generates the question 'how can theoretical sentences be true or false?' Would they not actually be sentence functions, and no more true or false than 'x is a man'? His answer, of course, taken from Ramsey, is that the conjunction of theoretical postulates is to be construed as preceded by an existential quantification of the predicate variables. This desperate expedient runs up against the objection based on the above analysis that it is logically odd, given that 'f' is a *descriptive* predicate variable, to suppose that the theoretical statement

(Ef) (x) fx

might be true even though no substitution instance of the function 'fx' *could* be true. But, on Nagel's view, the only descriptive predicate constants which are available to the empirical hierarchy and no empirical attributes satisfies the conditions laid down by micro-theoretical postulates.[10] Only a neo-pythagorean–such as Carnap, on occasion, tends to be,–would be happy with substitution instances involving purely arithmetical expressions.

18. Furthermore, it should be noted that the Ramsey sentence approach to theoretical sentences is inconsistent with the idea that theoretical statements can be factually true in the full sense in which statements in the observation framework are now conceived to be true. For the full sense of 'factually true' requires that there be a level of factual truth in which objects are directly 'mapped' or 'pictured' by atomic statements as linguistic events *in rerum natura.* Truth functional combinations of atomic statements and

[10] Cf. Grover Maxwell, "The Ontological Status of Theoretical Entities" *Minnesota Studies,* Vol. III, pp. 16-7; also my "The Language of Theories," in *current Issues in the Philosophy of Science,* edited by Herbert Feigl and Grover Maxwell, New York, 1961, p. 68. The latter essay is reprinted, with minor revisions as Chapter 4 in my *Science, Perception and Reality*. The corresponding reference is to p. 117.

[11] For an elaboration of this thesis concerning factual truth, which stems from Wittgenstein's *Tractatus, see* my "Truth and 'Correspondence'," *The Journal of Philosophy, 59,* 1962–reprinted with minor revisions as Chapter 6 in *Science, Perception and Reality*. See also paragraphs 89 ff below.

quantified statements are to be construed as recipes for constructing such maps or pictures, which latter do not themselves contain logical connectives or operators.[11] On Nagel's approach, it is *in principle* impossible for theoretical language to contain linguistic events to perform this role—for it can contain no sentences free of quantification. Such mapping or picturing as could be done by theoretical sentences must, in his view, be *indirect,* essentially mediated by correspondence rules correlating theoretical statements involving predicates in the empirical hierarchy.

III

19. But before I explore the topic of the factual truth of theoretical sentences, I want to call attention to another and more subtle way in which the idea that only predicate constants in an antecedent observation framework actually stand for properties or attributes can lead to a mistaken account of the meaning of theoretical predicates. It is more subtle, because it claims to be free from the myth of the given. I have in mind Miss Hesse's conception of the role of models and analogies in theory construction.[12]

20. It will be remembered that while Nagel gives models and analogies an important heuristic role in theory construction, and stresses their role as imaginative aids to theoretical thought, he emphasizes that the scientific *content* of the theory derives not from models or analogies but from the implicit definition of theoretical predicates by the postulates in which they occur, together with the correspondence rules which connect theoretical terms with expressions in the empirical hierarchy.[13] This is at best a half-truth.

21. It is a half-truth because theoretical postulates are often specified in a way which *logically* involves the use of the model. And even when a set of postulates is explicitly given in the form prescribed by contemporary logical theory, it turns out, in actual practice, (although *ideally* it *need* not) that the conceptual texture of theoretical terms in scientific use is far richer and more finely grained than the texture generated by the explicitly listed postu-

[12] Mary B. Hesse, *Models and Analogies in Science,* London, 1963.
[13] Nagel, *op. cit.* pp. 95ff.

lates. This is particularly true with respect to the categorial structure of what are lumped together from an abstractly logical point of view as 'predicates' and 'functors.' The thingish or quasi-thingish character of theoretical objects, their conditions of identity, the dispositional or non-dispositional character of attributes, are some of the more familiar categorial features conveyed by the use of model and analogies.[14]

22. Notice that I am *not* saying that these features *could not* be fully captured by the explicit working of a logistically contrived deductive system. To suppose otherwise is to return to some form of abstractionism or givenness. I simply call attention to the fact that the ability to capture these features in this logistical manner presupposes a degree of success in analyzing the categorial framework of scientific concepts which has to date been at best approximated in some limited areas. The familiar controversy over the logic of field concepts is a case in point. And who would be so rash as to say that the categorial structure of spatial and temporal concepts have been conclusively settled?

23. The notion of a micro-theory the conceptual texture of which is completely captured by an explicit set of postulates is, perhaps, a realizable ideal, but it is not realized in fact. The principal model of a theory does not play the *external* role of being an alternative realization of a *conceptually independent* set of postulates construed as propositional functions.

24. But how is the conceptual dependence of the theory on the model to be construed? Miss Hesse, if I understand her aright, construes theoretical concepts pertaining to micro-theoretical entities as *identical with* concepts pertaining to the entities of the model, once these entities has been purged of those features which constitute the known negative analogy. This I regard as a serious mistake which rests on a tacit extension to all analogies of a logical feature of the simple case with which her analysis begins. She asks us [15] to consider the case of the earth and the moon. Here the known positive analogy consists of attributes to be shared by those

[14] See also my remarks on the role of models in theory construction on pages 70-1 of "Theoretical Explanation," in *Philosophy of Science,* Vol. II, edited by Bernard Baumrin (New York, 1963).

[15] *Op. cit.,* pp. 64ff.

two objects; the known negative analogy consists of attributes known to be possessed by one but not the other; while the neutral analogy consists of attributes known to be possessed by one but whose possession by the other is problematical. The key feature of this account is the fact that it is in terms of the identity and difference of attributes of particulars.[16] And while one may sympathize with Miss Hesse's attempt[17] to analyze similarity in terms of identity and difference of attributes, it must be pointed out that not all similarity is similarity of *particulars.* And once one takes into account the fact that attributes themselves as well as particulars may be similar, the way is open for distinction between what might be called 'first order' and 'second order' similarity between particulars. In a simple case of first order similarity, two particulars would share an *identical* attribute. In a simple case of second order similarity, two particulars would have *similar* attributes.

25. Put in these terms, the question with which I am concerned becomes: Is the analogy between molecules and billiard balls a matter of particular molecules and billiard balls sharing *identical* attributes, or is it, at least in part, a matter of their having *similar* attributes. As I see it, it is the concept of similar or analogous attributes which is the key to the understanding of theoretical explanation.

26. But in what, it will be asked does the similarity of attributes consist? Without attempting a complete analysis of the ways in which attributes can be similar, enough can be said, I believe, to bring out the importance of the idea for the philosophical analysis of theoretical explanation. A simple example is the similarity of moments in a temporal series with points in a line. For, to nail down a logical point, the analogy between moments and points is not a matter of moments and points having certain attributes in common, but for moments on one hand and points on the other to have properties which, though they have no *first order* properties as common constituents do have certain *second order* properties in common, e.g. transitivity.

27. The fact that what is *said* by characterizing *to the left of* and *before* as transitive can be *shown* without the use of the predicates

[16] P. 65.
[17] P. 78.

standing for a second order property by exhibiting the postulates which hold of these relations, should not lead one to underestimate the role of reference to second level properties in science. For in the above example the ability to *exhibit* second order similarities presupposes the separate existence of the two conceptual frameworks of space and time. But the distinctive feature of the use of models and analogies in theory construction is that the conceptual framework of the theory is *generated* by specifying the analogies which are to obtain. Thus, as a first approximation, it can be said that models are used in theory construction to specify new attributes as *the attributes which* share certain higher order attributes with attributes belonging to the model, fail to share certain others (the negative analogy)—and which satisfy, in addition, the conditions laid down by the relevant correspondence rules.

28. But, it may be said, if the higher order attributes you have in mind are such 'formal' properties as transitivity and asymmetry, your answer can explain at most some very skeletal features of theoretical attributes. Furthermore, the move would be unnecessary, since it is exactly these formal properties of attributes and relations which can readily be exhibited by explicit postulates. This objection overlooks, in the first place, the indefinitely large number of such formal attributes of attributes and the complexity of which they are capable. Once again, what is possible *in principle* is mistakenly used in an explication of scientific practice. Thus we may not in fact be able to pick out from a model and formulate in tidy fashion the propositional functions which would serve as the positive and neutral analogies in a completely explicit theory. And while, by the same token, we would not have tidy expressions in actual use corresponding to these propositional functions, as 'transitive' and 'transitivity' correspond to

$$(x)\ (y)\ (z)\ xRy \cdot yRz \rightarrow xRz$$

our language *does* contain adequate resources for referring to second order attributes by more complex locations.

29. To see this, we must note that to refer to a specific second order property is, in first approximation, to refer to the corresponding propositional function. Thus

Before is asymmetrical

has, again in first approximation, the sense of

'(x) (y) xRy → ~ yRx' is true of *before*
i.e. becomes true on the substitution of 'before' for 'R.'

30. Not only can we refer to specific second order properties (i.e. mention specific propositional functions of second order) we can make *general* reference to such properties. Thus, in a simple case, we can say of two dyadic relations that they share *some* functions involving a dyadic predicate variable together with appropriate quantification over particulars. Furthermore we can, by imposing additional conditions, specify to which subsets of second order properties those which are shared by the relations belong. Thus the logical space of second order attributes admits of all the techniques of indefinite reference and definite description. And when we take into account the open texture and vagueness with which reference can be made, we begin to see how models can be the *fundamenta* of open-textured reference to second order attributes.

31. It was pointed out above that there is no limit in principle to the number or complexity of formal second order attributes. It should now be pointed out that it is not necessary that all of the predicate expressions of the propositional functions to which reference is made be predicate *variables*. Thus, an explication of the analogy between color and sound involves picking out propositional functions which have predicate *constants* in common (e.g. 'perceptible') in addition to the variables corresponding to the color and sound predicates being compared.

32. As I see it, then, models provide a basis for a more or less vague and open-textured reference to a framework of propositional functions which the predicates of a theory are to satisfy. They are specified as *the functions which* hold, with certain qualifications, of the predicates which apply to the entities of the model. I say 'with certain qualifications' because the reference to a model is accompanied by what I have called [18] a 'commentary' which eliminates specific functions from the analogy and modifies others. The important point is that these relatively precise qualifications operate within the context provided by an intuitive grasp of the framework of the model which it would be difficult if not impossible to articulate in terms of an explicit list of postulates.

[18] In "Empiricism and the Philosophy of Mind," *Minnesota Studies in the Philosophy of Science,* Vol. I, p. 313.

33. But whereas in the above 'rational reconstruction' of the use of models in theory construction it is implied that different sign designs would be used for the introduced theoretical predicates and the predicates of the model on the basis of which they are introduced, it should be noted that in actual fact the same designs may be used. That the predicates involved in theoretical explanation may be subtly ambiguous, having different meaning in different contexts, could easily be overlooked and lead to a failure to distinguish the conceptual framework of a theory from the conceptual framework of the model on which it depends.

34. Notice also that the above explication involves a three-fold distinction between

(a) the objects of the unqualified model
(b) the objects of a revised model which preserves the perceptual and imaginative role of the predicates of the unqualified model, but modifies the internal structure of the model in accordance with the commentary. The revised model involves, therefore, a controlled incoherence, an example of which is the visualizability of the imperceptible particles of micro-physics.
(c) the objects of theory.

35. The importance of distinguishing between (b) and (c) is that it puts the finishing touch on the idea that the positive analogy between theoretical entities and empirical constructs is not a simple matter of the identity of some of their first level attributes with those of perceptible things. Only if we recognize that the analogy between micro-physical entities and the entities of the model [19] is not to be construed in terms of the identity or difference of first order attributes, can we appreciate how the use of models in theoretical explanation can generate *genuinely new* conceptual frame-

[19] I say *the* model because the concept of model here is not simply that of a model in the set-theoretical sense (in which any theory would have an indefinite number of models.) The relevant concept is a methodological one pertaining to theoretical concept formation. That even in this methodological sense two (or more) models can be jointly used as a basis for introducing theoretical concepts is granted—compare the role of wave and particle models in quantum mechanics. This joint use as a *fundamentum* does not require that they be jointly used in application, and, indeed, in the latter use the models are incompatible. This incompatibility is I believe, given its proper location by the above account.

works and justify the claim to have escaped from the myth of the given.

36. Having made this point, it must now be emphasized that it is only from the standpoint of a rational reconstruction that we can draw a tidy distinction between three sets of entities as in paragraph 34. This tidy threefold distinction is, I believe, philosophically clarifying as a framework on which to project the various ways in which theories are used. But it can be philosophically misleading, if taken as a report of explicitly drawn distinctions in the practice of science. Thus, instead of putting the point substantively in terms of a distinction between the entities of the model, the entities of the revised model and the corresponding theoretical entities, one can speak of a model and the various roles it performs in theoretical explanation. And while Miss Hesse's account fails as I see it, to do justice to the complexity of these roles, it does have the virtue of stressing the essential and intrinsic involvement of the conceptual framework of the model in the conceptual framework of the theory.

37. The temptation to identify the theory with the qualified or revised model (to which, I believe, Miss Hesse succumbs) would indeed be overpowering if, inclined to take a realistic view of theories, one thought that only by such identification could the theory be provided with predicate *constants*. But this motive has, I believe, been undercut by the above critique of Nagel's interpretation of the predicate expression of a micro-theory. We have seen that predicate variables *do* come into the picture, but they come in not as theoretical language predicate variables, but as implicit in the specification of the second order attributes which are conceived to be shared by first-order attributes in the theoretical domain of the model.

IV

38. I find lurking in Nagel's account the presupposition that for an expression to become a meaningful descriptive predicate *constant* it must be given a place—for example, by explicit definition—within the antecedent observation framework. This presupposition would simply rule out the idea that theory construction could be a technique for enriching or revising this framework—though not, of

course, the idea that theoretical considerations enable one to pick out constructs and hypotheses *within* the observation framework which are likely to prove fruitful.

39. It is this presupposition which, as I see it, insures the ultimate shipwreck of Nagel's higher synthesis of instrumentalism and realism. It ultimately rests, as I diagnose the situation, on an interpretation of a predicate's

> standing directly [20] for a property of things

with its

> being associated with a property of things

This interpretation makes the meaningfulness of fundamental descriptive predicates a matter of a psychological connection between predicates and properties *construed as extra-linguistic entities.* It is the supposedly extra-linguistic character of these properties or attributes which is thought to guarantee them a status independent of the vicissitudes of human concept formation.

40. This realistic [21] interpretation of the perceptible attributes of things is, as I see it, aided and abetted by a confusion between the 'of-ness' of sense impressions and the 'of-ness' of thought.

Thus

> Jones has a sense impression of a red triangle

tells us about a non-conceptual state of Jones. On the other hand,

> Jones is under the perceptual impression that there is a red triangle (in front of him)

i.e.

> There appears to Jones to be a red triangle in front of him

tells us about a conceptual state of Jones. There is, of course, a close connection between these two kinds of state. Roughly, we explain the latter in terms of the former. To confuse them, however, is to identify the *conceptual* awareness of an item *as* red and triangular with a non-conceptual state of affairs. There arises the

20 As contrasted with standing for a property *via* a definition chain.

21 The term 'realism' is here used in the sense in which it contrasts with 'conceptualism' and 'nominalism,' rather than that in which it contrasts with phenomenalism.

bastard notion of a *pre-conceptual* awareness of items as being red and triangular which is supposed to be the language independent foundation on which the meaningfulness of language rests.

V

41. Of the dogmas of classical empiricism, two, have been generally abandoned. These are the twin ideas that empirical knowledge rests on an absolute foundation of knowledge pertaining to sense-data, and that the content of genuine descriptive concepts is derived from sense data. These 'dogmas' have increasingly been replaced by an anthropologically oriented interpretation of the conceptual framework in terms of which we perceive the world as a form of group life which has evolved from rudimentary beginnings by a process of adaptation which can be compared—in ways which are not well understood—to natural selection. This conceptual framework, though *brought into play by* sense impressions is not, at least primarily, *about* sense impressions.

42. It is in the spirit of this approach that Feyerabend correctly argues that although the meaningfulness of observation predicates requires that they be *responses to* sense impressions, it does not require that they stand for properties which either common sense or theoretical psychology or even ideal neuro-physiology would attribute to these impressions.

43. From the pragmatic picture of language as a technique of behavioral orientation, the conclusion might be drawn that we are free to replace, segment by segment, the framework in terms of which we perceive the world, by scientifically contrived structures which enable a more subtle orientation and fewer surprises. Thus, one who is at home in the micro-physics of his day would be free to train himself to respond to his environment in terms which, though they externally resemble the vocabulary of his fellow man, have, as their descriptive conceptual content, highly derived constructs in this theoretical framework. We are free, so to speak, to pour new conceptual content into old bottles.

44. And, indeed, as Feyerabend develops his 'pragmatic theory of observation' this seems to be what he is telling us. According to

him, the perceptual predicates which lie at the base of the empirical hierarchy are simply those which are in *de facto* possession of this behavior guiding role, and are subject to dispossession by any system of predicates which does the job better. As he sees it, it is simply untrue that an expression can become a descriptively meaningful predicate constant only if given a place in the *antecedent* observation framework. There is, indeed, no such thing as *the* observation framework. Observation frameworks come and go. For a descriptive predicate to be factually meaningful, it is sufficient that it belong to the current empirical hierarchy–i.e. to a framework in terms of which the relevant language using organisms–in this case scientists –currently orient themselves to, and cope with, their internal and external environment.

45. But though I am in general sympathy with Feyerabend's pragmatic theory of observation, I do not think that it will do as it stands. There is a core of truth in the concept of '*the* observation framework' and, indeed, of the abstractionist approach to basic empirical concepts which survives the exorcising of givenness. And, in my opinion, although Feyerabend arrives at the ontological truth that the world is in principle what scientific theory says it is, he does so by chopping the structure of science with a cleaver rather than carving it at its conceptual and methodological joints. As I see it, only someone who is unaware of the subtle interdependence of the various dimensions of the framework of empirical knowledge would speak cavalierly of the piecemeal replacement of part by scientifically better part.

46. We have come to issues in which the philosophy of science shows its organic connection with less exotic branches of the philosophical enterprise. And the first point to be made is that to reject the myth of the given is not to commit oneself to the idea that empirical knowledge as it is now constituted has no rock bottom level of observation predicates proper. It is to commit oneself rather to the idea that even if it does have a rock bottom level, it is *still* in principle replaceable by another conceptual framework in which these predicates do not, *strictly speaking*, occur. It is in this sense, and in this sense *only* that I have rejected the dogma of givenness with respect to observation predicates.

47. The second point to be made is that it is, in my opinion, a

serious mistake to move, as Feyerabend seems to do,[22] from the propositions:

(a) It is *possible* to train micro-theorists and their experimental colleagues to respond directly to visual stimulation by the use of sentences in the framework of micro- theory.
(b) Every physical process can *reasonably* be described in micro-physical terms.
(c) It is *possible and reasonable* to train physicists and chemists to tell by looking that certain specific micro-physical processes are going on in certain specific contexts—i.e. to respond directly, knowing that one is in such a context, to visual stimulation with sentences in the language of micro-theory, in a way which is reliable and known to be reliable. (for example, it is possible and reasonable to train a physicist to tell what a particle is doing in a cloud chamber i.e. to know what it is doing without inferring this from non-theoretical sentences *via* correspondence rules.)

to even the first—let alone the second—of the following propositions:

(d) It is reasonable to train scientists to respond in all kinds of circumstance to visual stimulation with sentences in the framework of micro-physics (e.g. when chopping wood or drawing water).
(e) It is reasonable to train scientists to respond to visual stimulation in all circumstances with sentences in the framework of micro-physical theory *only*—i.e. to *abandon* the perceptual use of the framework of common sense.

48. Let me emphasize that I am as firmly convinced as Feyerabend of the truth of the following propositions:

[22] Cf. Smart, p. 167; also pp. 159, 162.

(α) Micro-physical entities do not have the second class existence of mere 'conceptual devices.'

(β) The framework of common sense is radically false—i.e there *really* are no such things as the physical objects and processes of the common sense framework.[23]

(γ) Propositions (α) (β) are to be clarified in terms of the concept of its being reasonable *at some stage* to abandon the framework of common sense and use only the framework of theoretical science, suitably enriched by the dimension of practical discourse.[24]

But, as I see it, the ontological theses formulated by (α) and (β) involve no commitment to the idea that the time for scientists to abandon the framework of common sense is *now*.

49. Indeed, all the methodological gain which *exclusive* use of theoretical constructs would bring can be achieved by the use of correspondence rules as in the classical account, together with the direct use in experimental situations referred to in 47 (c) above. On the other hand, taken seriously, as we shall see, the *abandonment* of the common sense framework would result in serious methodological and conceptual loss.

50. To develop this point, I must return to the topic of what is meant in this connection by 'the framework of common sense'. In paragraph 5 I distinguish between common sense *beliefs* and common sense *principles*. The former are in no way binding on the scientist. Nor are the conceptual *constructs* of common sense binding on the scientist. It is the rock bottom concepts and principles of common sense which are binding until a total structure which can do the job better is actually at hand—rather than a 'regulative ideal.'

23 This does not, of course, mean that there are no tables or elephants, but rather that tables and elephants as conceived by common sense do not really exist, i.e., that these concepts are *in principle* to be replaced by a counterpart built on a theoretical foundation. For an explication of what is meant by 'really exist' in this context, see "Empiricism and the Philosophy of Mind," 39ff.

24 This aspect of the situation, which is not stressed by Feyerabend, is illustrated by the practical dimension of such common sense concepts as that of what it is to be a hammer.

51. That this bindingness is *methodological* and is incompatible with neither the *ontological* thesis that the world as conceived by common sense is, in the Kantian sense, phenomenal, nor with the *epistemological* thesis that it is the task of theoretical science to give us an accurate picture of things, is the fundamental contention of this essay. To which it is anti-climax to add that no one who stresses the methodological primacy of the framework of common sense is committed, as Feyerabend seems to think he is, to preserving either outmoded common sense beliefs (which is obvious) or common sense constructs (e.g. our everyday concepts of *up* and *down*) in the corpus of science.

52. But it will be asked, is not this talk of the 'methodological primacy' of the common sense framework, this talk of 'foundation' and 'super-structure,' 'basic concepts' and 'constructs,' 'principles' and 'beliefs,' simply an attenuated form of the myth of the given, a ghostly survival of abstractionism and self-evidence? I cannot hope to do justice to this challenge on the present occasion.[25] I shall limit myself to present one central line of thought which illustrates the broad strategy by which I would meet it.

53. The idea that the framework of common sense has a rock bottom does not require that this rock bottom consists of sense impressions. The framework of common sense is a framework of (among other things) colored physical objects extended in space and enduring through time. And while objects which are red on the facing surface have the power to cause normal observers in the standard conditions to have sense impressions of red, this is incorrectly taken to mean that the physical property of being red on the facing side is to be analyzed in terms of this causal power.

54. As Berkeley, Kant and Whitehead, among others, have pointed out, physical objects cannot have primary qualities only—for structural and mathematical properties presuppose what might be called 'content qualities'. And unless one falls into the trap of thinking of the framework of physical objects as a common sense theory evolved with unconscious wisdom to explain the manner in which sense data occur, it will scarcely do to say that the content

[25] It is met, I believe, by the argument of Chapters I, III, and V of *Science, Perception and Reality* (London, 1963), and applied specifically to the mind-body also "The Identity Approach to the Mind-Body Problem," this volume, Part 3.

qualities of physical objects are conceived, by a common sense use of analogy, to be the physical counterparts of the qualities of data—i.e. to play in the realm of physical things the content-role played in sense data by sense qualities. For if the conceptual space of common sense physical objects is *underived,* their content qualities must be *directly* rather than *analogically* conceived, for it is only in terms of perceived, and therefore conceptualized, qualitative difference that form and structure can be distinguished.

55. Thus the rejection of a phenomenalistic analysis of the framework of common sense requires that the physical objects of this framework have perceptible qualitative content. And once one realizes this, one sees that there is no alternative to construing these physical objects as colored in the literal occurrent sense. One might wish to say that this framework—which has as its central constituents items which are in *this* sense colored through and through[26]—is, from the standpoint of theoretically oriented science, *false,* although enabling a behavioral adjustment of sufficient accuracy for the practical purposes of life. But false or not, such is the framework of common sense.

56. Again, the rejection of givenness is compatible with a distinction between the attributes which physical objects and processes can *strictu sensu* be perceived to have, and those which they can be perceived to have in less stringent senses. Thus, I can perceive that this is a pad of writing paper—indeed perceive it *as* a pad of writing paper— but *strictu sensu* I perceive that it is, for example, a physical object which, on the facing side, is rectangular, yellow and lined with parallel blue lines. It should not be necessary to point out that the distinction in question does not require that a perception of the pad *as a pad* consists in perceiving it as an object which is rectangular and yellow lined with blue on the facing side and *inferring* that it is a writing pad.

57. If the framework of common sense physical objects is analyzed along the preceding lines it is easy to see that while the fundamental concepts of the framework are not concepts of sense impressions, they are concepts which, so to speak, project or transpose the attributes of sense impressions into the categorial framework of physical things and processes. But although the conceptual

26 As a paradigm case, think of a pink ice cube.

framework of physical color is in this sense ontologically grounded in visual impressions, the conceptual framework in terms of which common sense conceives these impressions is itself an analogical offshoot from the conceptual framework of physical color and shape.[27] To put the matter in Aristotelian terminology, visual impressions are prior *in the order of being* to concepts pertaining to physical color, whereas the latter are prior *in the order of knowing* to concepts pertaining to visual impressions.[28] This is what I referred to above as the kernel of truth in the classical empiricist doctrine that concepts of the qualities of sense are formed by 'abstraction' from sense impressions.

58. We saw above that the logical status of color as qualitative content is an essential part of the common sense framework of physical objects. We have also seen that this logical space is a conceptual projection of certain attributes of visual sense impressions. We are therefore in a position to make two key points:

> (a) The abandonment by scientists of the conceptual framework of common sense physical objects would involve either the abandonment of the conceptual space of color *tout court,* or the retention of this conceptual space as space of sense impressions. The latter would be cut off from its foundation and left to wither on the vine. In either case, the conceptual space of the qualities of sense ("secondary qualities" in one use of this phrase) would disappear from the public observation base of science. It would enter science only in linguistics, in the study of the structure of the language of non-scientists—and of scientists only to the extent that their sense im-

[27] For an elaboration of this point see the concluding paragraphs of "Empiricism and the Philosophy of Mind;" also *Science, Perception and Reality,* pp. 47-9; 91-5; also "The Identity Approach to the Mind-Body Problem."

[28] That common sense concepts of sense impressions are *in principle* as subject to replacement by concepts belonging to an adequate neuro-physiology as the common sense framework of physical objects by an adequate micro-physics, is the key to the solution of the mind-body problem. But any solution along these lines which fails to preserve the conceptual space of color, sound, taste, etc., suitably transposed, has lost touch with rather than solved the problem. See the essay referred to at the end of the preceding note, particularly sections 49 ff.

pression talk continued to reflect the pre-revolutionary framework of common sense physical objects.

(b) Only when the conceptual space of sense impressions has acquired a status which is not parasitical on the framework of common sense physical objects—in other words, only with the development of an adequate scientific theory of the sensory capacities of the central nervous system—could the framework of common sense be abandoned without losing conceptual contact with a key dimension of the world.

VI

59. I have said that to a first approximation I agree with Nagel that the framework of common sense is methodologically indispensable, but with Feyerabend that it is in principle false. I have sketched an argument for the first half of this thesis. Let me now attempt to bring the entire thesis to a second degree of approximation by taking a closer look at the second half.

60. The key to the ontological issue is the concept of a correspondence rule and its place in the classical account of theoretical explanation. Since I have already made clear my sympathy with Nagel's qualified endorsement of the classical dualism of observation and theoretical frameworks, the problem will be to show how this dualism can be reconciled with the assertion that in principle it is the framework of theory *rather than* the observation framework which is real.

61. The crux of the matter concerns the correspondence rules which connect theoretical concepts with observation framework counterparts, and, in particular those which can informally be said, in a comfortable but problematic sense, to 'identify' macro-objects and -processes with systems of micro-objects and -processes.[29]

62. These correspondence rules, by correlating concepts in the two frameworks, also correlate theorems in the deductive system of the theory with general propositions in the observation framework. Let us refer to these general propositions as the observation frame-

[29] In other words with what I have elsewhere (Baumrin, pp. 71-3) called 'substantive' as contrasted with 'methodological' correspondence rules.

work counterparts of the theorems. Here it is vital to note that there are two importantly different types of candidate which satisfy the general criterion for being an observation framework counterpart, i.e. which can be characterized as lawlike statements in the observation framework, correlated with theorems in the theory by correspondence rules. There are

> (a) lawlike observation framework counterparts which are *compatible* with the observational evidence.
> (b) lawlike observation framework counterparts which not only satisfy condition (a), but which were accepted, and would still be accepted, on purely inductive grounds —i.e. in the absence of theoretical considerations.

63. The distinction is crucial, for it is clear that a counterpart which satisfies condition (a) need not satisfy condition (b), i.e. need not be the lawlike statement which it would be reasonable to accept on purely inductive grounds, nor even contain the empirical concepts which it would be reasonable to construct and use in the absence of theoretical considerations. It is characteristic of good theories to show their observational counterparts in sense (b) to be false. On this point I am in substantial agreement with Feyerabend.

64. It should therefore be clear that the role of substantive correspondence rules is not to tie theoretical propositions to antecedently established observational counterparts in sense (b), but to tie them to observational counterparts in the weaker sense specified in (a). By so doing, the theory, taken in conjunction with other knowledge, typically provides an explanation *not* of the antecedent generalizations which were based on purely inductive grounds, but rather of why the inductive use of observational evidence came as close as it did to yielding the observational counterpart in sense (a).[30]

[30] As I put it in "The Language of Theories" in *Current Issues of the Philosophy of Science,* edited by Herbert Feigl and Grover Maxwell, (New York, 1961,) ". . . [micro-] theories about observable things do not explain empirical laws, they explain why observable thing obey, to the extent that they do, these empirical laws; that is, they explain why individual objects of various kinds and in various circumstances in the observation framework behave in those ways in which it has been inductively established that they do behave." (p. 71).

65. Nor is the role of substantive correspondence rules in the reduction of one theory to another that of correlating theorems in the reducing theory with the postulates of the theory to be reduced *as it existed before the possibilities of reduction were considered in detail.*[31] The theory to be reduced is typically modified in subtle ways, a fact which may well lead to a revision of the correspondence rules which connect the theory to be reduced with observation framework counterparts, thus generating revised lawlike statements which are inductively indiscriminable both from those to which unaided induction would point and those which were accepted in the light of the unmodified theory.

VII

66. There are no puzzles in principle about the status of substantive correspondence rules which connect theory with theory. One can envisage with equanimity the replacement of two correlated theoretical structures by one structure of constructs and theorems in which the correspondence rules have (*pace* Nagel) been replaced by straight-forward definitions. This unified structure would, of course, be methodologically tied to the observation framework by two different systems of methodological correspondence rules—although the systems would tend to overlap—because the reduction of the theoretical framework of one science to that of another must not be thought to be the reduction of one *science* to another. The latter truth is, as I see it, misinterpreted by Nagel to demand that the substantive correspondence rules cannot, in principle be replaced by definitions.

67. It is rather in the case of those substantive correspondence rules which connect micro-theoretical with empirical constructs, and which, if one were to take seriously their analogy with the rules which connect one theory with another, would tempt one to

[31] This point is not clearly distinguished by Feyerabend from the foregoing since he tends to run together problems pertaining to the relation of one theory to another theory with problems pertaining to the relation of theory to empirical generalizations. (Nor does Nagel always keep these two types of problem sufficiently apart.)

speak of 'reducing' the observation framework to the framework of the micro-theory, which generate philosophical perplexity. One can understand how primitives in chemical theory might come to be defined predicates in a deductive structure shared by physics with chemistry. After all, as *theoretical* concepts they were in jeopardy to begin with; they were the fruit of postulational concept formation, and could be modified or abandoned at the drop of a good reason. But basic observation framework attributes *seem,* as we have noted, to have an absolute status independent of thought and language. The corresponding predicates seem, therefore, to lack the vulnerability of theoretical primitives and to be immune to replacement by micro-theoretical concepts.

68. This, however, as I have already pointed out in section 40, is an illusion which has its source in the fact that the perception of an object as red and triangular involves not only the *concept of a red triangle,* but also the *sense impression of a red triangle.* The latter is, as we have noted, a non-conceptual state of the perceiver. Thus, to put it bluntly, the phrase

sense impression of a red triangle

refers to items which would exist even if there were no conceptual frameworks, whereas the phrase

concept of a red triangle

obviously does not. If, therefore, one confuses the role of the phrase

of a red triangle

in the former [32] with its role in the latter, the object language character of the former role may be attributed to the latter, and the 'of-ness' of perceptual thoughts will be interpreted as a relation *in re* between thoughts and extra-conceptual items—in this case the property of being a red triangle.

69. The truth of the matter, however, is that in the phrase

concept of a red triangle

the sign designs 'red' and 'triangle' are being used in 'second intention', as the scholastics put it, to form expressions which classify conceptual items. And while it is true that the thought of a red triangle is related to redness and triangularity and, indeed, the property of being a red triangle, these latter items do not have the

[32] Cf. paragraph 57 above.

absolute status "independent of thought and language" which is supposed to render the observation framework predicates which express them immune to the fate of theoretical primitives.[33] To see that theoretical *particulars* are not doomed to a second class status, it is necessary to see that theoretical *attributes* are not, as a matter of principle, second class cousins of the perceptible attributes of perceptible things.

70. Once we recognize that to say of a predicate that it stands for a certain attribute—e.g. triangularity—is not to characterize it as related to an item in the extra-conceptual order, but to tell us that the predicate functions linguistically as does the predicate built into the attribute expression—thus, 'triangular'—our attention is focussed on *how* this predicate functions. And this takes us from 'stands for attribute' talk to talk about specific matter-of-fact relations between the predicate and the linguistic and non-linguistic situation in which it is used.

71. And here the key point is that whereas in

(1) 'φ' (in L) stands for triangularity

the term 'triangularity' refers to a conceptual item, in

(2) 'This is φ' (in L) is a perceptual response to triangular things

the phrase 'triangular things' does not. The latter formulates a necessary condition of 'φ' being an observation predicate of L. And whereas in (1) 'φ' must be the *translation* in L of 'triangular', so that both expressions must belong to the same conceptual framework, the same is not true in (2). It *could* be correct to say

(2′) 'This is red' (in E) is a perceptual response to systems of particles emitting such and such electromagnetic radiation

and

(2″) 'This is a system of particles etc.' (in scientese E) is a perceptual response to red objects.

72. These distinctions put us in a position to explain the feeling

[33] For a general defense of the thesis that the *esse* of attributes, not only theoretical attributes but even perceptual attributes is concipi, and a more detailed application of this thesis to the problem of the status of theoretical entities, see the essay on "The Language of Theories" referred to in note 27 above. A technically more adequate statement and defense of the thesis is given in "Abstract Entities," *Review of Metaphysics, 16,* pp. 627-71 (1962).

philosophers have had that theoretical predicates have a 'second class' meaningfulness. For to commit oneself to the idea that the connection of theoretical statements with the extra-linguistic order is in principle either *always* or *primarily* mediated by observation framework statements to which transition is made by correspondence rules, is to deny to theoretical expressions the direct language-entry transitions which make the observation framework more than a syntactical game.[34]

73. Clearly, a theoretical framework can achieve first-class status only if a proper subset of its expressions acquire a direct role in observation. This Feyerabend sees, and it is the central truth in his account. It is indeed important to note that theoretical expressions can acquire a reporting role—that we can 'tell by looking' that certain theoretical states of affairs obtain.

74. But Feyerabend seems to me to confuse the sound philosophical point that realism with respect to theoretical entities involves the idea that *in principle* theoretical expressions could take over and monopolize the direct response role which they now share, to a limited and subordinate extent, with expressions belonging to the framework of common sense, with the unsound idea that this 'take over' is a live option *hic et nunc*. He thus turns a sound ontological insight into methodological error.

75. Nagel, on the other hand, reifies methodological insight into ontological error. His concessions to realism are superficial and leave untouched the hard core of instrumentalism in his position.

76. Unimpressed by the fact that it can be methodologically useful to train oneself to respond directly to experimental (or clinical) situations in the language of a theory, and recognizing that the ability so to respond exists in an ambience of common sense, the extirpation of which would be folly, Nagel commits himself to the idea that the framework of common sense is *in principle* indispensable. By so doing he dooms theoretical frameworks to an *in principle* second class status as contrasted with the *de facto* (and

[34] I have used the phrase 'language-entry transition' which contrasts with both 'intra-linguistic move' and 'language-departure transition'— in "Some Reflections on Language Games," *Philosophy of Science, 21*, 1954, reprinted in a revised version as Chapter 11 in *Science, Perception and Reality.*

methodologically reasonable) second class status they currently possess.

77. Nagel avoids the appearance of doing this by mislocating the issue. He compares the instrumental character of theories to the instrumental character of empirical laws in their capacity as "material rules of inference," and argues that "there is, on the whole, only a verbal difference between asking whether a theory is satisfactory (as a technique of inference) and asking whether it is true (as a premise).[35]

78. Let us examine this parallel. Nagel begins by pointing out that instead of going from the premises

All A is B

and

Wellington is A

to the conclusion

Wellington is B

in accordance with a formal principle of inference, we can go directly from

Wellington is A

to

Wellington is B

in accordance with a material principle of inference which can be represented for our purposes as

from 'x is A' infer 'x is B'

He argues that the question as to the soundness of this principle is equivalent to the question as to the truth of 'All A is B.'

79. Nagel's concept of a material principle of inference is, as we shall see, the concept of what I shall call a 'dependent' principle of inference. To bring this out, let us consider another argument which can be constructed from the above materials. One can go directly from

All A is B

to

Wellington is B

in accordance with the principle

from 'All A is B' infer 'Wellington is B'

[35] *The Structure of Science,* p. 139.

Here the question as to the soundness of the principle is equivalent to the question as to the truth of 'Wellington is A.'

80. Let me represent the two 'material' (i.e. dependent) principles as follows:

(1) 'x is A' —(all A is B)→ 'x is B'

(2) 'All A is B' —(Wellington is A)→ 'Wellington is B'

and let me call the parenthetical adjuncts to the arrows the 'presuppositions' of the principles. Finally, let me represent a dependent principle of inference which leaves its presupposition unspecified, but indicates that there is one, by the form

'...' ⌊→ '———'

Corresponding to (1) and (2) we would then have

(1') 'x is A' ⌊→ 'x is B'

(2') 'All A is B ⌊→ 'Wellington is B'

81. With these distinctions in mind [36] let us explore Nagel's use of his concept of a material rule of inference to clarify his "irenic instrumentalism," his higher synthesis of the claims of instrumentalism and realism. His general idea is that realism stresses the use of theoretical sentences as *premises* in theoretical explanation and insists that theoretical sentences are either true or false; instrumentalism, on the other hand, stresses the use of theoretical sentences as *principles of inference* and insists that the relevant question is

[36] Material rules of inference in Nagel's sense (which is concerned, in more traditional terminology, with enthymemes) must be distinguished from material rules in the fullblooded sense which illuminates traditional puzzles concerning lawlike statements, inductive argument, and counterfactual conditionals. According to the latter conception, a lawlike statement is *as such* a principle of inference and is 'material' in that the validity of the argument it covers depends on the descriptive predicates involved, and not on its governing enthymemes in which the major premise—the supposed lawlike statement—is suppressed. Induction is correspondingly analyzed not as the establishing of descriptive statements to the effect, for example, that all objects of a certain kind A are also B, but as a reasoned *decision* to adopt a material principle of inference which authorizes reasonings which can be illustrated by: "Although this is not known by observation or testimony to be B, it is A and so, in all probability is B." For a defense of this approach to lawlike statements and induction see my "Counterfactual Conditionals, Dispositions, and the Causal Modalities," in *Minnesota Studies in the Philosophy of Science*, Vol. II, edited by Herbert Feigl, Michael Scriven and Grover Maxwell; and "Induction as Vindication" *Philosophy of Science, 31,* 1964.

not 'true or false' but 'satisfactory or not satisfactory?' All this (argues Nagel) is much ado about nothing, the two ways of putting the matter being equivalent. Is he right?

82. Let us first consider the case where a theory functions as a premise in a formally valid argument. It will be convenient to use expressions of the form 'θ_i' and 'O_j' to represent theoretical and observation framework predicates respectively, and the variable 'x' to range over whatever it is that pragmatists means by 'situations.' In these terms the formally valid argument in question would presumably look something like

$$(x)\ \theta_1 x \rightarrow \theta_2 x$$
$$(x)\ O_1 x \rightarrow \theta_1 x$$
$$(x)\ \theta_2 x \rightarrow O_2 x$$

Therefore,

$$(x)\ O_1 \rightarrow O_2 x$$

The first premise would be a theorem in the theory; the second premise the counterpart of a correspondence rule correlating 'O_1' with 'θ_1'; the third the counterpart of a correspondence rule correlating 'θ_2' with 'O_2'; and the conclusion a lawlike statement in the observation framework.

83. Let us now ask about the form of the material principle of inference which would be used when the theory is functioning as a 'technique of inference' rather than a premise. Clearly it must be the dependent principle

$$\text{'}O_1x\text{'} \xrightarrow{(T_1 . T_2 . T_3)} \text{'}O_2x\text{'}$$

where the premises of the formally valid argument are represented by the expression 'T_i.' [37]. If the presupposition of this dependent principle is left to be supplied by the context, it becomes

$$MP_T: \quad \text{'}O_1x\text{'} \rightarrow \text{'}O_2x\text{'}$$

84. It would then seem open to Nagel to say that the question

[37] The situation can alternatively (but equivalently) be represented by analyzing the use of the theory as a 'technique of inference' as a matter of *deriving* the material rule of inference

$$\text{'}O_1x\text{'} \rightarrow \text{'}O_2x\text{'}$$

from the material rules of inference corresponding to 'T_1,' 'T_2' and 'T_3.'

whether the material principle of inference MP_T is *satisfactory* is equivalent to the question whether the theory, or more specifically '$T_1 . T_2 . T_3$' is *true.*.

85. But what is meant by 'satisfactory' in connection with material principles of inference? One possible meaning is 'truth preserving.' And, indeed, in the original example one shows that

$$\text{'x is A'} \llcorner\!\!\longrightarrow \text{'x is B'}$$

is truth preserving by showing that the statement which is presupposed (i.e. 'All A is B') is true.

86. But that MP_T is truth preserving is shown *not* by showing that its theoretical presupposition ('$T_1 . T_2 . T_3$') is true, but rather by showing that the corresponding observation framework material rule of inference

$$MP_O: \quad \text{'}O_1x\text{'} \llcorner\!\!\longrightarrow \text{'}O_2x\text{'}$$

which is the indeterminate form of

$$\begin{array}{l} \quad\ ((x)\, O_1x \rightarrow O_2x) \\ \text{'}O_1x\text{'} \llcorner\!\!\longrightarrow \text{'}O_2x\text{'} \end{array}$$

is truth preserving. And this, on Nagel's analysis is equivalent to showing that

$$(x)\ O_1x \longrightarrow O_2x$$

is a true lawlike statement

87. Notice, therefore, that the expression

$$\text{'}O_1x\text{'} \llcorner\!\!\longrightarrow \text{'}O_2x\text{'}$$

is ambiguous: if the context is theoretical explanation, it is the indeterminate form of

$$\begin{array}{l} \quad\ (T_1 . T_2 . T_3) \\ \text{'}O_1x\text{'} \llcorner\!\!\longrightarrow \text{'}O_2x\text{'} \end{array}$$

and its indeterminate or context dependent form should be represented as

$$\begin{array}{l} \qquad\qquad\quad (\text{theory}) \\ MP'_T: \quad \text{'}O_1x\text{'} \llcorner\!\!\longrightarrow \text{'}O_2x\text{'} \end{array}$$

If, on the other hand, the context is induction and the explanation and prediction of particular empirical matters of fact, it is the indeterminate form of

$$((x)\ O_1x \rightarrow O_2x)$$

$$`O_1x' \longrightarrow `O_2x'$$

and its indeterminate and context dependent form should be represented as

$$(\text{observation})$$

$$MP'_O: \quad `O_1x' \longrightarrow `O_2x'$$

88. There is, however, another possible meaning for 'satisfactory' in connection with theories as 'techniques of inference.' It could mean 'leading to observation counterparts which are experimentally confirmed.' It would be only too easy to run these two senses of satisfactory together if one thinks of truth as warranted assertibility. All one would have to do is argue that the warranted assertibility of theoretical sentences is a matter of leading to experimentally confirmed lawlike sentences in the observation framework.

89. But to make *this* move is to drape an instrumentalist account of theories in the language of factual truth. Theoretical principles construed as the instrumentalist construes them could legitimately be said to have warranted assertibility and even to be 'factually true' in a weak sense which simply constrasts them with mathematical and logical truths, as depending for their warranted assertibility on experimental considerations. But in the full sense of 'factually true' in which we think of statements in the observation framework as factually true, no statement in a framework which is, *as a matter of principle,* limited to an indirect encounter with the world mediated by a correspondence rules can be said to be factually true.

90. And factual truth in the full sense involves, as Nagel seems to realize[38] without drawing its full consequences, a picturing or mapping of events in nature by linguistic or, more generally, conceptual episodes in their own capacity as natural events. 'Picturing,' in this sense, which involves 'semantic uniformities' of the 'thing → language', 'language → language' and 'language → action' types, is often confused (as in the classical correspondence theory of truth) with the 'equivalence' expressed by the schema

$$\text{that-p is true} \equiv p$$

[38] *The Structure of Science,* p. 139.

which is as relevant to mathematical, logical and ethical propositions as to statements of empirical fact.[39]

91. Thus, to say that theoretical statements are capable of factual truth in the full sense is to say that a stage in the development of scientific theory (including the theory of sentient organisms) is conceivable in which it would be reasonable to abandon mediation by substantive correspondence rules in favor of a direct commerce of the conceptual framework of theory with the world. Such direct commerce exists already in limited contexts, and to the extent that it does exist, theoretical frameworks enjoy in anticipation the first-class status which would be theirs in that "long run" in terms of which, according to Peirce, we conceive the scientific enterprise and the 'truth' about 'what really exists' which is its formal, final and efficient cause.

[39] For an elaboration of this point and a discussion of the picturing relation see "Truth and Correspondence," *Journal of Philosophy, 59,* 1962, pp. 29-56, reprinted with minor alterations as Chapter 6 of *Science, Perception and Reality.*

August 18, 1964

Dept. of Philosophy
University of Pittsburgh
Pittsburgh, Pennsylvania

HILARY PUTNAM

HOW NOT TO TALK ABOUT MEANING

Comments on J. J. C. Smart

Professor Smart is not only a philosopher whose work has stimulated us and provoked controversy among us; he speaks to us as a representative of Australian philosophy—of a philosophy which has ties to both British and American philosophy, as well as a distinctive flavor of its own. He is a kind of philosophical Ambassador—and would that all Ambassadors were as well liked! On the present occasion he speaks to us not of his own philosophical work, unfortunately, but rather gives his impressions of the scene here. And, being a good philosophical diplomat, his impressions are friendly ones. But even giving friendly impressions of his host country can land an Ambassador in hot water. So it is not surprising, even if it is regrettable, that I, as a native of this particular philosophical jungle, should feel compelled to rebuff compliments so nicely turned, and should sadly retort that what he praises as the beauties of our philosophical landscape seem to me to be merely weeds.

The views that Smart reports *in extenso* are the views of Paul Feyerabend. Smart also has a few words to say about some related views of Wilfrid Sellars. In my comment, I shall concentrate on Feyerabend's view, with which I am better acquainted than the papers by Sellars; and I shall rely not only on Smart's account of those views, but on the paper[1] in *Minnesota Studies in the Philosophy of Science* to which Smart refers. Until otherwise indicated, all references to Feyerabend will be to this paper.

Reduction. The first part of Smart's paper (roughly, the first seven

[1] Cf. [1].

paragraphs) deals with an issue which is subsidiary to the main one, and which I should like to discuss first. This is the question: whether *reduction* (of one scientific theory or discipline to another, e.g. geometrical optics to electromagnetic field theory) is best conceived of as *deduction* of the reduced theory from the reducing theory, with the aid of biconditionals expressing "coordinating definitions," or as mere *replacement*. Is the older theory *deduced* from the newer theory, or is it merely superseded by it? My comments on this issue will be two: one historical and one philosophical.

The historical remark is that Feyerabend and Smart are unaware of the literature. The Nagel view (reduction by means of biconditionals) was criticized along these lines and in detail by Kemeny and Oppenheim[2] in 1956. Moreover, Oppenheim and I, writing in the same series as Feyerabend but one volume earlier (*Minnesota Studies in the Philosophy of Science,* Vol. II) had summarized the Kemeny-Oppenheim view. These proprietary questions are not of much importance, to be sure, but it is necessary to keep the literature straight.

The philosophical remark is that the Nagel view does not appear to me to be *so* seriously false. Smart himself expresses some doubts, in this section of his paper, on the substantive character of this particular dispute.

The issue comes down to this: according to Smart, Nagel might "save" his position by saying that it is the *approximate* truth of the reduced theory (say, again, optics) that is deduced from the reducing theory (say, electromagnetic field theory). But, Smart seems to suggest, this would not *really* save Nagel, for Feyerabend would have a (presumably effective) rejoinder: "Perhaps Feyerabend would say that truth can not be approximate but only predictive value can be: that the old theory is useful he will not deny, but he could still claim that this is more a matter for engineers than for natural philosophers."

I must now object that this rejoinder is not effective at all. It is perfectly clear what it means to say that a theory is approximately true, as it is clear what it means to say that an equation is

[2] In [5].

approximately correct: it means that the relationships postulated by the theory hold not exactly, but with a certain specifiable degree of error. In short, it means that the theory is not true, but that a certain *logical consequence* of the theory, obtained, for example, by replacing "equals" with "equals plus or minus delta" *is* true.[3] Such a logical consequence may be called an *approximation theory*. Let us now ask:

(1) Suppose O (the theory of geometrical optics) is not deducible from E (electromagnetic field theory) with the aid of bridge laws, but a suitable approximation theory O′ is so deducible. How much of a change is this in Nagel's view? *Answer*. No Change. Nagel himself is quite aware that the *unchanged* reduced theory O is *inconsistent* with the reducing theory E.

(2) *Can* the concept "light ray" be introduced into E in such a way that a suitable O′ turns out to be correct? *Answer*. Yes. (Define a "light ray" as a normal to the wave front.)

In view of (1) and (2), I no longer see what the fuss is about. Feyerabend's failure to see such obvious points seems, at first blush, perverse. Later, however, we shall see that this is no accidental or localized failure of vision on Feyerabend's part; it is his whole strategy to minimize and deny the extent to which science is the *cumulative* acquisition of knowledge about Nature, in order to enhance the plausibility of his own curious view, which is that the best we can hope for (indeed, all it makes sense to hope for) is to arrive, not at correct explanations of phenomena, but at alternative explanations by means of false theories. That we must learn to aim at a plurality of theories, rather than

[3] This replacement is only given for the sake of an example. Other possible replacements would be (a) the replacement of "equals" by "equals with probability greater than .90," and (b) the replacement of "equals" by "equals in circumstances of kind C." Of course, all of these kinds of replacements (and still other kinds) may be made simultaneously. The result in each case is a *weakening* of the original theory, in the logical sense of 'weakening'—i.e., the resulting theory is always a *logical consequence* of the original theory. The term 'approximation theory' introduced in the text may naturally be extended to cover all these kinds of replacements. I contend that scientific knowledge *is* cumulative, notwithstanding the fact that scientific postulates are usually false as stated, in the sense that, even if a good scientific theory is not *exactly* correct, a reasonable approximation theory usually is correct, and will eventually become a permanent part of our "background knowledge." Indeed, if scientific knowledge were not cumulative in even this sense, it is hard to see why it would be of any theoretical (as opposed to engineering) interest.

at a single correct explanation, and that we must expect that all of these theories will later turn out to be false, and will be superseded by new batteries of false theories, with no "convergence" to what Feyerabend scornfully calls "One True Theory," is the central contention of Feyerabend's papers. (This might surprise the reader who has only Smart's account of Feyerabend's view, however; for Smart reports only those of Feyerabend's views which he finds especially sympathetic or interesting.)

Incidentally, Kemeny and Oppenheim, unlike Feyerabend, did recognize "reduction by means of biconditionals" (Nagelian reduction) as an important special type of reduction.

Meaning and common sense. The most interesting part of Feyerabend's view, as reported by Smart, rests on the identification of the *meaning* of a term with certain accepted *theory* containing the term. In order to make this view apply to terms in ordinary language, Feyerabend holds that our common sense conceptual scheme is a false theory. It should be emphasized that this identification of meaning with theory is implicit rather than explicit in Feyerabend's paper. Feyerabend does not *say* that the meaning of a term is a theory. What he does is to slide from the term 'meaning' to some such locution as 'accepted usage' (which is harmless enough, in one sense of 'accepted usage') and then cite empirical beliefs containing the term as examples of the accepted usage. This occurs again and again in the paper. For example, the "manner" in which the terms "up-down" (strictly speaking, ἀνα and κατα) were "used" by Thales is supposed to show that "cosmological assumptions" were "implicit in the common idiom" (*op. cit.*, p. 85). (These "cosmological assumptions," as stated by Feyerabend, involve anachronistically the Newtonian concept of a "force.") Feyerabend has simply *confused* "usage," in the linguistic sense, with *occurrence;* and his whole argument rests upon this crude mistake.

It should be clear how the different parts of the Feyerabendian doctrine hang together, once this identification is accepted. Feyerabend's main contention, as he is interpreted and presented by Smart, is that if the same term occurs in two different theories T_1 and T_2, it cannot be supposed to have the same meaning.

For Feyerabend this is virtually a tautology: "difference in theory implies difference in meaning" only says that "difference in theory implies difference in theory." Feyerabend himself puts it that "meanings are not invariant with respect to the process of explanation," i.e., the fact explained contains terms which change their meaning when the statement in question is deduced from a theory.

The most radical suggestion Feyerabend is led to make (pp. 90-91) is that since (on Feyerabend's view of "meaning") the meaning of ordinary language psychological terms (e.g., "sensation," "pain") is a theory, and the theory may be false, we do not have to worry about "linguistic" objections to materialism[4], "which have, I hope, been shown to be completely irrelevant." "The mental connotations of mental terms may be spurious."

Since Feyerabend does not distinguish between conceptually necessary propositions (which rest, according to him, upon empirical theories) and empirical propositions, the assertion that, for example, *if I assert that I am in pain, and I am clear on what I am asserting, and I am sincere, then I must be in pain,* would be for him an assertion which future empirical theory might just as well refute as any other. I am aware that there are difficulties both in establishing the character of the analytic-synthetic distinction, and in the contention that the italicized statement is analytic; but surely Feyerabend's short way with "linguistic" arguments is just *too* short!

If we reject Feyerabend's view of "meaning" (as we must, if we want to talk about meaning, in the customary sense, at all), then we may perfectly well say that even if *some* common sense beliefs about pain *are false,* still it is certainly not false that there *are* pains—not in some Pickwickian sense of 'pain,' but in the customary sense. And if some statements about pain are conceptually

[4] By 'materialism' I mean here the doctrine that pain, anger, etc., *are* neurophysiological states (or events). I have argued in [6] that this doctrine is incorrect (on "linguistic" grounds, which Feyerabend would surely regard as "irrelevant"). I do not reject "materialism" in the wider sense—the view that a whole human being is simply a physical-chemical system with a certain cybernetical organization. Indeed, I think that in the wider sense materialism is correct. And I would agree with Feyerabend in rejecting "linguistic" arguments against materialism in the wider sense. But I don't know of any such arguments that are worthy of serious consideration.

necessary (i.e., necessary by virtue of the rules of the language, assuming that these do *not* essentially presuppose a "false theory,") then any definition of pain, materialistic or otherwise, must be compatible with these.

In his turn, Feyerabend must either (a) abandon the analytic-synthetic distinction altogether; but then he has no business using the term 'meaning' at all! or (b) claim that the *rules of English* (not just some common sense empirical beliefs) in connection with 'pain' essentially presuppose a false theory. In this latter case, he must say what the theory is, and show both that it *is* empirical[5] and that it *is* presupposed. This, however, would involve discussing just the linguistic issues which he intended to prove to be irrelevant.

The Feyerabend-Smart examples. Let us notice just how unconvincing, from a linguistic point of view, are the examples cited by Feyerabend and Smart. According to Feyerabend, we do not mean by the word 'temperature' what Galileo meant (i.e., what Galileo meant by the synonymous Italian word). The reason Feyerabend gives is that we have abandoned the proposition that "the temperature shown by a thermometer is not dependent upon the chemical composition of the fluid used," which Feyerabend takes to be constitutive of the Galilean concept. If this was, indeed, "built into" Galileo's concept of temperature, then Galileo would not have been able to understand the denial of this statement. This is, of course, absurd. What Galileo meant was that intrinsic property of the body which the thermometer measures, and not the result of the measurement. And Galileo could *understand* the statement that measured temperature does not exactly correspond to true temperature, and that measured temperature depends to some extent on the fluid used, just as well as you or I can, independently of our degree of physical sophistication.

Nor are the other examples more convincing. If it was constitutive of the meaning of "down" that the earth would fall down if not

[5] Feyerabend appears to suggest that the false theory in question is that pains, etc., are *mental*. Now then, it is plausible that it is a semantical consequence of the rules of English (or, more simply, that it is analytic), that, say, *anger is a mental state*. But why is this a (false) *empirical theory?* Why isn't it a tautology, which simply illustrates how we use the locution 'mental state'?

supported (or that the people at the antipodes would), and that down is always the same direction, then the ancients could not have speculated that the earth is an unsupported sphere (and the antipodeans do not fall off). For we cannot speculate that "some bachelors may really be married," since we have built being unmarried into the concept of bachelorhood; and no more can one speculate that an unsupported object may not fall down, or that down may not always be the same direction, if the contrary is linguistically required. But there is not the slightest evidence that "an unsupported object will fall down" was ever linguistically required, in any sense. And there is no evidence that "down is always the same direction" was ever linguistically required. There have always been fairy tales in which unsupported objects did not fall down; and native speakers—even small children—have shown no difficulty in *understanding* these tales. And there have been spherical earth theories for millenia. It is precisely because Feyerabend is wrong that one can explain such theories to anyone, even to a child, in quite ordinary English.

Some of Smart's examples are even less convincing. That 'chimney pot' has changed its meaning as a result of modern physics is obviously false.

The misuse of the term 'meaning' by philosophers. It is evident that Feyerabend is misusing the term 'meaning.' He is not alone in such misuse: in the last thirty years, misusing the term 'meaning' has been one of the most common, if least successful, ways of "establishing" philosophical propositions. But how did this distressing state of affairs come to be?

The blame must be placed squarely upon the Logical Positivists. The "Verifiability Theory of Meaning" ("the meaning of a sentence is its method of verification") was, from the first, nothing but a persuasive redefinition. If to call metaphysical propositions "meaningless" were only to assert that these propositions are empirically untestable, it would be harmless (the metaphysicians always *said* that their assertions were neither empirically testable nor tautologies); but, of course, it is not harmless, because the Positivist hopes that we will accept his redefinition of the term 'meaning,' *while retaining the pejorative connotations of being "meaningless" in the customary (linguistic) sense, i.e., being literally without sense.* Since, for example, theses which are literally without sense cannot be debated or discussed, anyone who says that metaphysical philosophy is "meaningless" feels free to reject it *in toto* without constructing arguments against it. If 'God exists' is meaningless, then Hume's careful examination of the arguments pro and con in the *Dialogues on Natural Religion* was unnecessary; indeed, how could there *be* arguments pro and con a *meaningless sentence?*

The stock Positivist reply, that the Verifiability Theory of Meaning is an "explication," and that an explication need not exactly agree with the "pre-analytical use" of the explicandum is disingenuous. "Method of verification" is *not* an explication of the concept of meaning which is employed in linguistic theory and in everyday life, and it was not (as we have just seen) really intended to serve as such. In order to accomplish its purpose (ruling out metaphysics, normative ethics, theology, etc.), it was precisely necessary that the Verifiability Theory of Meaning should *fail* to explicate the customary conception of "meaning."

One of the most widely discussed objections to the Verifiability Theory of Meaning in its original form ("the meaning of a sentence is its method of verification") is due to Quine. This is the objection that it is not an isolated *sentence* that has a method of verification, but a whole *theory*. It is large groups of sentences, not single sentences, that figure in the construction of empirical tests. We may regard the test, indeed, as the test of one single sentence S when the *other* sentences $S_1, S_2 \ldots, S_n$ involved in the context are not (practically) subject to doubt; but still, the test is not a "test of S" just by virtue of the meaning of S, but by virtue of the meaning of S together with the meanings of $S_1, S_2, \ldots, S_n$, and the fact that $S_1, S_2, \ldots, S_n$ are regarded as empirically true.

The Logical Positivists have reacted to this difficulty in two different ways. Some have adopted Quine's suggestion, and abandoned the notions of sameness of meaning and change of meaning altogether. On this view, there is such a thing as being meaning*ful* (and the criterion is incorporability in an empirical theory) and questions of verification enter indirectly in determining whether or not a word or sentence is meaningful; but there is no sense at all to the question, whether two words have the same meaning, or whether a word has changed its meaning. We can, of course, ask if two words have the same *extension;* but this is just a factual question, to be decided by ordinary scientific procedures. (What ordinary scientific procedures are to be used to tell if all bachelors are unmarried or not, is a somewhat embarassing question for this view; the somewhat vague answer given seems to be that such statements are "justified by their place in a whole system.")

The other approach, taken by Carnap, is as follows: questions of verification (existence of testable consequences) enter into determining whether or not a set of sentences is an empirical theory. The role of terms in the formulation of empirical theories is what renders them meaningful (as in the preceding approach). Thus, again, questions of verifiability enter indirectly into questions of meaningfulness; but not into questions of *sameness* or *difference* of meaning. These concern the *synonymy class* to which a word belongs, and are settled by consulting the semantical rules of the language. Thus, in effect, Carnap operates with *two* conceptions of meaning; he employs the customary (linguistic) conception when *synonymy* is at issue; he employs the Verifiability Theory of Meaning, in its present much modified form, when the question "is a term in the language at all?" is at issue.

Feyerabend expresses considerable opposition to Logical Positivism, especially in its current form. Yet his misuse of the term 'meaning' seems to have been influenced by the Verifiability Theory of Meaning. Thus, at one point he argues that Galileo meant by temperature "what a thermometer measures," which has some germ of truth, but he takes the term 'measures' in a queer sense, which amounts to making Galileo an instrumentalist[6]; and at another point, he uses Nagel's premise, that the "procedures" of phenomenological thermodynamics fix the meaning of the

[6] What Feyerabend asserts is that the thermometer "measures" not temperature, but a complicated function of many physical variables (including temperature). What he means, of course, is that the *exact reading* depends upon such a complicated function of many variables. On this usage of 'measures,' every measuring instrument is *perfect;* only we sometimes make mistakes about what is *measured.* And to attribute to Galileo the view "temperature is what the thermometer measures," where 'measures' is taken in *this* sense is just to turn Galileo into an extreme operationalist. Needless to say, this is an unreasonable use of the word 'measures': and it is unreasonable to say, as Feyerabend does, that we do not today believe that thermometers measure temperature.

term "temperature." Whether I am correct in conjecturing that it is Positivism that has seduced Feyerabend into playing fast and loose with the term 'meaning' or not, the fact remains that Feyerabend cannot escape the very difficulties that have bedevilled the Positivists (or, rather, he could not escape them if he ever attempted to subject his own usage of the term 'meaning' in philosophy to some kind of responsible analysis and elaboration).

To see that this is the case, it suffices to recall that for Feyerabend the meaning of a term depends on a whole *theory* containing the term. Thus, he cannot escape being involved in the following questions: *Which* theories are constitutive of the meanings of terms? Since a theory contains many different terms, how do we tell which *terms* it determines the meaning of? What changes in theory, or, more broadly, empirical belief do *not* affect the meaning of terms? No answer to these questions is even suggested in Feyerabend's paper; and these are just the difficulties that have forced the proponents of the Verifiability Theory of Meaning to restrict the Verifiability Theory to questions of *significance,* and to look elsewhere (if anywhere) for an understanding of *synonymy.*

One might, of course, take the radical line that *any* change in theory is a change in the meaning of terms. (Which terms? All of them? Even the logical connectives? Why not?) But I expect Feyerabend would not want to take this line. For to say that *any* change in our empirical beliefs about Xs is a change in the meaning of the term X[7] would be to *abandon* the distinction between questions of meaning and questions of fact. To say that the semantical rules of English cannot at all be distinguished from the empirical beliefs of English speakers would just be to throw the notion of a semantical rule of English overboard.

What is curious is that Feyerabend does *not* follow this course. Indeed, many of his purposes would have been better served had he chosen to follow Quine in repudiating the theory of meaning altogether. He wishes to show that linguistic philosophy is irrelevant and misguided; clearly, if it all rests on a mistaken notion,

[7] The second occurrence of 'X' in the sentence in the text should be in quasi-quotes (Quine's "corners") to avoid a mention-use mistake. I have ignored such logical niceties in the present paper.

that there are such things as *rules of language,* then it is badly misguided. (So is the whole science of linguistics; an objection which Quine has attempted to turn, in *Word and Object,* by allowing some purpose to constructing linguistic descriptions of natural language, and some operational and theoretical constraints to be satisfied by such descriptions.) He wishes to show that it is not a valid objection to 'pain is an event in the brain,' that to say this is merely to change the meaning of 'pain.' If "change of meaning" is itself a *meaningless* notion, his work is done for him. He wishes to show that false theories are presupposed by ordinary language; if no distinction is to be drawn between "ordinary language" and "common sense" (i.e., the everyday beliefs of most speakers), then this is just to say that most people believe many false things, and who has ever doubted *this?*

However, all appearance of sensation would have vanished, had Feyerabend followed this course. For the "sensation" here depends on sliding back and forth between a non-customary conception of meaning and the customary conception. Consider, for example, the assertion that excites Smart; that even the term 'chimney pot' has changed its meaning, since we have abandoned the common sense view of the composition of chimney pots. If the "meaning" *is* the common sense view, then this is a tautology: "we have changed our view, so we have changed our view." It is only by sliding from the fact that the "meaning" *in the Feyerabendian sense* (if, indeed, there is one) has changed to the assertion that the meaning *in the customary sense* has changed that a "sensation" results!

Finally, one of Feyerabend's main conclusions—the one with which Smart's entire paper is, in a sense, concerned—simply does not follow without this "slide." This conclusion is that, since ordinary language presupposes false theories, we may have to *discard ordinary language* (p. 76). It is no surprise that we may have to change many of our common sense beliefs; we may have to say different things *in* ordinary English, French, German, etc.; these beliefs are the "meanings" of our terms, in the Feyerabendian sense of 'meaning'; only by supporting that they are also the "meaning" in the customary sense can one obtain the conclusion that we may have to discard everyday English, French, German, etc., altogether. Here Quine and Feyerabend would, of course, part company. If the "ordinary usage" of a term X depends on the false proposition that all X are P, then Quine would recommend that we start saying instead what is true, namely "some X are not P," and not that we necessarily drop the term X. If "common sense" is false, let us find out what is true. Many technical terms may have to be introduced; but that ordinary non-technical language cannot be used to say what is true, that it is somehow *essentially infected with falsity* is a conclusion that Quine would reject as he rejects "essentialism" in general. Indeed, Feyerabend's views are an unholy mixture of Quine's refusal to separate meaning from empirical theory and just the essentialism that Quine attacks.

What meaning is. I have repeatedly spoken of the "customary sense" of the word 'meaning.' What is this customary sense?

In one sense, we all know well enough what 'meaning' means. Thus what is wanted here is not a *synonym* for the word 'meaning' (e.g., 'significance'), but a conceptual analysis. Attempted conceptual analyses of the concept of meaning have sometimes taken the form of investigations into the foundations of linguistic theory (cf. the important book by Paul Ziff), and sometimes the form of formal models. I shall attempt nothing so ambitious here. What I say will, I hope, be acceptable to all philosophers of language who do not, like Quine, reject the theory of meaning altogether.

The theory of meaning depends upon the idea that a natural language has *rules*. (Ziff objects to the term 'rule,' but his "state regularities" and "state projections" *are* "rules" as *I* shall use the term.) These rules are sometimes syntactical and sometimes not. The syntactical rules determine what phones are allophones of what phonemes in a language; what sequences of phonemes are morphemes (more strictly, allomorphs of morphemes); and what sequences of morphemes are well-formed sentences. The non-syntactical rules distinguish linguistically *regular* from linguistically *deviant* uses of sentences. To call a married man a "bachelor," or to refer to a geometrical point as "graceful" is to violate semantical rules of English.

To discover the different sorts of non-syntactical rules that there are, and to succeed in stating them perspicuously, is the main task of semantic theory. Thus I shall not commit myself here, beyond giving a few examples, on the form and content of the semantical rules of a natural language. Even at the present early stage, however, certain things seem clear. It is generally agreed that a speaker's knowledge of the rules of his native language is implicit and not explicit; only the very sophisticated speaker can verbalize even (some of) the *syntactical* rules of his native language. The unconscious and obligatory character of linguistic rules has long been noted. The "meaning" of a word is a function of the rules governing its employment. These determine, among other things, which locutions are synonymous (coextensive by linguistic stipulation), which locutions have more than one meaning (are governed by alternative batteries of rules), which sentences are analytic on which readings, etc.

The unconscious character of linguistic rules is important in understanding what happens when someone asks for the "meaning" of a word. What the inquirer wishes to gain is a knowledge of the rules governing the employment of the word—so that *he* will be able to employ it too. But he does not wish for an explicit statement of these rules, but for the kind of implicit knowledge that I alluded to above. Thus it is the respondent's task, in such a situation, to say something from which the inquirer, employing his considerable (implicit) linguistic knowledge, can "pick up" the information he wishes—"information" which neither the inquirer nor the respondent can verbalize, and "pick up" by a process which *no one* today understands.

It is this that accounts for the hodgepodge of things that get counted as answers to the question "what is the meaning of such-and-such a word?" Traditional dictionary definitions, for example, contain a fantastic mixture of empirical and linguistic information. Nevertheless, they succeed often enough in enabling speakers to acquire the usage of words they did not previously understand.

Suppose, for example, that one asks a typical native speaker of

English for the meaning of the word 'gold.' He is likely to give one a mass of empirical information about gold (that it is precious, normally yellow, incorruptible, etc.), in addition to the essential linguistic information that 'gold' is the name of a metal. Yet, if gold became as "cheap as dirt," or began to rust, or turned green, the meaning of the *word* 'gold' would not change. Only if we stopped using 'gold' as the name of a metal, or used it to name a different metal, would the primary meaning change. (I say "the primary meaning," because the *connotations* of 'gold' *do* depend on the facts that gold is normally yellow, precious, etc.).

One should notice, also, that one may know the meaning of the word 'gold' without knowing *how to tell* whether or not a given thing is gold. (Of course, someone must be able to *identify* gold; otherwise the word couldn't be used as the name of a real metal; but it isn't true that only those people who can identify gold know the meaning of the word, or even that they know *more* of the meaning of the word. They simply know *more about gold.)*

Similarly with Feyerabend's example of the word 'temperature.' As long as we continue to use the word 'temperature' to refer to the same physical magnitude, we will not say that the "meaning" of the word has changed, even if we revise our beliefs many times about the exact laws obeyed by that magnitude, and no matter how sophisticated our instruments for measuring temperature may become. However, one "theory" *is* essential to the meaning of the word 'temperature'—that the magnitude we identify as "temperature," and quantify by means of thermometers, or however, is the magnitude whose greater and lower intensities are measured by the human sensorium as *warmer* and *colder* respectively. This does *not* mean that the human sensorium is never fooled, but that when it is not fooled, when the differences in felt warmness are accounted for by a difference in some property of the object rather than of the subject, it is generally a difference in "temperature" that is responsible. The use of the word "temperature' rests upon the empirical fact that there exists a single physical magnitude (in fact, molecular energy) which is normally responsible for differences in "felt warmness," because it is *analytic* that "if X has a higher temperature than Y, then X is warmer than Y"—i.e., the words 'temperature' and 'warmer' are semantically linked—and because the functioning of the human sensorium underlies the "stimulus meaning"[8] of the word 'warmer.' The word 'temperature' is "theory loaded"; and, fortunately, the theory is correct.

I asserted in a preceding section that Galileo used the word (or, rather, the corresponding word in Italian) to refer to the physical magnitude we call 'temperature' today. But what reason is there to say this? One reason, of course, is that all scientists accept the well known reduction of phenomenological thermodynamics to statistical mechanics (identifying "temperature" with mean kinetic energy of the molecules). This is a "reduction" in the Nagelian sense—reduction by means of biconditionals—and it leads to the deduction of a very good approximation theory to phenomenological thermodynamics from statistical mechanics. Accepting the reduction is precisely accepting that phenomenological thermodynamics and statistical mechanics (with the added postulate 'temperature is average molecular kinetic energy') are theories of "temperature" *in the same sense,* as Nagel correctly points out.

Two points of a more "linguistic" character are also relevant to this example:

(1) On the basis of *linguistic intuition,* it seems clearly linguistically regular to say that Galileo was measuring and theorizing about the magnitude we call 'temperature' in English, but that we have somewhat different beliefs concerning it than he did. The fact that native speakers have certain intuitions of regularity does not *prove* anything, of course, but it *is* part of the data available to the linguist. When he goes counter to strong and widespread linguistic intuitions, he must have a good reason to give.

Feyerabend, as might be expected from his general position, rejects this kind of appeal to linguistic intuition. For him the "linguistic intuitions" of native speakers simply reflect the false theories that they have deeply internalized; and conformity to linguistic intuitions is thus a sign of conformity to false theories, and is hence *undesirable.*[9] But there is a confusion here. Linguists

[8] The term 'stimulus meaning' comes from [7].

[9] Cf. [2], p. 36.

and philosophers of language rely on linguistic intuition, as well as on methodological considerations (e.g., simplicity of the over-all description of the syntax and semantics of the language), in order to discover a correct semantic characterization of a word X. In saying that such-and-such is a correct semantic characterization of a word X, they do not thereby commit themselves to the existence of Xs, much less to the correctness of whatever empirical beliefs may be entailed in calling something an X. If the use of a word X *does* presuppose some false empirical theory or other, than this is just what we would hope to find out by this procedure. If we reject linguistic intuitions to begin with as somehow "dirty" (essentially infected with falsity), then we will never discover the semantic characterization of any word, and hence never discover if such theses as those of Feyerabend—that most words presuppose false empirical theories; that various terms have changed their meanings; that meanings are not invariant under the processes of reduction and explanation—are true or false.

(2) Given two semantic description of part or all of a language; even if they are approximately equal in simplicity and factual adequacy, there is always a strong *methodological* reason to prefer the one that postulates fewer meaning changes. Namely, the fact that two uses conform to the *same lexical* characterization (have such-and-such features in common) is a fact about the language; whereas the fact that they can be given two separate "entries" is trivial—the meaning of any word can be "split up" under as many entries as one pleases. Thus one will not discover the semantic structure of the language unless one conforms to the maxim "differences of meaning are not to be postulated without necessity." (Ziff calls this "Occam's eraser.")

Thus we see again that given the operational and methodological constraints appropriate to linguistic theory, it is unreasonable to say that such words as 'temperature' change their meanings whenever we change our theory concerning the corresponding magnitudes. We have just given two examples of the kinds of constraints that are accepted by semantic theorists. Quine has expressed considerable scepticism that such constraints single out a *unique* semantic description of a natural language. However, this does not trouble us in the present context.[10] Our difficulty today is not that we are confronted with a plurality of "equally valid descriptions" of English, and do not know how to choose (or if it makes sense to choose at all), but that no one has constructed even *one* wholly adequate description of English. And certainly it is not the case that *every* description is adequate (as Quine admits, in *Word and Object).*

The customary sense of the word 'meaning' may be somewhat vague; but the great interest in semantic theory among linguists, as well as among philosophers of language, and the developing integration of semantic and syntactic studies in linguistics, affords hope that it may be clarified, and perhaps supplemented by a battery of more technical notions. We are well on our way to knowing a responsible and theoretically fruitful way of talking about "sameness of meaning" and "change of meaning" in the linguistic sense. Smart offers the suggestion, in defense of Feyerabend, that Feyerabend may be talking about "meaning" *in some other sense*. But what *exactly* is the other sense? What constraints are there upon this other way of talking about "meaning"? In another paper, Feyerabend says in one place that it is "(empirical) meaning" that does not remain invariant. The only existing theory of "empirical meaning" is the Verifiability Theory of Meaning, and, as we have seen, in its present form the theory affords only a criterion of empirical meaning*fulness,* and no criterion of empirical *synonymy* (sameness of "empirical meaning"). I cannot join Smart in hailing it as a great discovery that terms

[10] Incidentally, the constraints allowed by Quine are clearly much too weak. For example, on the formal side, Quine requires only that a translation scheme be general recursive. This allows wildly irregular mappings as "correct." In practice something approximating to word-by-word substitution is always used, at least as a first approximation. On the operational side, Quine rejects conformity to the linguistic intuitions of native speakers as a constraint, on the grounds that this constraint begs the philosophical questions. (Where did these "linguistic intuitions" come from, if not from an implicit theory of the language, Quine asks. And what were the constraints upon this implicit theory?) But this seems puristic. Quine is right, that the *foundations* of linguistic theory cannot be permanently rested upon appeals to linguistic intuition (as they are, for example, by Chomsky and his followers); but this does not mean that we cannot rely upon such appeals at all. Fortunately, we can proceed with sciences whose conceptual foundations are far from being in order!

are constantly changing their meaning in some completely unspecified sense of the word 'meaning'; nor can I accept the argument that, since ordinary language presupposes false theories in some completely unexplained way, ordinary language may have to be discarded.

Dept. of Humanities
Massachusetts Institute of Technology
Cambridge, Massachusetts

BIBLIOGRAPHY

(I have included certain works on semantic theory *not* referred to in the text, for the convenience of the reader.)

[1] P. Feyerabend, "Explanation, reduction, and empiricism," in *Scientific Explanation, Space and Time, (Minnesota Studies in the Philosophy of Science,* vol. III), ed. H. Feigl and G. Maxwell, pp. 28-97. (Minneapolis, 1962)

[2] P. Feyerabend, "How to be a good empiricist, etc.", in *Philosophy of Science, (The Delaware Seminar, volume 2*), ed. B. Baumrin, pp. 3-40. (New York, 1963.)

[3] J. Katz and J. Fodor, *The Structure of Language, Readings in the Philosophy of Language.* (New York and Englewood Cliffs, N.J., 1964.)

[4] J. Katz and P. Postal, *An Integrated Theory of Linguistic Descriptions,* (Cambridge, Mass., 1964.)

[5] J. Kemeny and P. Oppenheim, "On reduction," *Philosophical Studies* 7:6-19 (1956).

[6] H. Putnam, "The mental life of some machines," forthcoming in a volume edited by H. Castañeda. (Detroit.)

[7] W. V. O. Quine, *Word and Object,* (Cambridge, Mass., 1960.)

[8] P. Ziff, *Semantic Analysis.* (Ithaca, 1960.)

PAUL K. FEYERABEND

REPLY TO CRITICISM

Comments on Smart, Sellars and Putnam

---

(1) *Proliferation.* In writing the papers which are discussed, and criticized by Profs. Smart, Putnam, and Sellars, my aim has been to present an abstract model for the acquisition of knowledge,[1*] to develop its consequences, and to compare these consequences with science. With respect to the last point it may be expected that a comparison of historical phenomena with epistemological views or models will lead to new historical evidence, and to new ideas concerning the *actual* structure of science (or law, or philosophy, or commonsense). But it is also important to realize that one must not allow this structure to interfere with the models. Such models tell us how to proceed if a certain aim is to be achieved. In this way they form a basis for the *criticism* as well as for the *reform* of what exists. It is clear that a theory of knowledge which is built up with such a purpose of criticism and reform in mind will be fairly different from an analytic theory that relies on the maxim that in the battle between the actual and the idea the latter must be regarded as a mere fancy, as a flight from reality, as a castle in the air[2] which shrinks into insignificance when confronted with the hard facts of (scientific, legalistic, commonsense) life.[3]

The model which underlies my own discussion has as its *aim* maximum testability of our knowledge. No argument will be given for this aim in the present brief note.[4] All I intend to do is to sketch some consequences and to remove the air of paradox which, according to some critics, surrounds these consequences.

The main consequence is the *principle of proliferation: Invent, and elaborate theories*[5] *which are inconsistent with the accepted point of view, even if the latter should happen to be highly confirmed and generally accepted.* Any methodology which adopts the principle will be called a *pluralistic methodology.* The theories which the principle advises us to use in addition to the accepted point of view will be called the *alternatives* of this point of view.

Summarising earlier arguments for proliferation[6] we may say: (1) that no theory ever agrees (*outside* the domain of computed error) with the available evidence. Hence, if we do not want to live without any theory at all, we must have means of accentuating certain deviations, of lifting them out of the ocean of "deviational noise" surrounding each theory. Alternatives provide such means.[7] (2) Theories agree with the *facts* (as opposed to *observations*) only to a certain extent. Indeed, it would be a complete surprise for everyone if a theory were found that represents all the facts perfectly. Some disagreement with facts is mirrored in the evidence. But there are other cases where physical laws prevent the discrepancy from ever appearing in the evidence.[8] Now if we find a theory which asserts the discrepancy, which is able to repeat the past successes of the accepted point of view, which has new, and independent evidence in its favor, then we shall have good reason to abandon the accepted point of view despite its success. Alternatives are theories of the kind described. (3) One need not mention the *psychological* advantages flowing from the use of alternatives, the fact, that is, that a mind which is immersed in the contemplation of a single theory may not even notice its most striking weaknesses.[9]

The principle of proliferation not only recommends invention of *new* alternatives, it also prevents the elimination of *older* theories which have been refuted. The reason is that such theories contribute to the content of their victorious rivals. Thus certain contemporary attempts to give a classical account of typically relativistic effects[10] provide further tests for relativity, and their failure, further corroborating evidence. The same is true of some well-known but unpopular approaches in the quantum theory.[11] Professor Naess has made it clear[12] that such corrective use may well continue forever without our being able to indicate a point where it becomes preposterous. Knowledge so conceived is not a process that converges towards an ideal view; it is an ever increasing ocean of alternatives, each of them forcing the others into greater articulation, all of them contributing, via this process of competition, to the development of our mental faculties. All theories, even those which for the time being have receded into the background, may be said to possess a "utopian" component in the sense that they provide lasting, and steadily improving, measuring sticks of adequacy for the ideas which happen to be in the center of attention.[13] Hegel's assertion that theories emerge as aspects of the development of "the idea" is therefore essentially correct, except that it confounds cause and effect and strives to retain a semblance of consistency by weakening the laws of logic. In our presentation the competition of theories or, to speak in a less Platonist fashion, the discussion of alternatives by individual scientists, philosophers, politicians etc. is the cause, the gradual improvement of all of them as well as of the minds of the participants, the effect. Hegel, on the other hand, regards the idea as the subject of development, and human thought as derivative,

There are many thinkers who find such considerations strange and not to their taste.[14] From what I have read I gather that Professor Putnam is one of them. Now instead of further engaging in abstract argument, I would invite these opponents of pluralism to compare what I have just said with the actual history of thought. Are not the history of heliocentrism, and the history of atomism, splendid examples of the way in which antediluvian, preposterous ideas, ideas which are regarded with contempt and are ridiculed by the learned and by the common folk alike, may yet be turned against "modern" views *and may even succeed in overthrowing them?* Are we not warned by the still quite recent attacks against Boltzmann and against his "old-fashioned" atomic theory[15] that things can never be regarded as settled and that the picture of a steady accumulative progress towards an aim which can be reached, at least approximately, either assumes too much knowledge on our part, or puts undue restrictions upon our instruments of criticism (in which latter case the possibility of a "criticism from the past" will of course be excluded). It is of course somewhat difficult to find events like those just described in the customary historical accounts which favor monism. But more recent research has gone a long way towards showing the decisive role the principle of proliferation has played at almost all major stages of the history of thought.

(2) *Strong Alternatives.* Not all alternatives are equally suited for the purpose of criticism. Thus a theory which merely replaces a prediction P of T by another prediction, P′ that is inconsistent with P and experimentally indistinguishable from it, will not be satisfactory. It contradicts T, but it does not provide any reasons for removing T. Szilard's non-kinetic modification of the second law of phenomenological thermodynamics is of this kind.[16] It is assumed that deviations from the second law will occur which average out over many cycles. Of course, it is quite possible that such deviations do actually exist. But they will then refute the law directly, and without help from Szilard's account. A similar remark applies also, to a certain extent, to the older quantum theory which was largely a "transcription of the Balmer formulae".[17] The *first* condition to be satisfied by a theory that is fit for criticism going beyond the criticism provided by the observations themselves and strengthening their refuting power is therefore: it must contain assertions over and above the assertion (the prediction) which leads to the contradiction. *Secondly,* the additional assertions must be connected with the contradicting assertion more intimately than by mere conjunction. *Thirdly,* we shall not be inclined to eliminate T simply because there exists a different story, T′, which contradicts it, even if this second story should possess considerable interest. There ought to be some independent reason to accept T′, and one of the best reasons is, of course, evidence in favor of it. Such evidence need not be there from the beginning and an alternative T′ should not be eliminated simply because it does not yet provide it. But (a) it must be *possible* to produce it; and (b) such evidence must at some time be *actually* produced. *Finally,* the critic should also be able to account for the earlier success of the criticized theory for it is only in this case that a clear judgment of superiority can be made and that correspondingly clear reasons for temporary removal can be given. Points 2, 3, and 4 show that a "strong" connexion is needed between the refuting prediction and the remainder of the theory (which includes the independent evidence, and the evidence supporting the theory to be criticized).

Now it would certainly be desirable to possess a more detailed account of the "strength" of the connection between different parts of a theory, or of its *coherence.* There are some mathematical functions for which such an account is already available. Thus complex functions of a single variable which are defined in a domain D and differentiable everywhere in the domain, so-called *regular functions,* coincide everywhere in D if they coincide along a path of arbitrary length, and even if they coincide in a denumerable manifold of points, provided the manifold converges (identity theorem for regular functions). Conversely two regular functions which differ along a path of arbitrary length in D will differ everywhere else in D. We see here how the high degree of coherence of the regular functions is directly connected with their critical ability. This leads to a more general *conjecture,* to wit that two coherent theories ("coherent" in the as yet unspecified sense introduced at the beginning of the present paragraph) which approximately agree in a certain domain (and are confirmed here by the same evidence) will differ noticeably almost everywhere else if they differ in a finite, or in a denumerably infinite number of points. Theories of this kind (*strong alternatives* as I shall call them) would be especially well suited for the purpose of criticism (as their postulates make them differ everywhere, they are not *ad hoc* anywhere).

It is interesting to note that conditions 1-4 and the demand for coherence which characterize severe critics coincide with the conditions characterizing acceptable explanatory systems. This means that a strong alternative T′ of T which has received independent support will not only remove T but will also be a satisfactory successor of T: *the best criticism is provided by those theories which can replace the rivals they have removed.*

(3) *A Model for Progress.* Let us now consider the features of an activity which consists in the invention of theories for the purpose of explanation, in the criticism of these theories with the help of strong alternatives, and in the replacement of refuted theories by the rivals which brought about their downfall. Such an activity will lead to a succession of theories T, T′, T″, T‴,... We want to investigate the relation between these theories. This investigation clearly has nothing whatever to do with historical fact. It starts from a certain aim—to obtain testable knowledge—it indicates how the aim can be reached, and it then goes on to describe what will happen if people proceed accordingly.

It provides a model for possible action without claiming that actual science conforms to the model.

It is not unimportant to discuss the features of models of this kind. First, the model has great advantages and one may hope that scientists will try to act in accordance with it. Secondly, using the model in the course of historical research we may discover features unnoticed by those who favor monism and we may come to realize that actual science is much closer to pluralism than one would expect when consulting the (usually monistic) historians. Thirdly, discussion of the model and removal of apparent "paradoxes" will lend support to those who intuitively favor pluralism and strong alternatives, but who are prevented by the general philosophical climate from fully relying on their intuitions. I really hope that the occasional disparity between the model and actual scientific practice will be regarded as a criticism of the latter, not of the former. In the struggle between an ideal and actual reality the ideal must always be given the upper hand. After all, we do not want to leave the historical development of a discipline to chance (or to Oxford-commonsense); we want to shape it, and improve it in accordance with ideas we find reasonable. This was the way in which modern science and its philosophy started (cf. footnote 3). And this is also the way in which we may hope further to improve them both.

(4) *Consistency*. What is the relation between two theories, T and T′ (where T′ succeeds T as the result of criticism)? We see at once that

(i) T cannot be subsumed under T′ (T′ is the successor of T)

(ii) T cannot be explained on the basis of T′

(iii) T cannot be reduced to T′

if we retain the demand that subsumption, explanation, reduction must be by logical derivation. The reason is simply that T′, being a critic of T, is also inconsistent with T. It is gratifying to see that in actual science, too, the transition to a new theory is often the transition to a theory which contradicts what went on before. The so-called "layer model" which has been elaborated by Professor Nagel therefore agrees neither with actual science, nor with the arguments supporting proliferation.

Now it is of course always possible to preserve the layer model and the philosophy on which it rests. This can be done either by replacing strict rules of derivation with approximate rules,[18] or by replacing T with another theory, $\overline{T}$, which does follow from T′.[19]

The first suggestion is unacceptable; it eliminates the very same procedure of criticism by theories which has brought about the transition T → T′.[20] But assume that there exists strong refuting evidence of T that is independent of T′. Then the removal of the inconsistency between T and T′ will still lower the empirical content of T, will make T more "metaphysical" and why?—just in order to preserve a philosophical model that would otherwise have to go. Is it really necessary to point to the similarity between this move and the manoeuvres of the much maligned metaphysicians of the past?

The second suggestion is nothing but a concealed admission of defeat. For what is so attractive about a philosophical point of view concerning the relation between theories which can be upheld only if *real* theories, theories which have been discussed in the scientific literature, are replaced by emasculated caricatures of them? Consider Putnam's own example. He admits, and says Nagel admits, that geometrical optics (O) is inconsistent with wave optics (W).[21] Now he looks for a theory, Ō, which is vaguely similar to O and can therefore be called a geometrical optics, but which follows from W. I am quite prepared to admit the existence of such an Ō. It will not be the theory Putnam suggests (in his comments, p. 207): light rays are normal to the wave front only in isotropic media; and what about those cases where no wave front exists, such as inside a container with perfectly reflecting walls? Nor will it be possible to write the desired Ō down explicitly ("rays" must be replaced by sausage-like structures etc., etc.) But let us disregard these difficulties and let us be content with the fact that an Ō of the desired kind can be imagined abstractly. What has been shown? It has been shown that while almost all that was actually said about light before the advent of W contradicted W and could therefore be criticized by W (and criticize W in turn) [22] there are some logical creatures, such as $\bar{O}$, which do not contradict W. This is a great discovery indeed, especially in view of the trivial fact that all the logical consequences of W follow from it and provide a reservoir for the choice of some suitable (though perhaps not analytically presentable) $\bar{O}$. But surely it was not the intention of those who formulated the layer model to assert that the *logical consequences* of the theories in the top level follow from them, can be reduced to them, can be explained on the basis of them. What they asserted was that certain theories entail other theories *which have been formulated independently*. And *this* assertion is refuted by my argument and by the historical examples provided.[23] Result: recourse to "approximate rules of correspondence" (Smart) and to "approximate theories" *is out*.[24,25]

(4) *Meaning Invariance.* In his paper Professor Putnam creates the impression that I am mainly interested in meanings and that I am eager to find change where others see stability. This is not so. As far as I am concerned even the most detailed conversations about meanings belong in the gossip columns and have no place in the theory of knowledge. This is true even in those cases where meanings are invoked to force a decision about some different matter. For even here their only function is to conceal some dogmatic statement which would not be accepted, if presented by itself, and without the chatter of semantic discussion. (See the example in footnote 26).

Still, the fact that meanings occur so often in philosophical debate forces us to deal with them too, at least occasionally, and some terminological hints are therefore indicated (the hints are given in footnote 27). On the basis of these hints we now put the following question: assume that T and T′ are two strong alternatives. What relation may we expect to find between their primitive descriptive terms?

In order to answer the question we examine the following three theories: T = classical celestial mechanics, including the space-time framework; T′ = general relativity; $\bar{T}$ = T, except for some slight modification in the strength of the gravitational interaction. $\bar{T}$ is inconsistent with T. In our universe, which does not contain regions free from gravitational influence $\bar{T}$ and T will nowhere lead to the same predictions. Yet $\bar{T}$ cannot aspire to be a strong alternative of T. The difference between the predictions of the two theories can be accounted for by changing part of T. No modification is needed of other principles (the law of inertia, the spatiotemporal framework, can both be retained in their original form). A strong confirmation of $\overline{T}$ endangers only a small part of T, it does not necessitate a reconsideration of the "nature" of space, time, force, mass. Using the terminology of footnote 27 we may express this by saying that the transition $T \rightarrow \overline{T}$ is not of the fundamental kind and that the "meanings" of the primitive descriptive terms of T and $\overline{T}$ will largely coincide.

The transition $T \rightarrow T'$ is an altogether different matter. The fact that within T′ all events are embedded in a four-dimensional Riemannian continuum which acts not merely as a somewhat unusual background of coordinates, but is supposed to describe *intrinsic properties* of physical processes now necessitates the reformulation of all the basic laws of T and the redefinition of all its basic concepts (spatial distance; temporal distance; mass; force; etc.). The new laws will not only *read* differently, they will also conflict *in content* with the preceding classical laws. And this is not just a matter of words. The classical, or absolute idea of mass, or of distance, cannot be defined within T′. Any such definition must assume the absence of an upper limit for signal velocities and cannot therefore be given within T′. *Not a single primitive descriptive term of T can be incorporated into T′.* Using the terminology of footnote 27 we may express this by saying that the change of rules accompanying the transition $T \rightarrow T'$ is a fundamental change, and that the meanings of all descriptive terms of the two theories, primitive as well as defined terms, will be different: T and T′ are *incommensurable theories;* (other incommensurable theories have been discussed in earlier papers).

Now one of the most frequent objections against admitting incommensurable theories into the class of the strong alternatives of a given theory T is that they cannot contradict T and cannot, therefore, be used to criticize T. "For in order for two sentences to contradict one another" writes Professor Shapere, expressing most clearly what has been said by many critics[28] "one must be the denial of the other; and this is to say that what is denied by the one must be what the other asserts; and this in turn is to say that the theories must have some common meaning. On the other hand, two sentences which do not have any common meaning can neither contradict, nor not contradict, one another".

This remark is entirely correct and it indicates that the use of incommensurable theories for the purpose of criticism must be based on methods which do not depend on the comparison of statements with identical constituents.[29] Such methods are readily available.

One of the main properties of strong alternatives is that they disagree everywhere if they disagree in a finite number of points (see section 2). Moreover, the disagreement is due to basic rules, and not to *ad hoc* adaptations from point to point (strong alternatives are *coherent* theories—see again section 2). Now one may be interested either in the relation between different theories, or in the features of the world as seen from the point of view of a single theory. Questions of the first kind may be called, by analogy to terms which Carnap has used on a similar occasion,[30] *external questions;* questions of the second kind may be called *internal questions.* External questions can be decided without bringing in meanings at all. We simply compare two infinite sets of elements with respect to certain structural properties and inquire whether or not an isomorphism can be established. Considering a Euclidian plane and an arbitrarily bent surface with a Riemannian metric imposed upon it, we realize that there is nothing corresponding to a large-scale Pythagorean triangle on the latter which establishes that, in two dimensions at least, Newtonian mechanics and general relativity are different theories. The actual situation (three-dimensional Euclidian space and four-dimensional Riemannian manifolds) is of course much more drastic and shows even more clearly that appeal to meanings is not needed to prove difference. Adding the usual physical interpretation, we may conclude that the validity of the one is incompatible with the validity of the other (I omit the detailed argument for lack of space). Of course, there is still the question as to which theory is to be preferred. Usually, one tries to solve the question on the basis of *crucial experiments.* This is quite proper, but may again lead into trouble. For if one assumes that a crucial experiment consists in selecting, on the basis of observation, one of two contradictory statements then we seem again to be forced to use only such theories as have a certain minimum of similarity in meanings.[31] However there may be empirical evidence against one and for another theory without any need for similarity of meanings. This can be in various ways. *First,* we may introduce incompatibility of theories as above (lack of isomorphism with respect to certain basic relations plus consequences following therefrom) and then confirm one of two incompatible theories. If the confirmed theory and their rival are strong alternatives, then the rival will have to go. *Secondly,* both theories may be able to reproduce the "local grammar"[32] of sentences which are directly connected with observational procedures. In this case the utterance of one of the sentences in question in accordance with the rules of the local grammar, or the utterance of a local statement, as we shall say, can be connected with *two* "theoretical" statements, one of T, and one T′ respectively (theoretical statements, corresponding to a given local statement S will be called statements associated with S within the theory in question). We may now say that the empirical content of T′ > the empirical content of T, if for every associated statement of T there is an associated statement provided by T′, but not vice versa. And we may also say that T′ has been confirmed by the very same evidence that refutes T if there is a local statement S whose associated statement in T′ confirms T′ while its associated statement in T refutes T. Similar remarks apply to the question of independent evidence. *Thirdly,* we may construct a model of T within T′ and consider its fate. The proof of the irrationality of $\sqrt{2}$ may be seen in this way. To start with, rational and irrational numbers form two distinct domains, R and I, so that it does not seem to be possible to relate a single entity, $\sqrt{2}$, to both of them. However, we may select a subdomain, R′, of I which is isomorphic to R with respect to all the properties we think essential of the latter, and interpret the "experiment" as being one between I and R′.[33] Within physics the construction, in T′, of a model $\overline{T}$ of T usually is not possible without violation of highly confirmed laws. This, then, is sufficient evidence for the rejection of $\overline{T}$ and, via the isomorphism, of T.

To sum up: it *is* possible to use incommensurable theories for the purpose of mutual criticism. This removes, I think, one of the main "paradoxes" of the approach suggested by me.[34]

Now for the *internal questions.* The disagreement between strong alternatives is due to basic rules, and not to *ad hoc* adaptation from point to point. According to footnote 27 we must conclude that the difference will turn up in the descriptive concepts, and that the theories will be incommensurable. We see here quite distinctly the function of the descriptive concepts of a *coherent* theory. *They reflect, within each individual description and prediction the peculiarities of the fundamental principles of the theory.* That is, they will be applicable whenever the principles are adopted, and inapplicable otherwise. Hence, the concepts of theories which have been shown to be incompatible by the first method above cannot be applied simultaneously, provided, of course, that the basic principles *are* allowed to contribute to their content. This proviso leads at once to the next three questions which are

(i) *historical question*: *do* we allow our basic beliefs, principles, etc., to influence our concepts?

(ii) *methodological question*: *should* we allow our basic principles to influence our concepts?

(iii) *physiological question*: *can* all our basic beliefs, principles, etc., influence our concepts?

It is in the consideration of these three questions that the greatest confusions arise. Usually (ii) is rejected by appeal to the fact that crucial experiments would then become impossible. This point has already been dealt with. Or it is rejected because empirical statements are supposed to retain an invariable meaning. Why? In order to provide a neutral basis for our decision between different theories. This point can be reduced to the previous one and is removed with it. But this last point, which is a logical point (and an invalid one) is then often interpreted as establishing limits to our interpretation of concepts, and especially of observational concepts. Now it is one thing to say that some interpretations of observational statements make some theories irrefutable. But it is quite a different thing to say that such undesirable interpretations *can never be carried out.* We meet here again Kant's familiar transcendental argument, only it is now presented upside-down, not as an argument showing which theories are certain, but as an argument showing how certainty can be avoided at one place and how it therefore turns up at another. I am sure that Strawson's dictum (see footnote 26) has been arrived at in some such manner—if it is based upon argument, that is. But it is equally evident that while he and others move here in a logical fairyland the problem is a factual one, to wit (problem iii above) whether all principles, including the most esoteric ones, can be incorporated, *physically,* into our reaction patterns (with possible effect upon our perceptions), or whether some such principles are not just "too much" for our organism. It is the great merit of Professor Smart that he has drawn attention to this side of the problem, and of Professor Sellars that he has arrived at his conclusion with this feature of the problem in sight [cf. his reference to neurological facts in paragraph 58 (b)]. This is an important step beyond the pseudo-logical rigmaroles we are usually invited to consider. I hope that I shall be able at some time to give a more detailed account of the matter. At the present, with but two days to go for completing the paper (including, of all days, Christmas Eve!), removed from anyone who might provide inspiration, either by discussion, or by beauty, I can only offer the following brief considerations.

(5) *The Historical Question.* The answer to the historical question is that people very often *did* allow fundamental principles to influence the most concrete parts of their everyday lives (or the "local grammar" of the words used in this context). As proof I invite the reader to consult historical accounts of the way in which religious beliefs have influenced all aspects of human activity, the most pedestrian parts included, so that even the distinction between waking and dreaming changed its position and its nature. This applies also to such apparently "rational" people as the Greeks.[35] All this refutes the idea, held by Austin[36] and others, that such beliefs were metaphysical superstructures erected on a single unchangeable basis—Oxford-commonsense. In my previous papers I have given very few examples of this all-pervasive character of basic theory indicating that the historical and sociological literature is full of cases of this kind. Let me briefly repeat one of these examples and deal with Professor Put-

nam's objections to it. The example is the way the pair "up-down" was used in antiquity (and in the early middle ages). Now first of all this example deals with concepts, and not with words, so that Professor Putnam's two-word excursion into Greek scholarship [early in his paper] is irrelevant (Besides, his interest in the Greek *words* seems to indicate that it is he, and not I, who deals with occurrences, rather than with usages). Secondly the grammar of the pair is not at all dependent on physical assumptions concerning *forces*.[37] "Aut est quisquam tam ineptus qui credat esse homines quorum vestigia sint superiora quam capita. Aut ibi quae apud nos jacet inversa pendere? Fruges et arbores deorsum versus crescere?" [Lactantius, *Divinae Institutiones,* III, *de falsa sapientia;* apologies for the Latin–however in arguments with linguistic philosophers, it is better to use the original *words* rather than a translation, which would preserve the *concepts* only]. Here reference is made to unusual *positions,* and not to unusual *motions* (of course, if we start from unusual positions, then even the usual forces will bring about unusual motions!) From considerations like these it is then inferred that the heavens cannot be spherical, which is indeed evidence that "down is always the same direction" was linguistically required [Putnam, p. 211]. The fairy tales which Putnam mentions only apply unusual *forces*: they are still compatible with the absolute sense of "up-down". This pair was therefore used much in the same way in which Professor Max Black thinks the pair "sooner-later" is used today.[38] And *this* use has disappeared today, at least in some circles (we no longer talk of the *lower* side of the earth, but of the *other* side). The change admittedly is not a universal one. This now leads at once to the following question: should we not educate people in a manner that makes their everyday behavior agree with basic science as it agreed with basic theology before? This is question (ii).

(6) *The Methodological Question.* The power of a theory can be fully utilized only if it is not treated as an instrument for prediction, so that the local grammar is allowed completely to determine the "nature of things". It is doubtful whether Newton would ever have identified the Galilean orbit of a projectile with the Keplerian orbit of the moon,[39] had he restricted himself to

data of observation (cf. his very instructive drawing in the *System of the World.*) No data were available for C, D, E; the

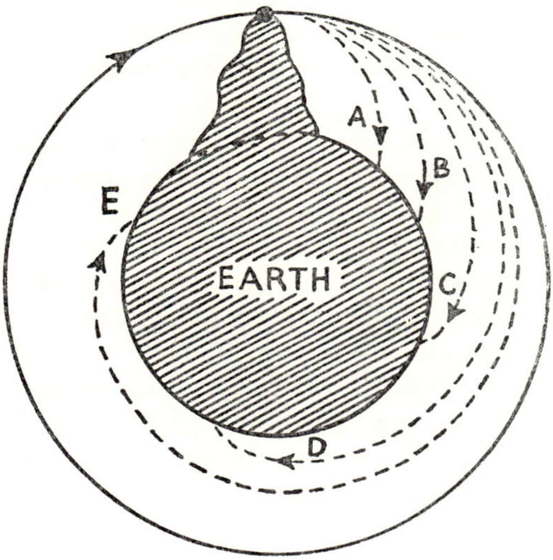

orbits of projectiles on the one side, and the orbit of the moon on the other were thereby shown to belong to different domains. However, quite apart from facts like these, it has never been explained why a new and satisfactory theory should be interpreted as an instrument, while the principles behind the established usage, including the local grammar, which are shown to be inadequate by the new theory, are not so interpreted. After all, the only advantage of the latter is that they are *familiar*, an advantage which is a psychological and historical accident and which should therefore have no influence upon questions of interpretation and of reality. One may try to answer this criticism by ascribing an instrumental function to *all* principles, old and new, and not only to those contained in the most recent theory. Such a procedure means acceptance of a sense-data account of knowledge. Having shown elsewhere[40] that such an account is impossible, I can now say that this consequence of a *universal* instrumentalism is tantamount to its refutation. Result: neither a restricted, nor a universal instrumentalism can be carried through in a satisfactory manner. This disposes of the instrumentalistic move.

It does not yet dispose of the arguments against the realistic interpretation of new, and successful theories, or against the *complete* replacement, in accordance with the principle enunciated at the end of section 2, of older theories by their victorious rivals. Of these arguments there are four:

Argument i: T, the older theory, or the corresponding principles of commonsense, express what a person *means* who describes events within the domain of overlap between T and the theory T′. Thus a layman talking about the temperature of his milk "surely" does not consider "the kinetic energy of the molecular constituents of the liquid" (argument against the universal use of statistical physics, discussed by Nagel in *The Structure of Science*). And "one can talk of trees, chimneypots, and telescopes in a way which is independent of the choice between Newtonian and Einsteinian physics" (argument against the universal use of relativity, discussed in Smart's paper). Or, "I *know* (pronounced indignantly, and emphatically) what pains are, and I also *know* (same as above) that a reference to material constituents is not only not implied, but is even nonsensical" (familiar argument against the universal use of materialism).

Argument ii: *It would be* practically impossible, too complicated, to use T′ inside D. "For example" writes Smart "it would be impossible in practice to use general relativity in celestial mechanics (though it may be needed to explain isolated anomalies, as with the perihelion of Mercury)." This point is sometimes given great weight in arguments against the feasibility of a purely physiologial account of the mind.

Argument iii: Scientists (or the common folk) *in fact* use T inside D.

Argument iv: T is our *observation language*. Eliminating *it* means eliminating the observational basis on which science rests, and thereby also T′ (this argument has been used by Feigl, Nagel, and others).

All these arguments have one thing in common: they would have provided excellent weapons for the Aristotelians against Galileo; for the defenders of witchcraft against their more enlightened opponents; and for the backward side of many other debates. As a matter of fact some of these arguments *have* been used for the purpose of defending an older doctrine which was in danger of being overthrown, and especially in the disputes which accompanied the emergence of modern science. It is therefore a little late to try to enforce them *now, after* the rash opponents of Aristotelianism (to mention the most popular example) have succeeded in eliminating it. Besides, if it is indeed so important to stick to what is commonly accepted, be it now part of commonsense,[41] or of some more esoteric doctrine, then consistency would seem to force us to retreat even further, at least to the conceptual system of *Pithecanthropus Oxoniensis,* and to start from there. On the other hand, if such a procedure is deemed too radical, if we *are* allowed to use the ideas of classical physics (for example) because they are accepted *now,* then what prevents us from voting now *for the future,* and from bringing about *its* general acceptance by proper methods of education (such as the use of educated mothers or nurses)? Moreover, the fact that comprehensive conceptual schemes have been replaced *in toto* (women having no longer the opportunity of betraying their ineffective husbands with the devil) would seem to indicate that the difficulties mentioned in the arguments are largely imaginary. A more detailed examination fully confirms this guess.

The *first argument is incomplete.* Different people, members of different language groups will "mean" different things when proceeding descriptively within D.[42] Whom shall we choose as a point of reference? Moreover, it cannot be denied at this stage of the inquiry that a practitioner of T′ may become accustomed to his theory to such an extent that in the end he will see D in its terms. The argument *even admits* this possibility and tries to prevent it from becoming an actuality [cf. the above quotation from Nagel with its *two* alternative descriptions of the same situation—the caloric properties of milk]. It *admits* the possibility of language groups whose members "mean" T′ when proceeding inside D. Why not take their meanings as a measure of adequacy? Why not try even to form such groups through education, especially in view of the fact that T′ has advantages over and above T (after all, we have assumed that T was removed by T′ and T′ is a strong alternative of T. According to the statement at the end of section 2 this means that T′ is superior to T). The argument in the form in which we have presented it so far cannot answer this question.[43]

We may now strengthen the argument by adding the assertion (which I shall call the *universality thesis*) that everybody, and not only a few laymen, thinks in terms of T. The universality thesis is silently assumed to be correct by many linguistic philosophers. Ryle's *Concept of Mind,* for example, is not intended as a field study of the conceptual system of *homo Oxoniensis;* it is supposed to possess universal validity. In section 5 we have referred to material indicating that the universality thesis is *false.*[44] It remains to point out that it is also irrelevant.[45] A language (a theory) is not only a vehicle for meanings, it is also supposed to inform us about what exists. The universally accepted idiom (if there is such a thing) may not satisfy the second demand. It may be a universal mistake. Universality of T therefore cannot convince us that T′ must not be universalised. Quite the contrary, it is only through the attempt to universalise a new theory that we can discover the mistakes of the accepted ideas [for this cf. again sections 1 and 2 of the present paper]. The universality thesis therefore is not only incorrect, and irrelevant. It is also desirable that it be false.

However—and this at last brings us up to a real difficulty—is it not plausible to assume that the (physiological) organization of the human mind puts limits to the replacement of theories and to scientific progress? Are there not very general theories, which cannot be replaced except by altogether destroying the functioning of the mind? This consideration, if I understand him correctly, is the basis of Professor Sellars' objections against me. It will be dealt with in the next section.

The *second argument* mentions formal difficulties: it may be impossible to develop the formalism of T′ in such a manner that everyday occurrences can be described in its terms. Usually a theory deals with fairly idealised elementary situations which are not accessible to direct observation. Concrete events, such as the processes occurring at the level of everyday observation, involve the interaction of many elementary processes and it is not always possible to calculate their interaction in terms of the most recent theory. This applies even to theoretical cases. For example, we do not yet see any way of completely replacing Newton by relativity in planetary astronomy (this has been observed by Professor Smart). However the universal application of a theory, or of some other general point of view does not require the universal use of its *syntactical apparatus;* all it requires is that the *categories* of the theory be regarded as basic and that these categories replace the categories of the preceding point of view.[46] Thus the adoption of a materialistic position with respect to the mind-body problem does not entail that from now on we shall have to employ density functions in phase space for the description of mental events and it is not exposed as impossible by the fact that the explicit form of these functions is not available, and will perhaps never be available. Nor does the universal application of general relativity entail that the *formalism* of classical mechanics must go and must be replaced by the formalism of general relativity (not even classical mechanics itself has as yet achieved coherence of expression in all domains of application, and it will perhaps never achieve it). In the first case all that is needed is the realization that we are dealing with complex patterns and sequences of physical events and that the appropriate categories will therefore be those of physical processes (localizability is one of them; mass is another). In the second case we may retain all the formulae of classical physics, but we must now interpret their key terms after the fashion of general relativity, either as four-dimensional invariants, or as projections into the space—or into the time—part of the four-dimensional Riemannian continuum. It would of course be desirable to add coherence of formal expression to coherence of content. The main thing is however the latter. Without it there would not even exist a motive for the attempt to achieve coherence of formal expression. Extension of a new theory to the domain in which an older point of view is still doing well is therefore much less difficult than might appear at first sight. We may retain the familiar words (such as "pain", "thought" etc.) or the familiar formulae; we may even retain the local grammar [see fn 32] of these words which is that part of their grammar that is restricted to the domain D of the overlap of T and T′. We may also continue to use all those reflexes which guide their automatic application in observational situations. Only, this local grammar cannot any longer determine *all* our linguistic reactions. It must be incorporated into a new framework (which corresponds to the takeover of a new level of neurological organization). Such integration may then retroactively lead to a partial reorganization even of the local grammar itself![47] It may lead to "seeing things differently" and it may in this way precipitate the discovery of quite obvious but still unnoticed weaknesses of T *inside D* (the realization that we can see the fixed stars only on a photon account of light is an example). All this must of course be examined in greater detail, expecially via a more thorough examination of physiological principles. Still, the difficulties insinuated (but never stated) in arguments i, ii, and iii seem to be largely imaginary. They seem to be due not to the existence of objective barriers, but rather to the non-existence of that amount of boldness that is necessary for overcoming the problems which arise in the course of our attempt to comprehend the world.

(7) *Observation.* Argument iv is not easy to evaluate. It may be given a physiological twist, but it also contains an element of logic. This dual character of the argument considerably improves the position of its defenders. They start with the stronger version and create the impression that their argument has *substance.* When attacked, they withdraw to the weaker version and can so pretend that their argument is *invulnerable.* A shifty procedure like this should be easy to look through—but this is not what happens. Quite the contrary, the procedure seems to possess some kind of intuitive appeal. Exhibition of the dubious character of this appeal is overdue.

The *logical* version of the argument assumes that there are statements, so-called basic statements, or observation statements, which can be absolutely confirmed by observational evidence. This property is thought to be connected with the "meaning" of their main descriptive terms (the observation terms). If this is true, then a radical enough change of the meanings of all observation terms will eliminate the property, and destroy the observational basis of our knowledge. One may therefore formulate the logical version by saying that a logical connexion is assumed to exist between the meaning of a term (apologies for again bringing in semantic matters!) and the fact that it is, or is not, an observational term. I shall call a theory of observation which is based upon such a connexion a *semantic theory of observation.*

A semantic theory of observation is usually supported by an analysis of such verbs as "to see", or "to observe", or "to hear", or else by an analysis of the nouns which form their objects. If such an analysis is correct (a matter that is open to considerable doubt) then the semantic theory forms an essential part of commonsense. There are also other, and less linguistic ways, of supporting it. Whatever the means of support, the semantic theory is unacceptable to anyone who rejects the synthetic apriori. For consider the principles P which guarantee that observational terms can be applied. These principles describe some general features of the world (or of the mind) and are therefore synthetic. On the other hand a theory implying their denial would eliminate its own observational basis, and would therefore be unacceptable. We can say in advance that such a theory must not be formulated–which guarantees the perennial correctness of P. P is synthetic apriori.

The undesirable consequences can be avoided by redefining observational terms so that the logical connection between meaning and observability is eliminated (as it is in any case within commonsense). Only if we admit that meaning and observability are two different and independent (though empirically related) dimensions of a term, only then shall we be sure that *this* source of a synthetic apriori has been destroyed. The new theory of observability which results from the described procedure (and which was formulated very clearly in the early thirties by Popper, Carnap, and Neurath) [48] may be called the *pragmatic theory of observation.* The choice between the pragmatic theory and the semantic theory is of course purely a matter of convention. If we want synthetic apriori statements, if we want to be able to derive eternal laws from facts of observation (as was done by Aristotle, and as is done again by the school of "pure phenomenology"), then the semantic theory will most certainly be a Splendid Thing. However, if it is our intention not to except any part of our knowledge from revision, then we shall have to choose the pragmatic theory. This choice eliminates argument iv.

For it is clear that the question whether or not a certain term, such as the term "electron" is, or can be, an observational term, is now a question of empirical fact. We have already dealt with this question in section 6 when asking to what extent a theory can be integrated into our behavior so that we can "see" the world in its terms. We have pointed out that so far no limit of integration has been found [for this cf. also the next section]. However physics provides also the opportunity of an *external integration,* as one might call it[49], that is of an integration which makes use of devices bridging the gap between the entities considered, and some of the more "basic" parts of human behavior. In this case the observation sentence is produced by the combined system [apparatus+observer]. There are no limits to a procedure of this kind. This refutes the "physiological" version of argument iv.

A variant of argument iv consists in pointing out that without some restriction in the use of new theories which allows for the definition of a neutral "experience", it may become impossible to obtain *impartial* evidence for a theory, or to stage a single *crucial experiment.* This has already been discussed in section 4. However I would like now to pursue the matter a little further as it is important to be quite clear about the situation in which we are here involved. As we have indicated above, it *is* possible to retain older ideas, and to allow modifications only at the outskirts of our knowledge. Professor Putnam's "approximate theories" would be one way of doing this. It is possible to obtain what many empiricists seem to regard as such an important feature of knowledge, viz. a relatively neutral "experience" that can decide between two universal theories without being influenced by either. But such a procedure introduces the synthetic apriori and may prevent discovery of weaknesses of the accepted point of view right in the center of its most successful domain of application. The empiricists therefore cannot have their cake and eat it too. *I* can [for this see again section 4 above].

It remains to make a few comments on *argument iii.* This argument, or rather an attitude corresponding to it is rather widespread among scientists. The attitude invites concentration on successful theories and deplores consideration of alternatives. It is the attitude of monism with which we are already familiar. Again, there are two kinds of arguments which may be used, and have been used to support the attitude. The first argument, which goes back to Newton (and was refuted by Goethe, among others), assumes that it is possible to prove a theory "directly, and positively" (Newton) and *not* only via the elimination of alternatives. This argument excludes alternatives as not being empirically adequate. The second argument admits that any theory may sooner or later break down, but adds that one should make the best of it while it is still successful. This latter argument [for which see footnote 14] excludes alternatives as bringing about an unwarranted dissipation of the valuable energy of the scientists. *In praxi* one usually meets a combination of argument one and argument two, argument one providing the punch, argument two a safe second line of retreat (this fluctuation between a more logical and a more practical position is quite characteristic of the "classical" empiricism of Newton and of his followers; it still appears in Niels Bohr). Neither the combination, nor the single arguments are acceptable, for which see above.

We may now summarize by presenting a scoreboard of arguments:

Argument i: incomplete without the universality thesis, irrelevant and leading to undesirable conclusions with it.
ii: irrelevant
iii: invalid
iv: invalid

As far as I can see these are the only arguments available against the idea that a good theory should replace a bad theory completely, and not only in isolated cases (the only argument not yet dealt with will be discussed in the next section). The arguments have also been presented here in a rather strong fashion. What one usually finds in the literature on this point is an attitude of conservatism struggling to become an argument. None of the arguments is satisfactory. This is the one side of the issue. On the other side, on our side, there exist very good *abstract reasons* why a new theory should be used everywhere: only this procedure will lead to the strongest possible criticism of the received point of view. These abstract reasons are fortified by *concrete examples* of extremely successful scientific research making use of the very procedure arguments i-iv want to forbid. Finally it is but *plain commonsense* (though perhaps not Oxford commonsense) to prefer total improvement to piecemeal tinkering and to accept such total improvement wherever it can be had. The case for a complete takeover therefore seems to be well established—but for one point to which we now at last turn.

(8) *The Physiological Question.* It is reported that Pavlov once criticized the early Bolshevists' theory of education as being incompatible with the laws of behavior and thinking which he himself had discovered. This criticism may be removed by the following consideration. Assume that the educational *aims* of the Bolshevists are attractive, and that their realization is therefore desirable. Then the attempt to realize them may either lead to success which is a good thing, and thereby to the refutation of Pavlov's theory which, being a step forward, is also a good thing. Or this attempt may fail and thereby confirm Pavlov's view, which is again a good thing. Hence, both scientific method and proper education make it desirable to attempt to carry out the Bolshevists' aim.

The very same argument can be applied to those who resist the complete replacement of commonsense by some scientific ideology on the grounds that the attempt is utopian. For if the attempt succeeds, then man will have succeeded in ridding himself of prejudices which would have prevented him from ever seeing the world as it is and from figuring out the details of its real structure, which is a good thing; at the same time the laws of the mind on which the opponents base their charge of utopianism will be refuted, which is also a good thing. If the attempt does not succeed, then this provides confirming evidence for these alleged laws, which is again a good thing.

These two arguments are precisely analogous except for the fact that Pavlov *could point to the laws* he thought would prevent certain attempts from succeeding whereas no one has as yet formulated the laws of the mind which allegedly prevent the universalization of scientific ideologies. Even worse, most arguments are nowadays based not on physiology, but on meanings, and are therefore entirely irrelevant. Among past arguments we may mention Kant's proposal to derive the structure of the mind from the structure of a theory then widely held. This proposal rested on the features of *one single* theory. The possibility that other theories might bend the forms of perception and the categories to accommodate *their* structure was either not considered, or was rejected outright, and without any experimental investigation. The opposite view viz. the view that we see things as we believe them to be—"you see a bird, I see an antelope; the physicist sees an X-ray tube, the child a complicated lamp-bulb; the microscopist sees coelenterate mesoglea, his new student sees only a gooey, formless stuff; Tycho and Simplicius see a mobile sun, Kepler and Galileo see a static sun"[50]—has again been held without experimental support. I strongly sympathize with this latter view, but I must now regretfully admit that it is incorrect. Experiments have shown that not every belief leaves its trace in the perceptual world and that some fundamental ideas may be held without any effect upon perception.[51] The laws of perceptual organization are then not wholly dependent on the information available to us even in those cases where the information has already become part of our automatic behavior. All this clearly cries for further research. Is there the slightest suspicion that such research may support the Strawsonian thesis of a "massive central core of human thinking that has no history"[52] and is never going to have a history?

Before turning to this question, let me first briefly deal with some points which were raised by Professor Sellars. Professor Sellars' paper is so rich in content and so full of interesting conjectures that it is perhaps somewhat impudent to deal with it in such a summary fashion. Yet, the editor beckons urgently! Here, then, the comments.

I start with a few remarks on Professor Sellars' attempt to distinguish between theories which express beliefs, and other principles which "are constitutive of the very concepts in terms of which we experience the world" [par. 5 of his essay]. It seems to me that this distinction does not separate science from commonsense. The space-time framework of classical physics and of the theory of relativity, to mention but one example, have quite a lot in common with frameworks of commonsense as described by Sellars. They have "no external subject matter" [6]; their concepts help "to project, or transpose the attributes of sense impressions into the categorial framework of physical things and processes" [57] so that they are (repeat) "constitutive of the very concepts in terms of which we experience the world" [5]. Now this similarity is strong enough to give the same treatment to both and to express this fact by a unified terminology ["theories"—see my footnote 5]. For example, we shall have to admit that a relativistic account of the spatio-temporal aspects of commonsense is possible *here and now* [48; 74]— after all, the theory of relativity is "already at hand", it is not "a 'regulative ideal'" [50]. Sellars' reluctance to admit this, his emphasis on the difference between science and commonsense, is due to his preoccupation with *microtheories* [47f; 67; 9. Putnam, too, seems only to consider microtheories and not framework theories such as the theory of relativity. Cf. his comment on the chimney-pots on page 215]. Microtheories indeed do not seem to contain any large-scale principles. They seem to permit a safe distinction between the inevitable external subject matter (gases, large-scale objects) and some internal subject-matter. One sees that both are necessary and one can at least envisage the possibility to *retain* the customary account of perceived objects, and to interpret the microaccount as *additional information,* filling in holes, or empty spots, not dealt with by the macroconcepts. True, Professor Sellars does not himself subscribe to this "hole theory" or "Swiss cheese theory" of commonsense (which Professor Smart has conjured up on another occasion in order to make a scientific materialism more palatable to the pundits of ordinary speech). Still, by likening scientific theories to microtheories, and by relating the latter to an external subject matter, independent of it, indispensable, and rooted in commonsense, he has created the impression that a particular shape of commonsense is "methodologically indispensable" [59], too. The fact that there are scientific theories without any external subject matter refutes this impression and imposes the task of *modifying* those aspects of commonsense which are incompatible with them. Now according to Sellars the framework of commonsense "is a framework of (among other things) colored physical objects extended in space and enduring through time" [53]. Question: in Euclidian space, and with time running down separately? If the answer is yes, that is, if the shape of the objects is fixed independently of their motion, then a revision is clearly needed. The intuitive content of this revision can be shown by large-scale models (such as the behavior of an image in a spherical mirror, or the behavior of the normal world when seen through distorting glasses) and may at some time become visible even in the real world. Four-dimensional invariants are not entirely beyond our faculty of perception (just consider our ability to perceive the *changing* character of a person behind the quite undescribable changes of his face, behavior, tone of voice). The question that remains is, of course: is it possible *always* to carry out such a modification even of the most primitive perceptual elements of commonsense?

The answer to this last question must of course remain hypothetical. No stronger claim can be made at the present time. There are some arguments in favor of an affirmative answer. To start with, we should consider that until now only two or three percent of the inbuilt circuits of the brain have been utilized. A large variety of further integrations is therefore possible. There may of course be difficulties: present education after all consists, according to an excellent remark of Professor Z. Young[53] in badly damaging our learning apparatus. "The moulding of the baby into a transmission receiver" writes Professor Waddington in the same vein[54] "seems a difficult, and complicated, and even slapdash process, and not at all what one might have thought out if one had set out to design this job. A frequent result of the process seems to be that people believe too much, and believe it much too strongly"; or, to use Gantt's even more drastic characterization,[55] they increasingly become "a museum of antiquities". (Note, that the main intention of our linguistic friends is to find "logical" reasons for continued infliction of such damage, and for leaving the museum unaired). It seems reasonable, then, to assume that a new mode of education whose principles have as yet to be found will make it easier to change points of view, to integrate theories into behavior, and in this way to see the world in terms of such theories. This education will have to be careful not to let the world of everyday objects come forth too impressively, and it will also have to keep the division between dream and reality somewhat open. The Greeks, whose perceptual world seems to have been much less settled than ours,[56] also produced thinkers with a range of imagination quite inconceivable today. For those already grown up, and messed up, a psychological procedure like the procedure of alienation suggested by Brecht[57] may do the trick. All these are problems which need investigating, especially through a more detailed study of the central nervous system. The mentalistic bias of much of contemporary philosophy is one of the most widely used means to prevent such investigation from becoming effective; and this is, and always has been, its ideological function.

So far we have been dealing with the higher levels of mental organization only. Now the structure of the central nervous system has been described by various authors as consisting of layers of organization, ranging from highly organized, involuntary levels of low complexity to lowly organized, voluntary and very complex higher levels. The higher levels control and overlay the activity of the lower levels. If they are put out of action, then the primitive levels break through and impose their own laws. This creates the "positive" components of "dissolution", first described by Hughlings Jackson. The whole system is somewhat precarious and always threatened by a "revolution from below". Now one might be inclined—though this is of course entirely hypothetical—to correlate certain basic features ("archetypes"?) of common languages with lower levels and the abstract notions of theoretical science with higher levels. If such a correlation does in fact exist, then the life of a theoretician who wants to "see" the world in terms of the most recent scientific ideology will not be an easy one. But neither was the transition from animal life to civilization with all its additional repressions and its precarious balance between sophistication and savagery. The argument from common usage, amplified by these physiological considerations, is therefore correct insofar as it maintains that certain modes of thinking, being physiologically more basic, are also more firmly integrated into human life, behavior, perception. The argument certainly goes too far when asserting that these modes of thinking are the only ones which can be integrated. After all, we do overcome our lower nature and nobody would be inclined to attack civilized morality by pointing out that we are apes with inbuilt constraints.

These are only some of the problems arising in connection with what we have above called the physiological question. It is evident that they are not even touched upon in the grammar-school philosophy of our friends, the contemporary school-philosophers of grammar.

Dept. of Philosophy
University of California
Berkeley, California

NOTES

1 *Historical Remarks.* In his paper Professor Smart concentrates on the views of Nagel and Sellars, and he also discusses some of my own publications. Perhaps unintentionally he creates the impression that a new philosophical position is in the making, a kind of neorealism, and that I have contributed to its development. Such an impression would be both incorrect, and unfortunate. It would be incorrect as the present discussion does not lead beyond the arguments which have been used in the traditional issue between realism and instrumentalism. I for one am not aware of having produced a single idea that is not already contained in the realistic tradition and especially in Professor Popper's account of it. For example, most of the points which I have made in "How to be a Good Empiricist" appear in Boltzmann. But such an impression of novelty would also be very unfortunate. It would increase the tendency to disregard the connexion between philosophy and the sciences which is so essential for the development of our problem, would further separate the most recent history of philosophy from its past and would thereby reinforce the provincialism and irrelevance that is a trademark of much of contemporary philosophy. Of course, many thinkers see the situation in a very different light. They believe that philosophy was born (by parthenogenesis?) in the 20th century and that reference to the past can be only of the didactic kind, showing the mistakes committed by those not yet in possession of the One True Method. Of course, they never succeed in completely eliminating traditions either. The old problems have the habit of appearing again; but separated from the context in which they arose they now look very strange indeed. Compared with them, even the most pedestrian part of commonsense seems to be full of virtue and even the most ridiculous argument derived from it a source of sun-like clarity. There arises then an activity where the correction of what was once a sensible hypothesis and what has now become a silly mistake can pose as a major philosophical advance (Austin against Ayer on sense data); and where a bumbling and tedious recapitulation of the less interesting parts of the phylogeny of our thought can be hailed as a splendid achievement of a "modern philosophy." We need the historical background both again to give substance to our arguments, to show that they are relevant to a much wider class of problems than a specialist treatment would suggest, and for raising the level of discussion to the heights already achieved by Democritus, Galileo, Descartes, Faraday, Kant, Boltzmann, Maxwell, Einstein, Popper. Considering such a wider background, Putnam's attempt "to keep the literature straight" [p. 206 of his essay] where by "literature" he means Kemeny, Oppenheim, and himself, is somewhat amusing.

2 For a more detailed criticism of this view cf. my review essay of *Scientific Change* ed. Crombie in *British Journal for Philosophy of Science* for November 1964.

3 Modern science and its methodology started as the rebellion of a few people against established modes of thought. These thinkers disagreed with the most fundamental beliefs of their time, which were held by scholars and by laymen alike and which also had impressive evidence in their favor. Yet it is only rarely that one finds in their work even a trace of hesitation and not once does the universal support for the established doctrine suggest to them that their own ideas might perhaps not be the whole truth. Commonsense, received opinion, the teachings of the schools count nothing for Descartes. He is prepared to reject them all. In defending his own pet ideas Galileo was forced to contradict experience and reasonable conjecture. An immense optimism is at work here, an optimism which is prepared to abandon the products of the combined efforts of many generations and "to begin the whole labor of the mind again" (Bacon). Ernst Mach who initiated the movement of modern positivism criticized the science of his time in the very same spirit. This is the attitude which has led to the sciences as we know them today. Every philosopher of science lives off the products of this great and distant past. But it is also clear that the contemporary tendency to dwell at length on *what is,* and either to oppose, or not to be concerned with, *what should be* would never have permitted such products to arise. It is time for the philosophers to recognize the calling of their profession, to free themselves from the exaggerated concern with the present (and the past) and to start again anticipating the future.

4 For such arguments cf. K. R. Popper, *Conjectures and Refutations* (New York, 1962).

5 When speaking of *theories* I shall include myths, political ideas, religious systems, and I shall demand that a point of view so named be applicable to at least some aspects of everything there is. The general theory of relativity is a theory in this sense, "all ravens are black" is not.

There are certain similarities between my use of "theory" and Quine's "ontology" ["On What There Is," *Proc. Arist. Soc.,* Suppl. Vol. 25 (1951)] Carnap's "linguistic framework" ["Empiricism, Semantics, and Onotology,' *Revue Intern. de Phil., 4* (1950)], Wittgenstein's "language game" [cf. my review "Patterns of Discovery," *Phil. Rev.,* Vol. 59 (1960)], Pareto's "theory" [*Treatise on General Sociology,* par. 7 ff], Whorf's "metaphysics" [*Language, Thought, and Reality,* passim], Kuhn's "paradigm" [*Structure of Scientific Revolutions,* passim] etc. I prefer this "accordion" use of the term because it provides a single name for problems which in my presentation are intimately related. However the "philosophical music" [Sellars, paragraph 4] I intend to make will not be extracted from the fact that the word is so used, but from the common features of the problems described. Cf. also section 8, i.

Terms such as (1) consistent, (2) incompatible, (3) follow from, are applied to pairs of theories, [T, T′] and they mean that (1) there do not, (2) there do, exist pairs of predictions, P and P′, one following from T, one from T′ (via corresponding initial conditions) which are incompatible. In a special case T′ taken together with the conditions of validity of T will be (1) consistent with, (2) incompatible with, (3) sufficient to derive, T. For modification of this explanation cf. section 4.

The *principle of proliferation* is an essential part both of Popper's critical rationalism and of the philosophy of possibilism as it has been developed by Professor Arne Naess. It was used, more or less consciously, by many scientists before. Boltzmann is one example [cf. the article *Boltzmann* in the *International Encyclopaedia of Philosophy,* ed. Paul Edwards]. Professor Popper shows ["Back to the Presocratics"; *Proc. Arist. Soc.,* Vol. 54 (1959)] that the Presocratics were the first thinkers to proceed in accordance with it.

6 Cf. e.g. "How to Be A Good Empiricist" Vol. II of the *Delaware Seminar in Philosophy of Science* (New York, 1963).

7 Not a single planet moves in the orbit calculated in accordance with Newton's celestial mechanics (this has nothing to do with relativistic effects). There exist other as yet unexplained discrepancies exceeding the error of measurement by a factor of 10. These discrepancies provide a background of "deviational noise" relative to which the 43″ of arc per century in the advance of the perihelion of Mercury lose much of their force. Moreover, there are many as yet unexplored possibilities of explaining this value within the framework of Newton's theory. An oblateness of the sun of magnitude 10^{-4} (corresponding to 0.09″ of arc) would increase the perihelion rotation by 7″. Owing to the turbulence of the atmosphere an oblateness of 0.05″ could be discovered only with great difficulty, so that at least 5″ of the famous 43″ depend on assumptions beyond the level of present observation. The as yet unexplained discrepancies mentioned above may be due to causes which are operative in the present case also. Is it not obvious that abandoning Newton's mechanics on the basis of so doubtful a number would be rather imprudent? And is it not also clear that only an account of the 43″ which explains them on the basis of new principles will lift them out of the background of deviational noise and turn them into an *effect* that is capable of *refuting* the Newtonian scheme? Naturally such an account will rest on a theory inconsistent with Newton—it will rest on an *alternative* to Newton's theory.

8 Cf. the discussion in section VI of my "Problems of Empiricism" *Beyond the Edge of Certainty,* ed. R. Colodny (Englewood Cliffs, N. J., 1965), as well as in "Realism and Instrumentalism" *The Critical Approach, Essays in Honor of Karl Popper,* ed. M. Bunge (Glencoe, 1964).

9 As an example one need only consider Galileo's intoxication with his preposterous theory of tides. For an evaluation cf. A. Koestler, *The Sleepwalkers,* (New York, 1959) pp. 464 ff.

10 Cf. R. H. Dicke "The Observational Basis of General Relativity" in Chiu H.-Y. and W. F. Hoffmann, ed., *Gravitation and Relativity* (New York, 1964), pp. 1-16.

11 Cf. my "Problems of Microphysics" *Frontiers of Science and Philosophy* ed. R. Colodny, (Pittsburgh, 1962)

12 In a manuscript which he kindly sent to me prior to publication.

13 In this respect scientific theories are much more similar to works of literature than one would be inclined to believe. "The domain of literature" writes Ingeborg Bachmann in her "Frankfurter Vorlesungen" [*Ingeborg Bachmann,* Munich, 1964, pp 298 ff; esp. pp 333f] "is an open domain—and this is true of ancient literature as well as of more recent works. It is less closed than any other domain—for example the sciences, where each new theory eliminates what has preceded [which is precisely the point at issue—P. F.]—it is an open domain as its whole past intrudes into the present. With the force of all times it presses towards us, towards the threshold of time where we reside . . . and we are made to realize that none of its works is dated, or can be made ineffective. It contains all those presuppositions

which resist final judgment and final categorization. These presuppositions which reside in the works themselves I would like to call 'utopian' presuppositions." A little less reliance on Wittgenstein whom Miss Bachmann seems to know quite well and a little more acquaintance with the history of science would have convinced her that the opposition between science and literature is much less pronounced than she seems to think.

14 Proliferation of views has been regarded with suspicion in religion, in philosophy and, finally, in the sciences. This tendency towards monism seems to be a most persistent inheritance from our savage past. The *philosophical* debate between monism and pluralism has a long and interesting history. It started when Plato, heavily relying on common prejudice, ridiculed the Presocratics because of the variety of theories they produced [cf. footnote 5]. Within empiricism the demand for monism was most clearly formulated by Newton [*Principia* Book iii, Rule iv; ed. Cajori, p. 400]: "In experimental philosophy we are to look upon propositions inferred by general induction from phenomena as accurately, or very nearly true, *notwithstanding any contrary hypothesis that may be imagined,* till such time as other phenomena occur by which they may either be made more accurate, or liable to exceptions." Today Newton's rule is frequently invoked in the defence of the (slowly disintegrating) Copenhagen orthodoxy. Even in the fields which have traditionally exhibited a healthy disregard for parsimony of theory construction, for example in cosmology, some thinkers object to the proliferation of theories as being inconsistent both with the history of science, and with a sound methodology. Thus Professor Heckmann, whose masterful analysis of Newton's cosmology and whose contributions to the relativistic theories have been a delight to all students of the subject, writes in an address delivered to the eleventh Solvay congress [E. Schuecking and O. Heckmann, "World Models," *Institut International de Physique Solvay, Onzième Conseil de Physique* Brussels 1958, pp. 1 ff]: "The steady progress of science, in our opinion was possible only because scientists thought it not permissible to put forward new theories unless new data forced them to abandon the older concepts." (This is historically incorrect. Galileo, Faraday, Boltzmann, Einstein brought about progress by the skillful application of alternatives. In many cases *no other way was possible* [cf. again my "Problems of Empiricism," esp. section VI].) "It would seem" so the authors continue "that some cosmologists have abandoned the aforementioned principle of methodology." But, so the passage concludes "it is sound policy to refrain from theorizing along the lines of Bondi, Gold, and Hoyle until there is strong empirical evidence for continuous creation of energy and momentum." (Quite the contrary. Cosmology, more than any other discipline offers such a close connexion between fact and theory, and such a vast amount of background noise that alternatives *will be quite indispensible* for the purpose of refutation.) Professor Kuhn, to whom we owe a clear account of the function of alternatives in the development of science still shows some sympathy for monism. It is true, he points out that what may seem to be a minor disturbance, still accessible to treatment, "can be seen, from another viewpoint, as a counterinstance, and thus as a source of crisis" [*Structure of Scientific Revolutions,* Chicago, 1962, p. 78]. He remarks that "Copernicus' astronomical proposals . . . *created* [my italics] an increasing crisis for . . . the paradigm from which it had sprung [74] and, more generally, that "paradigms [theories in our terminology] are not corrigible by normal science [science without alternatives] *at all"* [121]. However there is still the tendency to regard proliferation with suspicion, and perhaps even as a sign of immaturity [this comes forth very clearly in an earlier paper of Kuhn's which appeared in *Scientific Change* ed. Crombie; see above note 2]. Strictly speak-

ing, alternatives are allowed to *accentuate* trouble, they are not regarded as *giving rise to it.* "Invention of alternatives" writes Kuhn "is . . . what scientists seldom undertake . . . The reason is clear. As in manufacture, so in science retooling is an extravagance to be reserved for the occasions that demand it" [76]. The restriction to a single point of view "is a sign of maturity" [11, 21] and it is "essential to the development of science" [24]. Such restriction is not at all dangerous as science possesses "a built-in mechanism that ensures the relaxation of the restrictions that bound research when the paradigm from which they derive ceases to function effectively" [24]: Proliferation is admitted only "when the occasion demands it," only, "when the paradigm . . . ceases to function effectively." But my point is that trouble is either omnipresent [this is admitted by Kuhn, pp. 80 ff, 145] or arises by accentuation *only if alternatives are first used.* Proliferation *precedes trouble,* and not the other way around. One cannot say that Professor Kuhn has renounced monism completely and that he has given a satisfactory account of the function of alternatives.

15 "It is not easy today to recapture the attitude that prevailed towards the end of the last century" writes von Smoluchowski ["Experimentell nachweisbare, der üblichen Thermodynamik widersprechende Molekularphänomene", *Physikalische Zeitschrift, 13* (1912), p. 1070 f.] "At that time the scientific leaders of Germany and France were convinced, with only very few exceptions, that the atomistic-kinetic theory had ceased to be of importance." Boltzmann was then regarded as the "last pillar" of that imposing edifice, as a somewhat backward defender of a lost cause.

16 *Zs. f. Physik* Vol. 32 (1925), pp. 753 ff.

17 B. Hoffmann, *The Strange Story of the Quantum* (New York, 1959), p. 58.

18 Smart, page 161.

19 Putnam, p. 206ff.

20 The same applies to Professor Sellars' suggestion in paragraphs 62-64 of his essay insofar as we are here advised to weaken the relation between what Sellars calls "the theory" and the "antecedent generalizations."

21 This inconsistency was an important factor in the refutation of the particle theory of light which may be regarded as a wider theory, embracing O. Newton was acquainted with diffraction phenomena and he treated them extensively in his *Optics* [Book iii, part I]. He suggested an explanation in terms of an interaction between the particles and the bending edge. He was therefore able to accommodate these phenomena. It was only through the successful elaboration of the wave theory (which also gave an account of the rectilinear propagation of light—Newton's main objection against waves) that the inadequacy of the particle theory, and of its kinematics, viz, the geometrical optics, was finally demonstrated. The inconsistency between geometrical optics and wave optics was an important part of this demonstration.

22 Newton's objection against the wave theory (see the last footnote) can be construed as an indirect criticism of a primitive form of W via O.

23 This remark seems also to apply to Professor Achinstein's criticism in *Phil. Rev.*, for 1964.

24 The term "approximate theories" is due to Putnam, p. 206; see also Sellars, par. 65.

25 A word about *thermodynamics*. It is still far from clear how the relation between the *existing* phenomenological theories (P) and the *existing* statistical theories (S), is to be conceived. From the point of view of our model which *prescribes* how to construct theories a contradiction between S and P would be desirable. Historically one of the most decisive arguments against P, Einstein's theory of Brownian motion, did use an S that was incompatible with the second law of the phenomenological theory [for details cf. again my "Problems of Empiricism" *loc. cit.*]. But apart from this one case there reigns confusion. It seems to me that neither Nagel, nor Putnam, are aware of these facts. Nagel only discusses the schoolbook-relation $T \sim E_{kin}$, and Putnam speaks [p. 219] of the "well known reduction" of P to S. What is this "well known reduction" he refers to? Where in the literature is it to be found? As far as I can see there exists a great variety of different approaches, ranging from Boltzmann's original ideas, and investigations into the ergodic theorem arising therefrom, via Tolman's equal-probability approach, Khinchin's work (where we also find pertinent criticism of Nagel's "formula") to the very confused considerations which are based upon the papers of Szilard and upon von Neumann's book. There are of course also more popular accounts and I cannot get entirely rid of the suspicion that Professor Putnam, when speaking of the "well known reduction" is speaking mainly of them. But this is much too slender and too inarticulate material for philosophical generalizations! Does this mean that the case S vs. P is not yet ripe for discussion? Not at all! One of the basic results of a statistical physics that has not been emasculated, and turned into a Putnamian "approximate theory" is the existence of *fluctuations,* even between different levels of temperature. This phenomenon is in flat contradiction with the second law of P. The attempt, due to Szilard and others, to preserve the second law of P by corrective terms expressing information on part of the observer fails (a) because the fluctuations occur independently of any observer (sudden freezing of undercooled fluids); and (b) because we can easily imagine quite simple mechanisms which violate the second law [cf. my note "On the Possibility of a *Perpetuum Mobile* of the Second Kind" in *Mind and Matter, Essays in Honor of Herbert Feigl,* Minneapolis 1965]. Such a reinterpretation of S is also undesirable for methodological reasons for it robs the theory of its ability to criticize P and thereby reduces the empirical content of the latter. I agree not all is well with thermodynamics. However it would be very unwise to defend an unreasonable method (layer-monism) by reference to a mess in physics. And whatever the judgment on these matters is going to be, a third- or fourth-hand account of science is too little to be content with in basic discussions, even if such an account, because of its simplicity, should have managed to become "well known" in some circles.

26 The *materialistic position* with respect to the mind-body problem has often been criticized by the argument (the *first argument*) that there are things which can be said about material processes which cannot be said about mental processes and vice versa because such assertions would be meaningless. Thus for example a material process may consist of parts, it may contain elements moving on a curved path, whereas it would be utter nonsense (and not only false) to say that this sensation of red I am having now contains parts moving on a curved path (a simple thought such as Bühler's *Aha Erlebnis* would be an even more damaging counter example).

Now whether or not a statement makes sense depends on the grammatical rules governing the corresponding sentence (the reference to rules is criticized by some authors, but not by Putnam). The first argument appeals to such rules. It points out that the materialist, in stating his thesis, violates them. Note that the particular *words* employed are not the point at issue here. Whatever the words used by the materialist, the rules he introduces in stating his thesis would be incompatible with some fundamental rules of the idiom in which we usually describe pains and thoughts. This incompatibility is taken to refute the materialist.

It is evident that the argument is incomplete. An incompatibility between materialistic ideas and the more fundamental rules implicit in the mentalistic ideology will criticize the former only if the latter can be shown to possess certain advantages. Are there such advantages?

According to the *second argument* the mentalistic idiom has indeed advantages over and above materialism. Everybody uses it. Even the most ingenious materialist has not yet succeeded in making his theory complete and even he must frequently fall back upon a language which contains at least some mentalistic components.

Two things are involved in this argument. First, it is silently assumed that "there is a massive central core of human thinking which has no history [and that] there are categories and concepts which, in their most fundamental character, change not at all." [Strawson, *Individuals,* p. XIV] For if the mentalistic idiom is only one of many different languages which are all equally detailed and equally actual, then the appeal to actual use no longer leads to a single point of view.

There is no room, in the present brief note, to discuss this assumption in detail. Let me only point out that it is presupposed by many linguistic philosophers (Ryle's *Concept of Mind,* for example, is not intended as a field study of the conceptual system of *homo Oxoniensis*: it is supposed to possess universal validity); that it must be investigated empirically, by field work, not by considering "conceptual possibilities" (the "conceptual possibilities" occurring to Mr. Strawson will be those permitted by the fundamental rules of the language he speaks and their consideration will be irrelevant to the question whether there are other languages, with different fundamental rules; after all, one cannot expect a tapeworm to be aware of the possibility of three dimensional Euclidian space); *and that it is false.* (The last point will of course not be realised by those who project everything into their own language and thereby wipe out the very differences to which we are referring here). There is no one "massive central core" of mentalism contained in every language. I hope that one day I shall be able to publish part of the evidence in favor of this statement. For the time being the reader is invited to study Dodds's impressive *The Greeks and the Irrational* and to consider how large the differences must become when we leave the West and examine, for example. the conceptual background of Ancient Egypt, of the Hopi, of the Azande. Compared with the detailed empirical evidence which has been provided by many scholars, the evidence in favor of "archetypes" which might support Professor Strawson's general thesis can hardly be regarded as more than a "visionary rumour." * Of course, there still remains the question, raised by Professor Smart and, in somewhat different terms, by Professor Sellars whether the similarities in the physical organization of all men may not lead to a "massive core of human thinking." I shall comment on this problem in section 8. Suffice it to say at this point that the usual practice of meaning analysis does not even perceive its existence.

* For the notion of a "visionary rumour" see C. G. Jung, "Flying Saucers," Vol. 10 of the *Collected Works of C. G. Jung,* p. 314. The empirical character of the archetypes is emphasized by Jung *op. cit.,* 328, footnote 12, where we read: "I am an empiricist who keeps within the boundaries set for him by the theory of knowledge." Which of course reminds one at once of the many cases where association with the devil was proved empirically (dreams, hallucinations, phantom pregnancy).

Secondly, we must point out that even a universal mentalistic idiom is no sufficient reason for predicting that the attempt to construct an equally universal materialistic idiom is either bound to fail, or will lead to a very imperfect language. Universal acceptance may be due to a series of historical accidents (prevalence of theories often is; it depends on who dies when, who educates whom, and who is connected with the most powerful institutions). It does not show that (after the inevitable initial period of training) a different language would fare worse.

At this point recourse is made to the *third argument*: English, and the mentalistic idiom it allegedly contains, has been empirically successful. "Our common stock of words" writes J. L. Austin [*Phil. Papers,* p. 130] "embodies all the distinctions men have found worth drawing, and all the connexions they have found worth marking, in the lifetime of many generations. These surely are likely to be more numerous, more sound, since they have stood up to the long test of the survival of the fittest and more subtle . . . than any that you and I are likely to think up." Here at last it is realized that neither the existence, nor the popularity of a conceptual system can guarantee its adequacy, and that a *bridge* is needed between theories and the world. For Descartes who saw this problem quite clearly the bridge was the *veracitas dei.* Professor Austin, who is a little more modern, employs the survival of the fittest. However he does not seem to realise that a false theory may contribute to survival if acting in accordance with it improves the situation of the organism (towards the end of the Middle Ages, belief in the geocentric system had definite survival value which then contributed to the prolonged survival of the Ptolemaic system). Besides, the process of survival *presupposes* the existence of competing alternatives, and as it is still going on, it does so even today. The survival of ordinary English until now therefore cannot be used as an argument against building up different idioms, or as an argument to the effect that any such different idiom is bound to fail. Quite the contrary, it is to be expected that there are "mutations" which will be even more successful than ordinary English. It is therefore of importance to continue the creation of new and different languages such as the language of materialism. Appeal to survival value favors materialism, it does not eliminate it. The most decisive point is however next.

The idea that a language, or a theory, has advantages because it has *stood up to test* assumes that it was *subjected to test,* and that there existed the *possibility of failure.* Now we know from the history of thought that theories can be surrounded with safeguards so that failure becomes impossible. "Ordinary" idioms are very similar to such infallible theories—indeed, their omnipresence, the possibility to apply them under any circumstances, to adapt them to unusual situations by means of "adjuster words" [J. L. Austin, *Sense and Sensibilia,* p. 74] is usually regarded as one of the best arguments in their favor. They are connected with beliefs which permeate the whole life of the community, which are intimately connected with its fears and its hopes, which are defended and reinforced both from the inside, by prejudice, and from the outside, with the help of powerful institutions, and they are regarded as successful if they reflect these beliefs most adequately. All this raises the suspicion that even the simplest descriptive statement, such as "I am in pain" can never be really *confronted* with facts. and thereby *examined,* but merely functions as an *accompanying noise.* It is therefore as unwise to use existing idioms as a guide in ontological investigations as it would be to expect valuable information from an accomplished con-man, or from a soothsayer who is greatly respected by the society in which he lives, *unless* we are able to test, and to find out, the pretenses of either: questions of test turn out to be of crucial importance for the theory of knowledge. The point of the example was however this: the example shows that in philosophical arguments which are comprehensive, and which consider all aspects of the matter, appeal to meanings occupies only the most primitive initial stages and is quickly replaced by considerations of fact and of method.

27 In what follows I shall assume that the rules (assumptions, postulates) constituting a language (a "theory" in our terminology) form a *hierarchy* in the sense that some rules presuppose others without being presupposed by them. A rule R′ will be regarded as being more fundamental than another rule R″, if it presupposed by more rules of the theory, R″ included, each of them being at least as fundamental as the rules presupposing R″. It is clear that a change of fundamental rules will entail a major change of the theory, or of the language in which they occur. Thus a change in the spatio-temporal ideas of Newton's celestial mechanics makes it necessary to redefine almost every term, and to reformulate every law of the theory, whereas a change of the law of gravitation leaves the concepts, and all the remaining laws, unaltered. The former ideas therefore are more fundamental than the law of gravitation.

Now it seems reasonable to assume that the customary concept of meaning is closely connected, not with *definitions* which after all work only when a large part of a conceptual system is already available, but with the idea of a *fundamental rule,* or a *fundamental law.* Changes of fundamental laws are regarded as affecting meanings while changes in the upper layers of our theories are regarded as affecting beliefs only. There exists therefore a rather close connexion between meanings and certain parts of theories. Meanings in the sense just explained are also independent of the analytic-synthetic issue. Moreover, it is now quite plausible to assume that meaning-talk can be replaced, without residue, by theory-talk. For a more detailed account cf. my paper "On the 'Meaning' of Scientific Terms," *Journal of Philosophy,* 1965.

28 Letter of August 18, 1964. I am grateful to Professor Shapere for having provided me with some very detailed and clear criticisms of earlier papers.

29 A "constituent" of a statement containing logical constants is any statement that is part of it and does no longer contain any logical constants.

30 Cf. the references in footnote 5.

31 It was Professor Feigl whose incessant arguments on this point woke me from my dogmatic slumber and made me realize the need for a more detailed theory of test.

32 The local grammar of a statement is that part of its rules of usage which is connected with such direct operations as looking, uttering a sentence in accordance with ostensively taught (*not* defined) rules, etc.

33 For details of the relation between I, R and R′ cf. for example F. Waismann, *Introduction to Mathematical Thinking* (New York, 1951).

34 For a discussion of such paradoxes cf. also P. Achinstein, "On the Meaning of Scientific Terms," *Journal of Philosophy* 1964, pp. 497-509 as well as my reply in the 1965 Volume of the same journal. The argument in the text also invalidates one of my main criticisms of Professor N. R. Hanson's *Concept of the Positron* [the criticism was published in a rather unkind review, *Phil. Rev.* 1964, p. 266, lines 9 to 15]. Looking back upon this criticism I find it somewhat amusing that I should have used the very same argument against Hanson which Achinstein, Shapere, Putnam, and others are now using against me. It is sometimes difficult to recognize one's friends.

35 Cf. Dodds, *The Greeks and the Irrational* (Berkeley, 1956). See also the cases mentioned in note 26.

36 Private communication.

37 Talking of forces, by the way, is not as anachronistic as Professor Putnam makes it appear to be. There were of course no Newtonian forces as they are understood now. There were forces of a different kind, vitalistic entities, quite similar to some of the entities considered by Newton's contemporary, Henry More (who had a considerable influence upon Newton's thought).

38 "The Direction of Time," *Models and Metaphors* (Ithaca, 1962).

39 For this formulation cf. Koestler, *The Sleepwalkers,* p. 504.

40 "Das Problem der Existenz Theoretischer Entitaeten" in *Probleme der Wissenschaftstheorie* (Vienna, 1960).

41 Professor Sellars' objection against equating commonsense with theories such as the theory of relativity will be dealt with in section 8.

42 Cf. section 5 and footnote 26.

43 Reference to the empirical success of T, or of some common idiom mimicking T is clearly irrelevant, for we have admitted that T is successful in D. *But so is T'* which also has additional confirmations elsewhere (see section 2).

44 Cf. also footnote 26. The popular argument that a layman can function as an observer for the theoretician and that the latter is therefore committed to the categories of the layman's report, and perhaps even to his whole language, is easily refuted. The scientist cannot use the layman's report without some interpretation of his own (consider what would happen were he to take over the interpretation of a primitive Zulu's account of a comet). He very often uses not the *content* of this report, *but the fact that the sentence expressing it has been uttered.* After all, he might have employed a well conditioned dog without being thereby forced to include barks into his theory, and to "mean" with them what the dog "means."

45 This was of course well known to all those philosophers of the past who objected to regarding common consent as a basis for truth. For a more recent discussion with ample material cf. paragraphs 591 ff of V. Pareto's *Treatise on General Sociology.*

46 This, I hope, resolves the difficulties Professor Hanson has felt in connection with the principle of correspondence. Cf. also p. 251 of my review in *Phil. Rev., 69* (1960).

47 Physiologically speaking this reorganization may be quite minute (a point which was already made by F. A. Lange). This removes the objection (which has some real force in other contexts) that the necessity to stay alive and to function as an organism puts limits upon our ability to "see things in a new light." We are here confronted with some very interesting connexions between physiology on the one side, and the realism-instrumentalism issue on the other. For a more detailed discussion cf. section 8.

COMMENTS

48 R. Carnap, "Ueber Protokollsaetze' *Erkenntnis 3* (1933) expresses the theory thus: "It stipulates that . . . *any* concrete statement may be regarded as a protocol statement."

49 For details concerning this procedure cf. again my "Problem der Existenz theoretischer Entitaeten," *loc. cit.*

50 N. R. Hanson, *Patterns of Discovery* (Cambridge, 1958), p. 17.

51 Thus Professor Michotte in his epoch-making investigations of phenomenological causality has produced certain cases where "causal impressions appear in conditions which, from the point of view of commonsense mechanics, are paradoxical" *The Perception of Causality* (New York, 1963), p. 87. These cases refute the idea that the causal aspects of certain perceived processes are due to an "act of interpretation" on our part or that "under the influence of past experience or in some other way, we ourselves invest certain basic impressions of movement with a 'meaning' " [*loc. cit.*]. Conversely, he has shown that "it seems impossible to produce a visual causal impression of one object being attracted by another" [125].

52 *Individuals,* p. xlv.

53 *The Hitchcock Lectures,* Berkeley 1964.

54 *Issues in Evolution,* The University of Chicago Darwin Centennial Discussions Vol. III (Chicago, 1960), p. 172.

55 Darwin Centennial Vol. III, p. 199; Vol. II., p. 235.

56 "We have to take into consideration that man awake comes to know that he is awake only through the rigid and lawful net of the concepts he uses; this is why he sometimes believes that he is dreaming when this net of concepts has been torn to pieces . . . Pascal is quite correct when saying that a recurring dream would occupy us just as much as things we see every day . . . The waking day of a mythically excited people, such as the older Greeks is indeed closer to dreaming than the day of a sober thinker. The constantly acting miracle, as it is assumed by the myth, is responsible for this. If each tree can now talk as a nymph, if a god can rob a virgin under the cover of a bull, if Athene herself is suddenly seen, passing in a beautiful carriage on the side of Peisistratus through the marketplaces of Athens—and this every honest Athenian believed to be the case—then like in a dream everything is possible at every moment and nature surrounds man like a masquerade of the gods . . ." F. Nietzsche, "Ueber Wahrheit und Luege im aussermoralischen Sinn," *Werke* ed. Schlechta, Vol. III, pp. 319 f. Cf. also Dodds *op. cit.* and the literature quoted there.

57 Cf. the excellent analyses in Brecht, *Versuche 25/26/35* pp. 102-105 concerning the role of what he calls the V-effect in everyday life.

8

DAGFINN FOLLESDAL

QUANTIFICATION INTO CAUSAL CONTEXTS*

Presented January 7, 1964

In this paper I will discuss some problems connected with contrary-to-fact conditionals, scientific laws, confirmation, and some types of probability statements and disposition terms. These problems are analogous to those that Professor Quine, in several articles and books,[1] has pointed out for the logical modalities.

In *From a Logical Point of View,* after having shown that quantification into contexts of logical modality encounters serious difficulties, Quine continues:

> What has been said of modality in these pages relates only to strict modality. For other sorts, for example, physical necessity and possibility, the first problem would be to formulate the notions clearly and exactly. Afterward we could investigate whether such modalities, like the strict ones, cannot be quantified into without precipitating an ontological crisis. The question concerns intimately the practical use of language. It concerns, for example, the use of the contrary-to-fact conditional within a quantification; for it is reasonable to suppose that the contrary-to-fact conditional reduces to the form 'Necessarily, if *p* then *q*' in some sense of necessity. Upon the contrary-to-fact conditional depends in turn, for example, this definition of solubility in water: To say that an object is soluble in water is to say that it would dissolve if it were in water. In discussions of physics, naturally, we need quantifications containing the clause '*x* is soluble in water', or the equivalent in words; but, according to the definition suggested, we

* I am grateful to W. V. Quine for helpful remarks.

> should then have to admit within quantifications the expression 'if x were in water then x would dissolve', that is, 'necessarily if x is in water then x dissolves'. Yet we do not know whether there is a suitable sense of 'necessarily' into which we can so quantify.[2]

Obviously, much is at stake here; the usefulness of the notions of physical necessity, contrary-to-fact conditionals, etc., depends largely on whether one can quantify into them, as illustrated e.g. in Quine's example that I just quoted, concerning solubility in water. If one cannot find a quantifying into these contexts, then these notions, together with a wealth of related notions, will probably have to be abandoned, to the extent that they cannot be made clear without appeal to physical necessity, e.g. in the case of dispositions by conceiving them as enduring structural traits of objects.[3]

However, it seems to me that arguments analogous to those of Quine against quantification into logical modalities apply to the contexts of physical necessity and related contexts as well. The first part of my paper will be an attempt to show this. To make the parallelism between these arguments and those of Quine apparent, Quine's wording will be used wherever possible.

Adapting an example from Chisholm's "The contrary-to-fact conditional,"[4] let us suppose that there is a well which is such that anyone who drinks from it gets poisoned. Let us further suppose that just one man has drunk from that well (and of course got poisoned). Then, presumably, the following is true:

(1) it is causally necessary that the man who drank from that well got poisoned.

Let us now further suppose that this man was born in place p at time t, i.e. that the following is true:

(2) the man who drank from that well = the man who was born in p at t.

Now in spite of (1) and (2), presumably

(3) it is causally necessary that the man who was born in p at t got poisoned.

is false.

To see how causal contexts fare with respect to quantification, let us now try to apply existential generalization to (1). We get:

(4) $(\exists x)$ it is causally necessary that x got poisoned.

However, what is this object that got poisoned? The man who drank from the well, that is, the man who was born in p at t? But to suppose this would conflict with the fact that (3) is false.

This then is a straight causal parallel to one of Quine's arguments against quantification into contexts of logical necessity.[5]

Our problem relates to Nelson Goodman's problem in "A query on confirmation"[6] that the degree of confirmation may vary widely with the way the given evidence is described.

The trouble in our example clearly springs from the fact that the objects over which we quantify can be uniquely described in ways that are not causally equivalent. Hoping to make sense of quantification into causal contexts, one might therefore try to weed out from one's universe of discourse all such "stubborn" objects that can be uniquely specified in ways that fail of causal equivalence. That is, one may require that whenever each of two open sentences 'Fx' and 'Gx' uniquely determines one and the same object x, the sentences are causally equivalent. In symbols:

(5) If $(w)(Fw \equiv .w = x)$
and $(w)(Gw \equiv .w = x)$
then (it is causally necessary that $(w)(Fw \equiv Gw)$)

But this postulate annihilates causal distinctions, for as pointed out by Quine in an exactly parallel argument concerning the logical modalities,[7] let 'p' stand for any true sentence, let y be any object in our purified universe of discourse, and let $x=y$. Then

(6) $(w)(p.w = y. \equiv .w = x)$

(7) $(w)(w = y. \equiv .w = x)$

By (5), next, with its 'Fw' taken as '$p.w = y$' and its 'Gw' as '$w = y$', we can conclude from (6) and (7) that

(8) it is causally necessary that $(w)(p.w = y. \equiv .w = y)$

But the quantification in (8) implies in particular '$p.y = y. \equiv .y = y$' which in turn implies 'p', so from (8) we conclude

it is causally necessary that p

So, causal distinctions collapse.

But, one may object, at least one system of quantified causal logic has been proposed and appears to work, viz. that of Arthur W. Burks.[8]

This system consists of a standard system of truth-functional logic and quantification theory, supplemented with the following axioms and rule for the causal and modal notions:[9]

A.1 $\Box A \supset \boxed{C} A$
A.2 $\boxed{C} A \supset A$
A.3 $\Box(A \supset B) \supset . \Box A \supset \Box B$
A.4 $\boxed{C}(A \supset B) \supset . \boxed{C} A \supset \boxed{C} B$
A.5 $(x) \Box A \supset \Box (x) A$
A.6 $(x) \boxed{C} A \supset \boxed{C} (x) A$
RL If $\vdash A$ then $\vdash \Box A$

And surely, there is no collapse of causal distinctions in Burks's system. However, the Quinean argument that I just presented was a semantic argument, relating to the *interpretation* of quantified causal logic. And important semantic points may easily remain hidden without being reflected in the syntax.

In order to bring the difficulties of causal logic into the open let us, therefore, supplement Burks's system with a standard system of identity and descriptions. This is not done by Burks, and for good reasons, for see what happens: We now become able to carry through the following disastrous argument (I will use Quine's system of natural deduction, but of course any system of quantification theory will do):†

(1′) $x = x$ — axiom of identity
(2′) $\Box(x = x)$ — (1′) RL
(3′) $\boxed{C}(x = x)$ — (2′) by A.1

* (4′) $(\iota y)\ (y = x.p) = x$
* (5′) $(y)(y = x.p. \equiv .y = x)$ — (4′) by contextual definition of definite description
* (6′) $x = x.p. \equiv .x = x$ — (5′)
* (7′) p — (6′) by axiom of identity
(8′) $(\iota)\ (y = x.p) = x. \supset p$ — *(7′)
(9′) $\Box\ ((\iota y)\ (y = x.p) = x. \supset p)$ — (8′) RL
(10′) $\boxed{C}\ ((\iota y)\ (y = x.p) = x. \supset p)$ — (9′) by A.1

† EDITORS NOTE: Limitations of the Linotype font have made it necessary to use the regular instead of inverted iota.

*(11′) p

*(12′) $y = x.p. \equiv .y = x$ (11′)

*(13′) $(y)(y = x.p. \equiv .y = x)$ (12′) y

*(14′) $(\iota y)(y = x.p) = x$ (13′) by def. of description

*(15′) $\boxed{C}((\iota y)(y = x.p) = x)$ (3′) (14′) by substitutivity of identity

*(16′) $\boxed{C}((\iota y)(y = x.p) = x. \supset p)$
$\supset .\boxed{C}((\iota y)(y = x.p) = x) \supset \boxed{C}\,\mathrm{p}$ A.4

*(17′) $\boxed{C}p$ (10′)(15′)(16′)

(18′) $p \supset \boxed{C}p$ *(17′)

That is, modal distinctions collapse. What now are we to do? One thing would be to confine ourselves to quantification theory when we do causal logic and be silent about the rest. But, if the foundations of causal logic are so muddled as they now seem to be, this is not a position to rest content with.

If we look at our proof (1′)–(18′), however, another way out might seem a natural one to take:

Line (15′) was got from lines (3′) and (14′) by the principle of substitutivity of identity. By this principle one can prove, as we did in this proof, that all identities are causally or logically necessary, and this certainly does not seem to be the case, witness our example of the man who drank from the well. He was supposedly identical with the man who was born in p at t, yet this identity was apparently neither causally nor logically necessary. Why not, therefore, forbid substitutivity of identity in causal context in order to prevent inferences like the one leading to (15′) and thereby to the fatal (18′)? This is the way taken by many working on the logical and epistemic modalities, and it has probably been most ably argued for by Jaakko Hintikka.[10]

However, again it seems to me that the problems are being covered up rather than solved.

For one thing, tampering with the substitutivity of identity may easily make the notion of identity unintelligible. Thus, e.g., Neil Wilson: "If identity does not mean universal interchangeability, then I do not really understand identity at all."[11]

However, it seems to me that also the interpretation of quantifiers requires unrestricted substitutivity of identity; that is, *quantification and substitutivity of identity go hand in hand.*

To see this, let us try to get some insight into what actually happens when we quantify into causal contexts:

To start with statement logic, it can be given a helpful algebraic representation by adapting Leibniz's idea of possible worlds. '$\boxed{C}\ p$' will simply be said to be true if 'p' is true in every "physically possible world." This talk about worlds is of course a purely algebraic device. The notion of a "physically possible world" is just as much in need of clarification as the notions of physical possibility and necessity themselves. However, this device is helpful for one's intuitions and has been used frequently by modal logicians from Leibniz on. Leibniz's idea is inadequate for the representation of iterated modalities; however, in 1957 Hintikka got the idea of representing modalities algebraically by considering, instead of a set of worlds that are possible outright, a set of worlds that are, or are not, *possible with respect to* another.[12]

Iterated operators are needed in causal logic, e.g. to express that one dispositional property causes the appearance of another, or that if a certain causal law were false such-and-such would be the case, or that it is a causal law that a given kind of habit or resolution arises under certain circumstances (Burks's examples).[13] However, the problems of quantifying into causal contexts are basically the same whether these contexts are governed by several or just one causal operator. Let us therefore stick to Leibniz's worlds that are possible outright, and let us now leave statement logic and go on to examine quantified causal logic.

Consider the following simple example:

$$(\exists x)\boxed{C}Fx$$

Using the Leibnizean representation, this should mean that there is an object, let us call it 'a', such that not only is 'F' true of it in our actual world, but 'F' is true of it in every physically possible world, that is, if we use a diagram and let horizontal lines represent possible worlds:

(9)

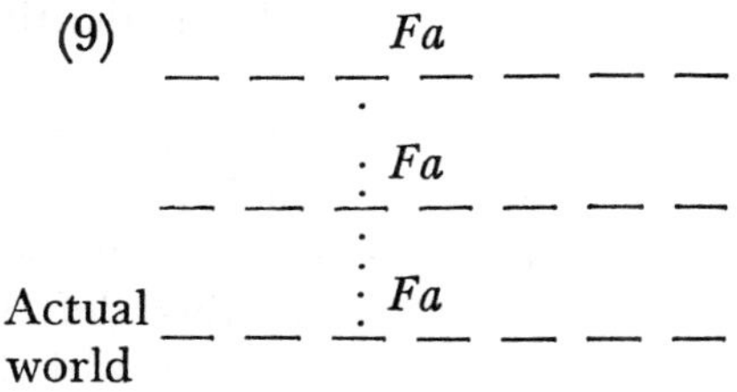

What now about an identity like

(10) $x = y$

Can (10) be true in our actual world and yet false in some physically possible world? That is, can we have a situation like this:

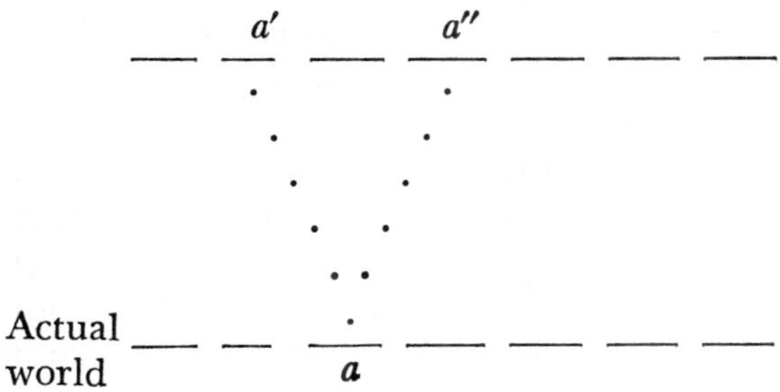

where to one and the same object a in our actual world there correspond two objects a' and a'' in some physically possible world: If so, what would it mean, when seeking to interpret e.g.

$$(x) \boxed{C} Fx$$

one picks the object a and says that 'F' is true of *it* in every physically possible world? Apparently, this makes no sense; the pronouns of quantification, in this case 'it,' make sense only when there is one and only one object that we are talking obout. So, if we want to make sense of quantification into causal contexts, we apparently have to require that *no identity statement which is true in our actual world is false in any physically possible world.*

What this amounts to, is that in order to make sense of quantification, identity should be *universally substitutive.* Tampering with the substitutivity of identity prevents the derivation of undesirable results, but only at the cost of making both identity and quantification unintelligible.

We have now explored two attempts to get out of the difficulties, restrictions on our universe of discourse, which led to a collapse of modal distinctions, and restrictions on the substitutivity of identity, which led to unintelligibility. Are we then left with a choice between unintelligibility and a collapse of modal distinctions, either one of which would make quantification into causal contexts pointless?

There are surely many other steps in the derivation (1′)–(18′) which we might attempt to forbid in our efforts to prevent this fatal proof from going through.

However, not any *ad hoc* restriction will do, as we just saw in the case of identity. There are, however, restrictions which prevent derivation of undesirable results and yet avoid the difficulties we just had in making sense of the pronouns of quantification. I will now describe one of these, which appears fairly natural.

Clearly, as we saw in our Quinean argument, the root of our trouble is that objects of our universe can be described uniquely in ways that are not causally equivalent. This, however, is not a peculiarity of the objects; as we just saw, no choice of universe will prevent this. Whether we restrict our universe or not is irrelevant, nothing was presupposed about the universe of discourse in the derivation (1′)–(18′) (except that it be non-empty). Rather than looking for a solution by doing something to our objects, let us therefore direct our attention to the expressions we use to refer to these objects:

To make sense of quantification, the variables have to *keep their references as we pass from the actual world to another which is physically possible*. Now, in order to avoid trouble, all we have to require, is that *all* referring expressions do the same. But definite descriptions obviously do not do this, for if we consider our diagram (9) again, we see easily that a general term, e.g. 'man who drank from the well', which is true of a unique object in our actual world, may be true of another object in another physically possible world, of several objects in a third such world, of none at all in a fourth, and so on.

That is, definite descriptions that have a reference in our actual world may fail to have one in other worlds, and even if they have one in some other world(s), their references may differ from one world to the next. Definite descriptions do therefore not in general satisfy the requirement we just found it necessary to put on our referring expressions in causal logic. In order to avoid trouble, we should admit into our stock of singular terms *only those descriptions which keep the same descriptum in all physically possible worlds.* That is, we can admit a description '$(\iota x)Fx$' as a singular term if and only if the following condition is satisfied:

$$(\exists y) \boxed{C} (x)(Fx \equiv .x = y)$$

(Note the order of the existential quantifier and the causal operator.)

All singular terms must satisfy a condition to this effect. That is, every admissible singular term must name one and the same object in every physically possible world.

For quantification into iterated modalities, parallel reasoning leads to similar, but slightly stronger conditions.

If, with Quine, we insist on the "primacy of predicates," an even simpler way out is possible. We simply eliminate *all* definite singular terms in favor of general terms and variables.[14]

Looking back to our fatal derivation, (1′)–(18′), we now see how our strictures on singular terms make the derivation break down. The substitution of '$(\iota y)\ (y = x.p)$' for 'x,' which gave us (15′) from (3′) and (14′), is no longer legitimate, since line (14′) is not a genuine identity statement. A genuine identity sign can be flanked only by genuine singular terms, and hence not by '$(\iota y)\ (y = x.p)$'. In (14′) and the other lines in our proof where '$=$' is flanked by '$(\iota y)\ (y = x.p)$', the '$=$' indicates merely that the two expressions which flank it relate to the same object in our actual world. This relation of "co-extensiveness" will of course not be universally substitutive in non-extensional contexts, like causal contexts. To avoid ambiguity, we should perhaps not use '$=$' but a different symbol for this relation.

That undesirable results disappear from modal logic when descriptions are not treated as names, but contextually defined, was first observed by Church more than twenty years ago.[15] However, modal logicians have frequently preferred other measures to avoid difficulties, especially restrictions on the substitutivity of identity. The semantic considerations which we have just been through, indicate, however, that restrictions on singular terms, and not restrictions on the substitutivity of identity, are what we need in order to make sense of quantification into causal contexts.

At last, a few words about the intelligibility of causal contexts and of the logical modalities.

The arguments which I have given concerning causal contexts can of course be paralleled for the logical modalities. It seems therefore that one can give no formal argument against quantification into either type of context. However, we found that substitutivity of identity and quantification go hand in hand, and this has certain philosophical consequences which many champions of quantified modal logic and quantification into causal contexts may find hard to digest.

To take the logical modalities first: In quantified modal logic one will affirm statements of the form

$$\Box Fa$$

Yet Carnap, Lewis, and others who are opposed to so-called Aristotelian essentialism, will deny that this commits them to the view that 'F' is essential to a. '$\Box F$' is true of the object a, yes, but only of the object *qua a*. Which properties are necessary of an object will depend on how we refer to it. If $a = b$, '$\Box F$' may be true of this object *qua a*, but false of this object *qua b*. These philosophers will reject as unintelligible the view attributed to Aristotle, that certain properties, the essential ones, are in a way inherent in the object itself, necessary of it regardless of the way we refer to it.

Now we have seen, however, that if one wants to quantify into modal contexts, identity must be universally substitutive in these contexts. But that means that if '$\Box F$' is true of an object, '$\Box F$' is true of it *regardless of the way in which this object is referred to.* That is, quantification into modal contexts commits us to essentialism.

Carnap and Lewis, who on the one side quantify into modal

contexts and on the other denounce essentialism, therefore appear to be in a rather awkward position.

What now about causal contexts? Depending on exactly what one means by essentialism and what position one takes to it,[16] quantification into causal contexts leads to a similar quandary. If one rejects as incomprehensible the view that causally necessary properties are inherent in objects, that is that '$\boxed{C}F$', if true of an object, is true of it regardless of the way in which it is referred to, then one should bettter not quantify into causal contexts; one should avoid contrary-to-fact conditionals, scientific law-statements, confirmation statements, and many types of probability statements and disposition terms—if one wants to make sense.

Dept. of Philosophy
Harvard University
Cambridge, Mass.
and
Norwegian
Research Council
for Science and
the Humanities

NOTES

1 W. V. Quine: "Notes on existence and necessity," *Journal of Philosophy 40* (1943), pp. 113-127. Reprinted in Leonard Linsky (ed.), *Semantics and the Philosophy of Language* (Urbana, Ill., 1952); "The Problem of interpreting modal logic," *Journal of Symbolic Logic 12* (1947), pp. 43-48; *From a Logical Point of View* (Cambridge, Mass., 1953, 1961); "Three grades of modal involvement," *Proceedings of XIth International Congress of Philosophy* (Brussels, 1953), vol. 14, pp. 65-81; "Quantifiers and propositional attitudes," *Journal of Philosophy 53* (1956), pp. 177-187; *Word and Object* (Cambridge, Mass. and New York, 1960).

2 Quine, *From a Logical Point of View,* pp. 158-159.

3 Cf. Quine, *Word and Object,* § 46.

4 Roderick M. Chisholm, "The contrary-to-fact conditional," *Mind 55* (1946), pp. 289-307. Reprinted in Feigl, Herbert, and Wilfrid Sellars (eds.), *Readings in Philosophical Analysis* (New York, 1949).

5 One might try to avoid this dilemma by arguing that (3) is perhaps true after all, if we understand it properly. However, this can easily be shown to lead to a collapse of causal distinctions. In fact, the weeding out of "stubborn" objects, which will be discussed shortly, amounts to this.

6 Nelson Goodman, "A query on confirmation" *Journal of Philosophy 43* (1946), pp. 383-385.

7 Quine, *Word and Object,* pp. 197-198.

[8] Arthur W. Burks, "The logic of causal propositions," *Mind N. S. 60* (1951), pp. 363-382.

[9] Here, '$\Box$' stands for 'it is logically necessary that'. For convenience, Burks lets his '$\boxed{C}$' stand for 'it is causally or logically necessary that.' 'It is causally necessary that' is thus perhaps best paraphrased by '$\boxed{C}p.-\Box p$' (Burks, *op. cit.*, pp. 374-375, 380).

[10] E.g., in "Modality as referential multiplicity," *Ajatus 20* (1957), pp. 49-64; *Knowledge and Belief* (Ithaca, N. Y., 1962); and "The modes of modality," *Acta philosophica fennica 16* (1963), pp. 65-81.

[11] Neil L. Wilson *The Concept of Language* (Toronto, 1959), p. 39.

[12] Jaakko Hintikka, "Quantifiers in deontic logic," *Societas Scientiarum Fennica,* Commentationes Humanarum Literarum 23, No. 4, Helsinki, 1957, esp. p. 13; "Modality as referential multiplicity," *Ajatus 20* (1957) esp. pp. 61-62. The idea was anticipated in work by Jónsson and Tarski on Boolean algebra in 1951: B. Jónsson and A. Tarski, "Boolean algebras with operators," *American Journal of Mathematics 73* (1951), pp. 891-939, *74* (1952), pp. 127-162; and was used by Stig Kanger in modal logic in 1957: Stig Kanger, *Provability in Logic,* Stockholm Studies in Philosophy I (Stockholm, 1957). Later, Saul Kripke got the idea independently, cf. his "Semantical analysis of modal logic I: Normal modal propositional calculi," *Zeitschrift für mathematische Logik und Grundlagen der Mathematik 9* (1963), p. 69, n.2.

[13] Arthur W. Burks, *op. cit.*, p. 380.

[14] Cf. e.g. Quine, *Methods of Logic* (New York, 1950, 1959), pp. 218-219. By insisting on the "primacy of predicates" and the eliminability of *all* singular terms in favor of general terms and variables, Quine thus can be said to have levelled the road for modal logic.

[15] Alonzo Church, Review of Quine, "Whitehead and the rise of modern logic," *Journal of Symbolic Logic* 7 (1942), pp. 100-101.

[16] This qualification is important. It seems that "causal essentialism" is better off than "logical essentialism," since many arguments directed against the notion of logical necessity (and thereby also against logical essentialism) e.g. in Quine's "Two dogmas of empiricism" *(From a Logical Point of View,* pp. 20-46), have no counterparts for the notion of causal necessity. In fact, treating dispositions as inhering structural traits of objects (cf. note 3 above) is a form of causal essentialism; dispositional properties remain properties of the object regardless of the way the object is referred to.

COMMENTS

R. M. CHISHOLM

QUERY ON SUBSTITUTIVITY

Comment on Follesdal

Mr. Follesdal has given us an admirably clear account of what he believes to be the consequences of quantifying into contexts of logical and causal modality. These consequences, he suggests, present us with a number of unattractive and extreme alternatives; e.g., abandoning these modalities; forbidding substitutivity, an alternative which (I agree) "leads to unintelligibility"; accepting Aristotelian "essentialism," possibly along with another metaphysical doctrine (unintelligible to me) implying that an object named "De Gaulle" in this world might be *identical* with an object named "De Gaulle" in some other possible world despite the fact that the first-named De Gaulle also *differs* in various respects (there is a slight difference in weight, say) from the second. I should like to ask whether we might not be able to avoid such extremes.

Let us think of "F" and "G" as being abbreviations for certain substantival expressions (e.g., "Frenchman" and "gourmet") and of "H" as being an abbreviation for a certain adjective (e.g., "happy") and let us assume:

The F is identical with the G.

Using only simple English, let us now distinguish three relevant types of case in which there is either genuine failure or only apparent failure of substitutivity of identity.[1] The cases are:

(1) (a) The F is necessarily an F, but (b) it is false that the G is necessarily an F.

(2) (a) It is causally necessary that the F is H, but (b) it is not causally necessary that the G is H.

(3) (a) A certain event caused him to be the F, but (b) that event did not cause him to be the G.

I suggest that no two of these three types of case should be treated alike.

(1) Our first statement, then, is: "(a) The F is necessarily an F, but (b) it is false that the G is necessarily an F." Why not say, in this case, that we do not have genuine failure of substitutivity—in other words, that (1) is false? There seems to me to be no good reason for affirming that the F is necessarily an F. Hence if we deny it, as I suggest that we do, we will be saying that, though (b) is true, (a) is false.[2]

Most contemporary logicians, apparently, would accept the (a) of statement (1). Yet I wonder whether they may not have been misled—either as a result of confusing (a) with a slightly different statement, which *is* a true statement, or as a result of committing an ancient fallacy. If, following Quine, we turn substantives into verbs or adjectives (and if, now, we use lower-case letters for verbs and adjectives), we may say, not only of the F, but of everything else as well, that it is necessarily *f-if-f*. Though nothing is necessarily french, everything is necessarily french-if-french (just as everything necessarily has the property of being either red or non-red). But though everything—and therefore the F—may be said to be necessarily *f-if-f*, this is not what is asserted by the (a) of (1), and it does not commit us to "essentialism." [3] We must not be misled into essentialism by committing the fallacy of inferring *necessitate consequentis* from *necessitate consequentiae*: from "Socrates is necessarily rational-if-human" and "Socrates is human," we should not infer "Socrates is necessarily rational." And from the fact that the evening star is necessarily self-identical, as is everything else, we must not infer that there is anything that is necessarily identical with the evening star.

(2) Our second example is: "(a) It is causally necessary that the F is H, but (b) it is not causally necessary that the G is H." We dealt with our first example by saying that, although (b) is true, (a) is false. I suggest (contrary to my earlier paper, to which Mr. Follesdal refers) that we deal with the second example by saying that, although (b) is false, (a) is true. Here, too, then, we are saying that the statement of the example is false and hence that it does not exhibit genuine failure of substitutivity. And why not say this? If it *is* causally necessary that the F is H (that the man

who drank from the well was poisoned), and if the F is identical with the G (if the man who drank from the well is the man from Omaha), then it is also causally necessary that the G is H (that the man from Omaha was poisoned). This way of putting the matter certainly accords with ordinary usage ("Why did the man from Omaha die? It was causally necessary, for he was poisoned") and, so far as I can see, it leads to no consequences as extreme as those that Mr. Follesdal presents to us.

(3) The third example is: "(a) A certain event caused him to be the F, but (b) that event did not cause him to be the G." (The assassination caused the former Vice-President to become the President, but it did not cause him to become Mr. Johnson, nor did it cause him to become the 1959 majority leader in the Senate.) Here there is a genuine failure of substitutivity of identity, and we cannot plausibly deny either (a) or (b). If we apply the apparatus of quantification theory to such statements as (3), it may well be that we should introduce certain restrictions—for example, with respect to existential generalization. But are we led in this case to any of the extremes to which Mr. Follesdal refers, and to which we might well have been led had we countenanced (1) and (2) as instances of genuine failure of substitutivity of identity? It seems to me that we are not, for we can countenance this type of case without committing ourselves to essentialism, and without being tempted to say that the object in question has a property which is such that its *having* that property is "a function of the way in which the object is referred to." What could the property be? It certainly could not be the property of being caused by a certain event to be the F, or the more general property of being affected by that event, for *these* properties belong to the thing in question no matter *how* it is described or referred to.

Dept. of Philosophy
Brown University
Providence, Rhode Island

NOTES

[1] I shall assume with Quine that statements of logical necessity *de dicto* (e.g., "It is logically necessary that either all men are mortal or some men are not

mortal") may be thought of as being statements about statements ("'Either all men are mortal or some men are not mortal' is logically necessary") and thus as presenting no special problem in the present context. Psychological statements in which there is failure of substitutivity require a somewhat different treatment than what is proposed here, but these, too, I believe, can be handled without making any far-reaching commitments; my own view of them is set forth in "Leibniz's Law in Belief Contexts," in *Contributions to Logic and Methodology in Honor of J. M. Bochenski* (Amsterdam, 1965), edited by A. Tymieniecka and C. Parsons, pp. 243-250.

[2] And thus we would reject the second step, among others, in Mr. Follesdal's derivation (1′) – (18′).

[3] In *Word and Object* (p. 199), Quine puts the puzzle of essentialism as follows: "Mathematicians may conceivably be said to be necessarily rational and not necessarily two-legged; and cyclists necessarily two-legged and not necessarily rational. But what of an individual who counts among his eccentricities both mathematics and cycling? Is this concrete individual necessarily rational and contingently two-legged or vice versa?" I would suggest that whatever holds necessarily of mathematicians also holds necessarily of cyclists and of everything else as well: everything is necessarily rational-if-mathematic and also necessarily bipedal-if-cyclistic.

9

HERBERT MARCUSE

ON SCIENCE AND PHENOMENOLOGY

Presented February 13, 1964

The Crisis of European Science and Transcendental Phenomenology is Husserl's last work. Written in the thirties, the first part was published in 1936, the second part only after Husserl's death.

I would like to indicate first where I see the general historical locus of this work. It seems to me that we have to place it into the context of the radical reexamination of the Western concept of Reason, of Western rationality that begins in the last decades of the nineteenth century and to which so essentially different thinkers as Bergson, Dilthey, Max Weber, Spengler, Piaget, and Bachelard belong. All of them have in common this questioning of the very idea which has guided Western thought since its Greek origins, i.e., the rationality typical of the occident. It seems to me that Husserl is the last in this group, and in a certain sense, (which may strike you as strange) the most radical of these re-examiners. In Husserl, it is modern science itself, this most sacrosanct child of Western rationality, that is questioned. In this reexamination, modern science appears as the end of a fateful development which begins with Greek thought, that is, with the origins of Western thought itself—as the "end" of this development in the twofold sense of termination and of fulfilling the *telos,* the purpose, the objective of this thought.

According to Husserl, science,—modern science, Galilean as well as post-Galilean,—originates in the Greek idea of knowledge and truth and comes to rest in a scientific rationality in which truth and validity contain in themselves illusion and repression. Before I try to present Husserl's radical thesis, I have to stress that it is not the result of a sociological analysis or of a sociology of knowledge. It is precisely the fascinating aspect of Husserl's work that it is a *philosophical* analysis within the academic framework of intellectual history, even within the academic division of labor. Husserl emphasizes philosophy as *Beruf,* as calling, and that philosophy is done in the *Berufszeit,* that is to say, in the time reserved, in the academic division, for such investigations. Husserl adds (and this is important: I come back to it at the end) that the calling of the philosopher is a unique calling because (and I quote him)

> this calling is linked with the "possibility of a radical transformation of humanity," and not only a radical transformation of humanity but also a "liberation," and this possibility makes the calling of the philosopher unique within the division of labor.[1]

In the course of such a philosophical undertaking (philosophical also in the sense of a discipline!), in the course of its own inner development Husserl's analysis transcends itself, or rather it descends from the pure theoretical to the impure pre-theoretical, practical dimension. Better—the pure theoretical analysis discovers its own internal impurity, but only to return from this impure sphere to the still pure theoretical dimension of transcendental phenomenology as constituent of the practical, pre-theoretical dimension, the *Lebenswelt.* (I use the German term *Lebenswelt.* The literal translation "life-world" is too large and too vague in this context; what Husserl means is our own empirical day-to-day world as it is given in immediate experience, practical and other—the world of life and death, in our empirical reality. So I will use either '*Lebenswelt*' or 'empirical reality').

I will now devote some time to presenting Husserl's own thesis (the work is not fully translated; we only have Gurwitsch's excellent abstract of it), but I shall focus it in such a way that the critical problems stand out. Husserl begins with a very brief description of what he considers the Greek concept of Reason, namely the idea of human being as self-determination and determination of its world by virtue of man's intellectual faculties, the concept of Reason, according to which man's intellectual faculties are at the same time capable of determining his own life and of determining, defining, and changing the universe. This conception presupposes that the universe itself which is thus rationally comprehended is in its very structure a rational system and therefore accessible to knowledge and change on the grounds of man's own rational knowledge. In other words, Reason for the Greeks, is objective and subjective at one and the same time, and on this basis, Reason is the subjective as well as objective instrument for changing the world in accord with man's rational faculties and ends. But in this process, Reason itself as *theoria,* is and remains the basis of the transformation of the world. Philosophy is thus established as *science,* and as first, most excellent and general science, which must give direction and the end to all other sciences.

What are the implications of this original concept of Reason? First, it implies a supra-factual, supra-temporal validity of Reason, so that the really real as discovered and defined by Reason is rational as *against* the immediately given fact. Reason establishes an authority and reality which is in this way antagonistic to the immediately given facts. Secondly, true being is ideational being (a conclusion from the first implication), not being as we experience it immediately in the flux of our empirical, practical world. Thus "Platonism" is the basis of all scientific knowledge. Thirdly, objectivity is necessarily correlated with subjectivity, again the subjective as well as objective structure of Reason. Husserl here gives a formulation which, in an entirely different context, recaptures the very question and thesis with with Western philosophy began, namely, the final identity of Being and Reason. He says:

> Can Being and Reason be separated if cognitive Reason determines (the essence of being?) [2]

So we find at the very beginning and at the late stage of western philosophy this almost literal identity in the formulation of the basic problem, the mysterious union and even identity of Reason and Being, Knowing and Being. Now this concept of Reason, which is theoretical and practical Reason in one, is understood by Husserl as a *project.* I use the term here as it was elaborated in the philosophy of Sartre: "project" in the sense that this idea of rationality and its application is a specific way of experiencing, interpreting, organizing and changing the world, a specific historical project among other possible ones, not the only, necessary project. This project, according to Husserl, came to fulfillment with the foundation of modern science, namely, in Galilei's mathematization of nature. Galilei's mathematization of nature established that purely rational, ideational system which was the dream of all Platonism; Galilei established the ideational world mathematically as the true reality, Substituting this scientific universe for the only given reality, namely, our empirical *Lebsenswelt.* But the very fulfillment of this project was also its collapse, according to Husserl. For this scientific rationality, this idea of Reason and its application proved successful only in the positive sciences and in the technological conquest of Nature, while the original foundation of this entire science, that which originally was supposed to constitute the very structure, content and end of science, namely, philosophy, remained an impotent, abstract, meaningless metaphysical sphere of knowledge and continued in this impotent form a hopeless academic existence which, in addition, was more and more dissolved into psychology. Thus separated from the basic philosophy which, according to the original ideas of Reason, was supposed to give the ends, the objectives, the meaning of science, separated from this basic philosophy which was supposed to provide the truly universal concepts, Reason was at the same time divorced—and this is decisive for Husserl—from that rational *humanitas* envisaged in the original philosophical project. Scientific, technological rationality became reason *kath' exochen.* Divorced from the validating "ends" set by philosophy, the rationale set by science and the rationale of its development and progress became that of the *Lebenswelt* itself, in which and for which this science developed.[3] Instead of rationally transcending the *Lebenswelt,* science comprehended, expressed, and extended the specific rationale of the *Lebenswelt,* namely, the ever more effective mastery of the environment *(Herrschaft über die praktische Umwelt),* including the ever more effective mastery of man.[4] But that was not the inherent *telos* of science, which was first and foremost, and not only in a chronological sense, the *telos* defined by the empirical reality in which science developed. Thus theoretical Reason, pure Reason, without losing its scientific character as theory, becomes practical Reason. Theory, by virtue of its internal dynamic rather than on external grounds, becomes a specific, historical practice. But (and this is decisive for Husserl and the justification of his own subsequent phenomenological reduction) this entire development, this entire transformation of Reason, this essential, structural, internal commitment of pure Reason, pure theory and pure science to the empirical reality in which they originated, this entire transformation *remains hidden to science itself,* hidden and unquestioned. The new science does not elucidate the conditions and the limits of its evidence, validity, and method; it does not elucidate its inherent historical denominator. It remains unaware of its own foundation, and it is therefore unable to recognize its servitude; unable to free itself from the ends set and given to science by the pre-given empirical reality.—I should like to stress again, because these formulations can be easily misunderstood, that it is not a sociological relation which is here established between an empirical reality and the pure science which develops in this empirical reality. Husserl's concept goes much farther. He maintains that the empirical reality is the framework, and dimension in which the pure scientific concepts develop. In other words, the empirical reality constitutes, in a specific sense, the very concepts which science believes are pure theoretical concepts.

Before I go on with Husserl's interpretation of this development, I would like to reformulate and to extend his thesis in a way which may bring out its provocative implications. What happens in the developing relation between science and the empirical reality is the abrogation of the transcendence of Reason. Reason loses its philosophical power and its scientific right to define and project ideas and modes of Being beyond and against those established by the prevailing reality. I say: "beyond" the empirical reality, not in any metaphysical but in a historical sense, namely, in the sense of projecting essentially different, historical alternatives.

Now back to Husserl's interpretation.

The new science (by which he understands mainly Galilean science) establishes a rational "infinite" universe of Being (I follow his words here literally), systematically organized and defined by science itself. Within this universe, every object becomes accessible to knowledge, not incidentally, in its contingent, particular occurrence, but necessarily and in its very essence.[5] Thus, it becomes object of scientific knowledge, not as this individual object but as exemplification of general objectivity (the falling feather as *res extensa* in motion).[6] That is to say, the concrete and particular object, the Aristotelian totality is no longer the *Wesen,* the essence; Platonism supersedes Aristotelianism, not only in physics, but in the very concept of scientific rationality. And concomitant with this de-individualization, which is the pre-requisite for the quantification of the scientific universe, is the familiar reduction of secondary to primary qualities; devaluation of the inexorably individual sense experience as nonrational.[7]

As a result of this twofold process, reality is now idealized into a "mathematical manifold": everything which is mathematically demonstrated with the evidence of universal validity as a pure form (*reine Gestalt*) now belongs to the true reality of nature.[8] But (and here is the great gap which separates the new science from its classical original) in contrast to the ideational forms of Plato, the ideational forms of mathematical physics are freed from any substantive connection with other than mathematical ends. The ideational realm of Galilean science no longer includes the moral, esthetic, political Forms, the *Ideas* of Plato. And separated from this realm, science develops now as an "absolute" in the literal sense no matter how relative within its own realm it may be, absolved from its own, pre-scientific and nonscientific conditions and foundations. According to Husserl, the absolute evidence of mathematics (which as we shall see we question), was for Galilei so self-evident that he never asked for the actual foundation of its validity, for the validating ground of this evidence, and of its extension to the whole of nature. Thus, the validation of the new science remained in the dark; its own basis never became the theme of scientific inquiry; science contained an unmastered, unscientific foundation. This is of the utmost importance for *the validity of science* itself, because the relation between science and the pre-scientific empirical reality is for Husserl not an external one but one which affects the very structure and meaning of the scientific concepts themselves.

Now according to Husserl, where is this pre-scientific validating ground of mathematical science? It is originally in geometry as the art of measuring *(Messkunst)* with its specific means and possibilities.[9] This art of measuring in the empirical reality promised and indeed achieved the progressive calculability of nature, subjecting nature to the ever more exact "foresight" in mastering and using nature. (Foresight—*Voraussicht,* perhaps better translated as projection and valid, rational anticipation). Foresight and anticipation, rational anticipation can then guide the practical orientation in and the transformation of the empirical *Lebenswelt,* without however (and this is decisive) setting or defining or changing the goals and ends of this transformation. Geometry can and does furnish (and the same holds true for the extension of geometry, mathematics) the methods and ever more exact, ever more calculable approaches for the transformation and extension of the established *Lebenswelt,* but remains forever incapable of defining, anticipating, or changing, by its own concepts, the ends and objectives of this transformation. In its method and concepts, the new science is essentially non-transcendent. This is what I consider as Husserl's key sentence: Science "leaves the *Lebenswelt* in its essential structure in its own concrete causality unchanged."[10]

As to the interpretation of this paradoxical and provocative thesis (so obviously paradoxical since we are used to seeing in science one of the most dynamic forces in the world): In my view, what is at stake is not the more or less external relation between science and society, but the internal conceptual structure of science itself, its pure theory and method which Husserl now reveals in their essential historicity *(Geschichtlichkeit),* in their commitment to the specific historical project in which they originated.[11] Pure science retains, *aufgehoben* (to use Hegel's term now) the practice out of which it arose, and it contains the ends and values established by this practice. The empirical reality thus performs the *sinngebende Leistung* (constituent act): It is constitutive of scientific truth and validity. Science is *Aufhebung der Lebenswelt*

(1) inasmuch as science cancels the data and truth of immediate experience,
(2) inasmuch as science preserves the data and truth of experience, but
(3) preserves them in a higher form, namely in the ideational, idealized form of universal validity.

And this threefold process takes place in the scientific abstraction. The quantified ideational forms are abstracted from the concrete qualities of the empirical reality, but the latter remains operative in the very concepts and in the direction in which the scientific abstraction moves.

In this way, the pre-scientific, pregiven empirical reality enters the scientific enterprise itself and makes it a specific project within the preestablished general project of the empirical reality. However, the abstract, ideational, mathematical form into which science transforms the empirical conceals this historical relation:

> The *Ideenkleid* (the ideational veil) of mathematics and mathematical physics represents and [at the same time] disguises the empirical reality and leads us to take for True Being that which is only a method.[12]

This is perhaps the most effective and lasting mystification in the history of Western thought! What is actually only one method appears as the true reality, but a reality with a *telos* of its own. The mathematical ideation, with all its exactness, calculability, foresight, leaves a void *(Leerstelle)* because the objectives and ends of this calculability and anticipation are not scientifically determined. This void can thus be filled by whatever specific end the empirical reality provides, the only condition being that it is within the range of scientific method. This is the famous neutrality of pure science which here reveals itself as an illusion, because the neutrality disguises, in the mathematical-ideational form, the essential relation to the pregiven empirical reality.

In Husserl's terms: The objective a priori of science itself stands under a hidden empirical a priori, the so-called *lebensweltliche* a priori.[13] Moreover, as long as this empirical a priori remains hidden and unexamined, scientific rationality itself contains its inner and own irrational core which it cannot master. According to Husserl, modern science thus operates like a machine which everyone can learn to handle without necessarily understanding the inner necessity and possibility of its operation.[14] In other words, pure science has an inherently instrumental character prior to all specific application; the Logos of pure science is technology and is thus essentially dependent on external ends. This introduces the irrational into science, and science cannot overcome its irrationality as long as it remains hidden from science. In Husserl's words: Reason is Reason only as manifest Reason *(offenbare Vernunft)*, and Reason "knows itself as Reason only if it has become manifest."[15] In as much as Reason remains non-manifest in science, scientific rationality is not yet the full rationality of science. How can Reason become conscious of itself?

Husserl proposes to break the mystification inherent in modern science by a phenomenological analysis which is in a literal sense a *therapeutic* method. Therapeutic in the sense that it is to get behind the mystifying concepts and methods of science and to uncover the constitutive *lebensweltliche* a priori under which all scientific a priori stands. This is to Husserl first a methodological problem. The pregiven empirical reality as a whole must become the object of the philosophical analysis, otherwise the a priori prior to the scientific a priori could never come to light. But obviously philosophy itself is part of this empirical reality and philosophy itself stands under the a priori of the empirical reality. The circle is to be broken by a dual phenomenological reduction (suspension, *epoche*) : first the suspension of the objective a priori; the suspension of scientific truth and validity; secondly the suspension of the *lebensweltliche* a priori, of the *doxa* and its validity.

Now what do we retain, what remains as the residuum of this twofold suspension? In the first *epoche*, "we put in brackets" (that is to say, we do not deny but simply suspend judgment on) scientific truth and scientific validity. What remains as the residuum is (a) the entire general structure of the empirical reality,[16] the infinite manifold of things in time and space, the orta, and (b) the world itself in which all these things necessarily appear—the world as the universal, unsurpassable horizon of all particular objects. But this first *epoche* is not sufficient: it cannot do what it is supposed to do, namely, break through the mystification and uncover the ultimate foundation of scientific truth. It cannot do this because with this first "bracketing" we are still on the basis *(auf dem Boden)* of the empirical reality, within the "natural position" of our day-to-day experience. A second *epoche* is necessary which "at one stroke" leads to a total alteration of the "natural position" of experience, to the suspension of the natural validation of everything that we naturally accept as valid in our immediate experience.[17] Once we have suspended these judgments too, we reflect no longer on the pregiven world and the particular objects appearing in it, but on *how* these objects appear, on the *modes* in which this entire world is given to us. The residuum of this *epoche* is thus the world as correlate of a totality of *modes of consciousness,* as a "synthetic totality." What we have now as residuum is the *transcendental* subjectivity,[18] and to this transcendental subjectivity the world is now given as phenomenon of and for an absolute subjectivity.[19] This transcendental subjectivity is no longer any particular or individual or group subjectivity. It is "absolute" because whatever object or object-relation may appear, now appears as necessarily constituted in specific acts of synthesis which inseparably link objectivity and subjectivity. In other words, we have now what we might call the absolute original experience: the experience which is at the origin of and is constitutive of any possible objectivity that can ever become the object of scientific and of any other thought. The phenomenological reduction has now opened the dimension in which the original and most general structure of all objectivity is constituted.

I shall add only a few critical remarks. The breakthrough to the transcendental subjectivity is supposed to be the road to uncover the foundation on which all scientific validity rests. I ask the question: can the reductive phenomenological analysis ever attain its goal, namely, to go behind scientific, and pre-scientific, validity and mystification? I shall offer three suggestions.

First: The phenomenological analysis is confronted with the fact of reification (Husserl does not use this term). Reification is a form which is usually not examined. Scientific as well as pre-scientific experience are false, incomplete inasmuch as they experience as *objective* (material or ideational) what in reality is subject—object, objectivation of subjectivity. In founding the analysis on the constitutive subject—object correlation, Husserl's dual *epoche* does go behind the reification—but so does all transcendental idealism. Thus far we are, in my view, in no way beyond Kant. I know Husserl's own interpretation of the difference between phenomenology and Kant; I think that in the context of my criticism this difference is not very relevant. My point is that the phenomenological breakthrough stops short of the *actual* constituent subjectivity. Husserl transcends the objective a priori of science in the first *epoche* and the empirical a priori in the second *epoche*. He thus creates a conceptual metalanguage for the critical analysis of the empirical reality. But my question is: does this conceptual metalanguage really come to grips with the constituent subjectivity? I think not.

Second: The phenomenological reduction arrives at a subjectivity which constitutes only the most general forms of objectivity, for example, the general form of appearing as object, changing as object, being related to other objects. But does this subjectivity give us "manifest Reason" behind the disguising Reason, the validation of scientific truth? Can this transcendental subjectivity ever explain–and solve–the crisis of European science? Husserl's transcendental subjectivity is again a pure cognitive subjectivity. One does not have to be a Marxist in order to insist that the empirical reality is constituted by the subject of thought *and* of *action,* theory and practice. Husserl recognizes the historical subject in its *sinngebende Leistung;* but then, by suspending, bracketing it, the phenomenological analysis creates its own a priori, its own ideation, and its own ideological veil. Pure philosophy now replaces pure science, as the ultimate cognitive lawgiver, establishing objectivity. This is the *hubris* inherent in all critical transcendentalism which in turn must be cancelled. Husserl himself seems to have been aware of this *hubris.* He speaks of the philosopher as *"urquellend fungierende Subjektivität"*: the philosopher functions as the primordial source of what can rationally be claimed as objective reality.

I come to the conclusion and leave it as a question. Husserl recognizes the fetishism of scientific universality and rationality by uncovering the specific historical-practical foundations of pure science. He sees that pure science is in its very structure technological–at least potentially applied science. The scientific method itself remains dependent on a specific *Lebenswelt.* This is the hidden irrational element in scientific rationality. Husserl finds the reason for this dependence in the loss of the philosophical dimension, which was originally the basic dimension of science. Classical philosophy defined the method and function of science in terms of an idea of Reason which claimed higher truth and validity than those embodied in, and established by, the given empirical reality. This validating idea of Reason was that of the *telos* of man as man, the realization of *humanitas*. According to Husserl, the humanistic structure of Reason collapses with the release of science from this philosophical foundation. This would imply that humanism becomes an ideology at the very time when modern humanism is born. In other words, the birth hour of humanism itself would be the degradation of humanism to a mere ideology. Apparently there must be something wrong with this formulation. The fact remains that humanism is still today an ideology, a higher value which little affects the inhuman character of reality. The question with which I would like to conclude is this: Is philosophy entirely innocent of this development, or does it perhaps share the *hubris* of science? Does it share the reluctance to examine its own real foundation and function and is it therefore equally guilty of failing in the task of *Theoria,* of Reason—to promote the realization of *humanitas?*

Dept. of Philosophy
Brandeis University
Waltham, Massachusetts

NOTES

1 Husserl, *Gesammelte Werke,* vol. VI (Den Haag 1954), ed. W. Biemel, p. 154.
2 Ibid., p. 9, 12.
3 Ibid. p. 49 f.
4 Ibid. p. 67.
5 Ibid. p. 19.
6 Ibid. p. 40.
7 Ibid. p. 54.
8 Ibid. p. 20-21.
9 Ibid., p. 27, 30.
10 Ibid., p. 51.
11 Ibid., p. 152.
12 Ibid., p. 52.
13 p. 49, 143 f.
14 p. 52.
15 p. 53.
16 p. 143 f.
17 p. 151 f.
18 p. 147 f.
19 p. 155.

COMMENTS

COMMENT ON THE PAPER BY H. MARCUSE

Aron Gurwitsch

Rather than commenting point by point on the interesting analysis which my friend, Professor Marcuse, has presented to us to-night of Husserl's *Die Krisis der Europäischen Wissenschaften und die transzendentale Phänomenologie* (to be referred to henceforth as *Crisis),* I wish to raise the question as to the perspective under which this work of Husserl's is to be seen. Professor Marcuse started by placing *Crisis* within the framework of contemporary philosophical thought, of course construing "contemporary" broadly enough so as not to exclude the more or less recent past. Thus he referred to the work of Dilthey, Bergson, Spengler, Gaston Bachelard, Piaget, and others. Far from denying the legitimacy of Professor Marcuse's approach, it must be pointed out that his is not the only possible one. One must, moreover, raise the question as to whether his approach is adequate, that is to say, does full justice both to Husserl's intentions and—much more important—to the philosophical significance of *Crisis.* A change in perspective and approach might entail revisions of some of Professor Marcuse's formulations and interpretations.

Crisis is the last work of Husserl. After the first two parts, whose central piece is the analysis of "Galileo's mathematization of nature," had appeared in 1936, that is, still in Husserl's life time, in *Philosophia* vol. I (a long-extinct periodical edited by A. Liebert in Belgrade), Husserl continued to work on the planned subsequent parts until the onset of his final illness in August 1937. This is mentioned to recall that at the time of his preparatory studies for that book (some of the studies date from the late 20's, some of the relevant ideas can be found, at least in germinal form, as early as 1913), phenomenological philosophy was already in existence. Husserl had already done whatever work in phenomenology it had been given him to do. *Crisis* has the sub-title "An Introduction to

Phenomenological Philosophy." Of course, "Introduction" is not to be understood in the sense of an elementary exposition for beginners. Nor does it mean, in view of what has just been said, that phenomenology is here presented for the first time. Rather it means opening up a *new* avenue of approach to an already existing body of phenomenological thought. The novelty of this approach, in contradistinction to two earlier books by Husserl which likewise have the sub-title "Introduction" (about those two books a few words will have to be said at the end), consists in choosing modern science, especially modern physics, as point of departure.

It is possible and perfectly legitimate to isolate Husserl's analysis of Galilean science and to concentrate upon it almost exclusively, as Professor Marcuse has done, provided one does not lose sight of the contextual linkage between that analysis and the late parts of *Crisis,* as well as Husserl's earlier writings, and is prepared to re-insert it into that context at some appropriate point. To begin with, I shall follow Professor Marcuse's example in isolating the analysis of Galilean science, that is to say I shall consider it under a perspective other than the whole of Husserl's phenomenology. That perspective, however, must be much wider than the one Professor Marcuse has chosen. It can be none other than the history of modern philosophy of science.

Not being an historian in the conventional and generally accepted sense, Husserl, when he speaks of Galileo, does not mean the historical figure of that name who lived at a certain time, any more than by Galilean science he means the scientific work actually done by that historical figure. Rather Galilean science denotes the science inaugurated by Galileo. The name is used as a symbol for the historical development of modern science from, roughly speaking, 1600 to 1700, that is the constitution of classical physics, and even beyond. Within the meaning of Husserl's definition, Einsteinian physics and quantum physics are sciences of the Galilean style.

What characterizes that style? To say it in one sentence: the cleavage between the world as it presents itself in the perceptual experience of every day life and the world as it is in scientific truth and "in reality." The world as it appears in direct perceptual experience is the world of common sense, called by Husserl the life-world (*Lebenswelt*). Its description would require too lengthy an exposition to be tackled here. Some aspects will be mentioned later; to a few points, attention must be called now, however briefly. The things encountered in the world of common experience exhibit chromatic qualities as properties belonging to them in their own right. Human voices, musical notes, and noises of all kinds are accepted as occurrences being in reality as they are given in auditory experience. Finally, there are observed regularities and causal connections, but certainly not of such a nature as to be expressible in the form of functional dependencies in the mathematical sense. Modern science of the Galilean style starts by refusing to accept the perceptual world at face value. Reality contains, embodies, and conceals a mathematical structure. As to its true and real condition, in contradistinction to perceptual appearances, the world (to express it in modern terms) is a mathematical manifold. To pierce through the veil of appearances, to discover the mathematical structure of the universe, to disclose reality as a mathematical manifold, is precisely the task of Galilean science. It is the task Galileo had set to himself and passed on to his successors. Galileo, the founder of physics, as Husserl points out, is not a physicist himself in the same sense as his successors who, as heirs to a scientific tradition, are already in the possession of the "correct" methods of physics; and here methods are not understood in the sense of techniques and technical procedures, but rather in the etymological sense of access and way of approach. Being first to conceive the idea of nature as a mathematical manifold, Galileo had to develop the methods of physics, understood in the just mentioned sense, so as to substantiate that idea. This is what justifies the use of his name as a symbol. The point requires some emphasis, because the contention that the world is not in reality as it looks but that its true condition and constitution has to be disclosed and discovered by means of mathematical construction is not retrospectively formulated on the basis of results attained. On the contrary, it is the guiding principle of the science of physics still to be developed, and it institutes its very development.

For that reason, the contention in question is in need of a justification which is the main concern of the first period of the modern philosophy of science, as represented chiefly by Descartes and Malebranche, in some sense also by Spinoza, and certainly by Leibniz. To illustrate the style of the *period of validation* by a significant example, let me merely recall that for the justification of the incipient new science and its underlying principle of the mathematical, especially geometrical, structure of the external world, that is, of corporeal nature, Descartes resorts to divine veracity. The first period comes to a close in 1748 when Euler submitted his *Réflexions sur l'espace et le temps* to the Royal Academy in Berlin. In this memoir, Euler maintains that, although it appertains to philosophy to provide a clarification of the fundamental concepts of physics (the question at issue concerns absolute motion, absolute space and time) the decision as to whether or not the mentioned concepts are to be admitted, and more generally, the decision as to which the legitimate fundamental concepts are, no longer falls under the jurisdiction of philosophy, but of physics. It is in terms of the theoretical exigencies of physics, and in these terms alone, that the foundation problems of physics have to be discussed. Euler's memoir is a "declaration of independence" on the part of physics with regard to philosophy, in so far as *science must be accepted and respected as a fact*. The task of philosophy is to account for the very fact of science; the goal which philosophy has to reach is set for it from outside of itself. Kant's *Critique of Pure Reason* follows the orientation initiated by Euler, which also prevails in the several varieties of Neo-Kantianism in the late 19th and the early 20th centuries. The same holds for all contemporary attempts at a logical analysis of science as unquestionably accepted, however they might differ from one another in questions of detail.

It is the historical significance of Husserl's Galileo analysis to challenge and even to abandon the acceptance of science as an ultimate fact and rather to see in it a problem. Husserl is far from questioning the technical, or, more precisely, the intrinsic validity of science, and nothing could have been further from his mind than dismissing it in any sense. What is in question is not science itself, nor any particular scientific theory, but rather the interpretation of science. Husserl concerns himself with the problem of the very existence and the sense of science of the Galilean style, that is the conception of nature as in reality possessing a mathematical structure. Galilean science rests on presuppositions which make it possible, orient scientific methods and procedures, and determine the sense of the scientific explanation of the universe in the Galilean style. Because the presuppositions in question partly were not made explicit and partly were taken for granted, the sense of science had been obfuscated since the earliest phases of its history. Explication of those presuppositions does not mean their denial nor their nullification. It means their elucidation and the clarification of what is built upon them, an account of the grounds on which the validity of science rests and, by that very token, a delimitation of its legitimate validity, that is to say a clarification of the sense of its validity and legitimacy. Husserl's analysis of Galilean science is a *critique,* but in a sense significantly different from that of Kant's *Critique of Pure Reason.* Kant confined the fundamental concepts and principles of Newtonian physics to the realm of possible experience, and he precluded their application beyond that realm. As to the phenomenal realm, Kant, along the line of Euler's orientation, endeavored to demonstrate the unrestricted validity and necessary applicability of the mentioned concepts and principles. Husserl, on the contrary, in calling attention to the presuppositions of Galilean science, raises the problem of its sense, and of the sense and limits of its validity, precisely with regard to what in Kantian parlance is called the phenomenal world.

Among the presuppositions of Galilean science and of the historical Galileo himself, we have to mention, in the first place, the conception that the model and standard of knowledge, knowledge in the genuine and emphatic sense denoted by the Greek term *epistémé,* is mathematical knowledge, which in Galileo's historical situation meant the geometry of Euclid. In that respect Galileo was himself an heir to an existing and accepted tradition. The disclosure of Galileo's presuppositions, therefore, necessitates going beyond himself and inquiring into the origin of Euclid's geometry. "Origin" is not meant to be understood in historical or psychological terms; what Husserl means is rather origin of sense, or genesis of sense (*Sinnesursprung, Sinnesgenese*).

In the life-world in which we find ourselves, we encounter things of a circular contour, or, to express it more properly, things whose contour presents a circular physiognomy. However, these contours are not circles in the strict sense in which the term is understood in geometry, any more than trees whose shape presents a cylindrical physiognomy are cylinders in the geometrical sense. As the etymology of its name indicates, geometry was originally an art of measurement, especially of measuring the earth, demarcating adjoining fields, and the like. Measurement requires a technique and admits of varying degrees of accuracy. The degree of accuracy to be attained in a given case depends upon both the available technique and the purpose at hand. Practical exigencies may, and do, require an increasingly higher degree of accuracy, and this may lead not only to a "better" use of the available technique but also to an improvement of the technique itself. All such improvements are, and remain, in the service of practical purposes. The same holds for the practice of the craftsman who, e.g., working on a wooden plank tries to make its surface smoother and smoother. Here again, the exigencies of the practical situation may demand perfections of the available technique, like the invention of new tools. Here as before, all improvements and perfections are motivated and guided by practical ends and purposes.

In the course of such improvements and perfections, notions of *ideal limits* may arise, like that of the plane whose surface is of "absolute" smoothness and cannot be made still smoother, that of the straight line which has no width, that of the circle whose points are equidistant from the center, "absolutely" and no longer more or less so as in the case of perceptual configurations exhibiting physiognomical circularity. That is to say, geometrical figures in the strict sense can come to be conceived. In the geometry of Euclid, the geometrical notions still have intuitive content; the figures can be visualized, though they cannot be perceived because they cannot be encountered in reality. One can speak of them as ideal limit-poles located at infinity in so far as they indicate a direction for the process of progressive perfection, but they can never actually be reached in the course of this process. In this sense, geometrical notions designate models of perfection. *Geometrical concepts arise on the basis of perceptual and other experience in the life-world by means of idealization.* Once the ideal limit-poles have been conceived, the spatial configurations encountered in perceptual experience, like the trees of cylindrical shape, the lines actually drawn, can be seen in their light and under their perspective, that is to say, the configurations acquire the sense of approximations, more or less close, to the ideal figures which because of their ideality can never be actually realized. Plato compared the relation between the configurations drawn in sand and the geometrical figures which the mathematician has in mind even when looking at those configurations with the relation between the shadows which objects cast on water and the objects themselves.

Ideal geometrical figures lend themselves to exact and definitive determination, valid once forever and for everyone. Geometrical determination is independent of the circumstances, conditions and contingencies of actual observation and measurement. For the same reason, the reference of perceptual configurations to ideal geometrical figures which the former more or less closely approximate is emancipated from all consideration of practical exigency. Of equal importance is the momentous discovery that on the basis of comparatively few fundamental propositions, the axioms, all possible geometrical figures can be constructed and all properties of these figures demonstrated. Geometrical proofs are conclusive, and not a matter of opinion. It can be proved that in Euclidean geometry the sum of the interior angles of a plane triangle equals two right angles and that a denial of this proposition entails a contradiction. Assuming the character of a demonstrative deductive science, geometry proves to correspond to, and to realize the idea of *epistémé;* it appears as the model and the standard of knowledge worthy of that name.

Idealization is only the first step in the constitution and development of geometry. Soon after Galileo's time begins the *algebraization of geometry* by means of Fermat's and Descartes' analytical geometry and, later, the differential calculus invented by Leibniz and Newton. The path is opened for a progressive algebraization and even logification of geometry, culminating in contemporary axiomatics and the geometry of abstract spaces. This development is not confined to geometry alone. All of modern mathematics develops in the direction of increasing *formalization.* In a formalized discipline, of which common algebra besides, incidentally, traditional Aristotelian logic, are the oldest and the most elementary examples, the terms are divested of all intuitive content and are defined merely by the relations obtaining between them and the operations which can be performed upon them. This development culminates in group theory, in which even the relations and operations are left undetermined as to their intuitive meaning and are solely defined by their formal properties like symmetry and transitivity in the case of relations, commutativity and associativity in that of operations. Formalization is tantamount to the development of *algorithms,* that is to say the establishment of systems of symbols and rules for operations upon those symbols. The operations can be performed "blindly" and mechanically, the only requirement being conformity with the operational rules. Because of their formal nature, algorithmic systems lend themselves to multiple interpretations so that the results of blind and mechanical operations on symbols can receive the appropriate intuitive meaning, according to their interpretation in a given case.

For two reasons it appeared necessary to dwell upon this at some length. In the first place, Professor Marcuse has repeatedly spoken of calculation and calculability of nature. Perhaps he meant prediction and predictability, notions which require too lengthy commentary to be tackled here. At any event, the term calculability betrays too narrow a view of mathematics which is not the science of numbers and magnitudes, but rather of order as such and of the possible forms of order, numbers and magnitudes representing but a special case of order, the most elementary of its possible forms. In the second place, in speaking of the increasing technization of both mathematics itself and the sciences relying on mathematics, that is especially physics, Husserl, in contradistinction to what Professor Marcuse has intimated, does not have in view the practical use to which the results of physics and mathematics can be put, that is technology. By technization, Husserl rather means formalization and algorithmization.

The entities which arise by means of idealization and formalization acquire a semblance of independence, self-containedness, and self-sufficiency. Once established, geometrical and algorithmic methods can be taught and learned; they can be transmitted from generation to generation. In using and perfecting them, one proceeds from invention to invention, the accomplishments of later generations surpassing those of earlier ones. Such use and perfection of methods, the gradual development of the several mathematical disciplines, the ever increasing growth of mathematical knowledge as witnessed in the last three centuries, does not require reference to, and explicit disclosure of, the foundations upon which the edifice of mathematics rests. The presuppositions underlying mathematical thought and construction, namely the experience of things as encountered in the life-world, on the one hand, and, on the other hand, the processes of idealization and formalization may, and do, fall into oblivion. Severed from the sources from which they spring, the products resulting from idealization and formalization, that is to say, all the formal disciplines like geometry, the whole of mathematics, and also contemporary mathematized logic, come to be considered as autonomous in the sense of seeming to have no presuppositions outside of themselves. Such proceeding upon the basis of obfuscated and even forgotten presuppositions is what Husserl means by *traditionality*. It is the right of the positive scientist, the logician, mathematician, and physicist, to remain within this scientific tradition and to abstain from concerning himself with its origin and institution. It is the duty of the philosopher to raise precisely that question in order to clarify and account for the very sense of modern science.

Let us return to Galileo who was an heir to Euclid's geometry and stood within its tradition. That also includes the unquestioned acceptance of geometry as the standard of knowledge in the true and genuine sense. Consequently, if there is to be an authentic science of nature, it has to be fashioned after the model of geometry. Galileo attempts to mathematize nature and he succeeds: nature does lend itself to mathematization. The mathematization of nature can be regarded, and has been regarded by Husserl, as a successful venture; its success assumes the form of an historical process, the history of modern physics, physics of the Galilean style, from its beginnings to its present phase. That means that the venture has thus far proved successful and is expected to prove successful in the future.

It is neither possible nor even necessary to enter here into an analysis of the several steps involved in the application of mathematics to nature, the direct mathematization of the phenomena pertaining to motion and dynamics in general, the indirect mathematization of what has come to be called "secondary qualities." Instead we have to dwell upon an interpretation of physics, the "realistic" interpretation, which prevailed until recent times and even in our day has found an advocate in the person of Max Planck.

Because of the obfuscation of the presuppositions underlying the elaboration of the science of physics and determining its sense, this elaboration is not seen for what it actually is. It is the construction of an exact universe, consisting of entities defined and definable exclusively in mathematical terms. This construction is guided by certain principles and is meant to accomplish the specific purpose of providing an ever extending rationalization of observed phenomena and their unitary explanation. Previsions and predictions on a scale unheard of before are thus made possible, which cannot be obtained as long as one relies merely on the regularities as experienced and even as methodically observed in the life-world. Overlooking the processes of *idealization, formalization,* and whatever other forms of *conceptualization* might be involved, processes which are *mental operations in and through which the entities,* with which the physicist deals, are *conceived and constructed,* he or the interpreter of physics is wrapped up in the very products to the disregard of the producing activity from which those products spring. Failure to refer the accomplished products and results to the mental operations from which they derive and whose correlates they are, makes one the captive of those products and results, that is to say, the captive of one's own creations, and that is a further aspect of traditionality. Thus, as Husserl expressed it—Professor Marcuse referred hereto—a cloak or tissue of ideas (*Ideenkleid*), of mathematical ideas and symbols is cast upon the life-world so as to conceal it to the point of being substituted for it. What in truth is a method and the result of that method comes to be taken for reality. Thus we arrive at the conception of nature (mentioned in the beginning) as possessing a mathematical structure or being a mathematical manifold. This conception is expressed in the famous utterance of Galileo: "Whoever wants to read a book, must know the language in which that book is written. Nature is a book and the characters in which it is written are triangles, circles, and squares." Nature as it really is does not reveal itself in direct sense experience, but must be disclosed by means of the specific methods of mathematical physics. Such disclosure is interpreted as discovery of true reality behind and beneath the appearances. As to the latter, that is the life-world as it presents itself in perceptual experience, it is relegated to the domain of mere subjectivity. Hence a new science becomes necessary whose task is to account for the rise of the perceptual appearances on the basis of the true and real condition of nature as discovered by the science of physics. It is no historical accident but deeply rooted in the logic of the historical situation of the incipient physics of the Galilean style and its subsequent development, that psychology in the specific modern sense as the science of error has evolved along with that physics and in logico-historical continuity with it.

Let us approach our problem from another angle. At every period of the historical development, the distinction, according to Husserl, must be made between the state of the science of nature at that time, the conception of nature prevailing at that period, or nature as at that time it is believed to be, and nature as it really is. By the latter can only be meant the definitive conception of nature, or nature as it will finally come to be believed to be, when the historical process of the development of science will have reached its end. Nature as it really is denotes a goal towards which the historical process of the development of science is supposed to converge, which the successive conceptions of nature are supposed to approximate. In the interpretation under discussion, the goal is anticipated as attained, though not by us, and, therefore, it is as yet unknown to us. Moreover, the goal—if I may say so—is projected into nature as its true and real condition, waiting to be discovered; in a word, it is hypostatized. Another version, among several, is Leibniz' conception of the omniscient God who, as the supreme logician, is from eternity to eternity in the possession of all knowledge for which we humans have to strive, to the extent to which it is accessible to us.

By now the sense of Husserl's Galileo analysis and of his critique of modern science has become clear. It is overlooked that nature as it really is, the nature of the physicist, in contradistinction to nature as it presents itself in common perceptual experience, is a mental accomplishment, more precisely, the Idea of a goal towards which a sequence of mental accomplishments is converging, an Idea in the Kantian sense. No notice is taken of the mental processes of mathematization, idealization and formalization from which those accomplishments derive as their correlates and results. Finally, sight is lost of the life-world as given in common experience, upon which the mentioned processes operate in re-interpreting it in the light of idealized and formalized entities.

Thus far we have confined ourselves to Husserl's Galileo analysis contained in *Crisis* and to its immediate consequences. Now the perspective must be enlarged. *Crisis* is only one of three literary documents which together express what may be called the late phase of Husserl's thought. The two others are *Formale und transzendentale Logik* (*Formal and transcendental Logic*) and *Erfahrung und Urteil* (*Experience and Judgment*). Returning to, and continuing, his earlier work in *Logische Untersuchungen* (*Logical Investigations*), Husserl, in those two books, concerns himself with the foundations of formal logic and the genesis of its sense; *Erfahrung und Urteil* being given the significant sub-title "Investigations concerning the genealogy of logic." As in *Crisis,* he is brought before the life-world and led to emphasize the specific mental processes involved in the constitution of logic. The three books form a unitary group and should be seen as related to one another.

It results from these three works that logic as well as mathematics and physics have to be understood as correlates and products of activities of consciousness and mental life. Correspondingly the same can be shown with respect to the life-world itself. It is the world in which we find ourselves and pursue all our activities, which in our every day life we take for granted. It is permanently experienced by, and permanently present to, us; it is an experienced world and the world as given in direct and immediate experience, prior to all conceptualization, idealization, and formalization. Primarily, the term has a socio-historical meaning. Properly speaking, there is no life-world *per se*. Every concrete life-world refers to a certain social group at a certain phase of its history, such as the world of the ancient Babylonians, the ancient Egyptians, etc., and, of course, also our own life-world, the world of the Occidentals in the twentieth century. This is to say that every life-world is understood, conceived of, and interpreted in a specific way, by the social group whose life-world it is. The schemes of interpretation are transmitted from the older to the younger generation and so are the socially accepted or approved modes of conduct and ways of coming to terms with typical situations. For this aspect of the life-world as a social world, I wish to refer to the several writings of my late friend and predecessor in my professorial chair, Alfred Schutz. Though the life-world refers to a socio-historical group and, accordingly, changes from group to group, and also for the same group in the course of its history, the question arises as to an invariant structure pertaining to every possible life-world. Of special interest in the present context are spatiality, temporality, and a certain style of typical causality: what may be called the customs of nature, the habits which things have of typically behaving, with greater or lesser regularity, under given circumstances. All processes of idealization take their departure from those invariant structures.

The analysis of the foundations of both logic and physics leads to the discovery or rather rehabilitation of the life-world which can no longer be dismissed as a merely "subjective" phenomenon requiring an explanation on the grounds of nature as it really and truly is. On the contrary, it proves to underlie, and to be presupposed by, the elaboration of "objective" nature. Still, we have not yet reached the ultimate, but only the penultimate, presupposition and foundation. *The life-world,* in its turn, *refers to, and, in that sense, presupposes mental life, acts of consciousness, especially perceptual consciousness through which it is experienced and presents itself as that which it is,* that is to say, *as that which we take it.* It is not until we arrive at consciousness as the universal medium of access (in the sense of Descartes' *Second Meditation*) to whatever exists and is valid, including the life-world, that our search for foundations reaches its final destination. As far as the processes of conceptualization, idealization, and formalization are concerned, they now appear in their proper place as acts of consciousness of a higher order in so far as they presuppose the more elementary and more fundamental acts through which the life-world is given or, in Husserl's parlance, they are built upon pre-predicative experience. This is another expression of our previous result that the universe of physics, objective nature as conceived in physics, is a product of mental life constructed on the basis of the pre-predicative experience of the life-world which equally proves to be a correlate of consciousness.

It is a misconception, frequently encountered, that in re-instating the life-world in its right, Husserl meant it to be the final dwelling place for philosophical thought. On the contrary, in *Crisis* itself he emphasizes again the principle of a universal and thoroughgoing correlation between objects of any description whatever, objects in the broadest and most general sense, so as to include ideal constructs of every kind, on the one hand, and on the other hand, acts of consciousness and systematic interconcatenations of such acts through which the former present themselves. In other words, the Galileo analysis is intended to lead us to the threshold of phenomenology.

Once more we have to widen the perspective, if only for a brief moment. Two other books of Husserl's, usually considered as belonging to his middle period, have the sub-title "Introduction to phenomenology," neither, of course, in the textbook sense. They are *Ideen zu einer reinen Phänomenologie und phänomenologischen Philosophie* (*Ideas concerning a pure phenomenology and phenomenological philosophy*) and *Cartesianische Meditationen* (*Cartesian Meditations*). In these works Husserl follows what he calls the Cartesian way, taking his departure directly from consciousness and conscious experience. He endeavors to establish a general theory of consciousness whose central notion is that of *intentionality*. Especially *Ideas* presents a general outline of the theory of intentionality which, if fully developed in all directions and ramifications, would be coextensive with a complete phenomenological philosophy.

Beyond this point I cannot go. Embarking upon an exposition, however sketchy and superficial, of the concept of intentionality would lead us so far afield as to trangress the limits within which these remarks have to be confined. I can have no other purpose than indicating that and how the Galileo analysis merges into the mainstream of phenomenological thought. If, as mentioned in the beginning, Husserl's Galileo analysis inaugurates a new period in the philosophy of science, it is because this analysis developed within, and grew out of, the framework of general phenomenology. To be properly understood and not to be mistaken for an expression of an "anti-scientific" attitude on the part of Husserl, the Galileo analysis must be seen under this widest perspective.

Graduate Faculty of Political and Social Science
New School for Social Research
New York, N.Y.

BIBLIOGRAPHY

Only writings which directly or, at least, indirectly, bear upon the phenomenological theory of science are included in this selected bibliography for which even in this respect no completeness is claimed. Among the translations, only those into English or French are mentioned.

1. Suzanne Bachelard, *La logique de Husserl,* (Paris, 1957.)
2. ——, *La conscience de rationalité* (Paris, 1958.)
3. Oskar Becker, "Beiträge zur phänomenologischen Begründung der Geometrie und ihrer physikalischen Anwendungen" in *Jahrbuch für Philosophie und phänomenologische Forschung* vol. VI, 1923.
4. ——, "Mathematische Existenz" in *Jahrbuch für Philosophie und phänomenologische Forschung* vol. VIII, 1927.
5. Jean Cavaillès, *Méthode axiomatique et formalisme,* (Paris, 1938.)
6. ——, *Remarques sur la formation de la théorie abstraite des ensembles* (Paris, 1938).
7. ——, *Transfini et continu* (Paris, 1947).
8. ——, *Sur la logique et la théorie de la science* (Paris, 1947).
9. Aron Gurwitsch, "Présuppositions philosophiques de la logique" in *Revue de Métaphysique et de Morale* vol. LVI, 1951; reprinted in the collection of essays *Phénoménologie-Existence,* (Paris, 1953).
10. ——, "Sur la pensée conceptuelle" in *Edmund Husserl 1859-1959, Phaenomenologica vol. IV,* (The Hague, 1959) translated into English "On the conceptual consciousness" in *The Modeling of Mind* (Notre Dame, 1963).

11. Edmund Husserl, *Logische Untersuchungen,* Halle (M. Niemeyer) 1900/01, 2nd ed. 1913; translated into French by H. Elie, L. Kelkel, and R. Scherer as *Recherches logiques* (Paris) .
12. ——, *Ideen zu einer reinen Phänomenologie und phänomonologischen Philosophie I, Husserliana vol.* III, (The Hague, 1950) translated into English (not an adequate translation) by W. R. Boyce Gibson *Ideas* (London and New York, 1931) translated into French by P. Ricoeur *Idées directrices pour une phénoménologie* (Paris, 1950).
13. ——, *Ideen* II, *Husserliana* vol. IV (The Hague, 1952); reviewed by Alfred Schutz in *Philosophy and phenomenological Research* vol. XIII, 1953.
14. ——, *Ideen* III, (Die Phänomenologie und die Fundamente der Wissenschaften) , *Husserliana* vol. V, (The Hague, 1952 reviewed by Alfred Schutz in *Philosophy and phenomenological Research* vol. XIII, 1953.
15. ——, *Formale und transzendentale Logik,* (Halle, 1929) translated into French by Suzanne Bachelard *Logique formelle et logique transcendantale,* (Paris 1957; translation into English is in preparation).
16. ——, *Cartesianische Meditationen, Husserliana* vol. I, (The Hague 1950 and 1963); translated into French by G. Pfeiffer and E. Levinas *Méditations Cartésiennes* (Paris, 1931) translated into English by D. Cairns *Cartesian Meditations,* (The Hague, 1960).
17. ——, *Erfahrung und Urteil,* (Prag, 1939 and Hamburg, 1948) translation into English is planned.
18. ——, "Die Frage nach dem Ursprung der Geometrie als intentional-historisches Problem" in *Revue Internationale de Philosophie* vol. II, 1939; also contained in No. 19; reviewed by D. Cairns in *Philosophy and phenomenological Research* vol. I, 1940.
19. ——, *Die Krisis der Europäischen Wissenschaften und die transzendentale Phänomenologie, Husserliana* vol. VI (The Hague 1954); partly translated into French by E. Gerrer "La crise des sciences européenes et la phénoménologie transcendantale" in *Etudes Philosophiques* vol. IV, 1949; reviewed at length by Aron Gurwitsch in *Philosophy and Phenomenological Research* vol. XVI, 1956 and vol. XVII, 1957.
20. Alfred Schutz, *Collected Papers* I, *Phaenomenologica* vol. XI; II, *Phaenomenologica* vol. XV, (The Hague, 1962 and 1964); III is in preparation.

10

ABNER SHIMONY

QUANTUM PHYSICS AND THE PHILOSOPHY OF WHITEHEAD*

Presented March 3, 1964

I. *Introduction*

One of the virtues which Whitehead claims for his philosophy of organism is that it provides a conceptual framework for quantum theory (SMW chapter 8, PR 121-2 and 145).[1] The theory which he has in mind is the 'old' quantum theory, consisting of the hypothesis of Planck (1901) and of Einstein (1905) that electromagnetic energy is emitted and absorbed in quanta, together with Bohr's model of the atom (1913) in which discontinuous transitions were supposed to occur between discrete electronic orbits. The philosophy of organism was presented in a preliminary form in the Lowell Lectures of 1925 (published in the same year as SMW) and in its most systematic form in the Gifford Lectures of 1927-8 (published as PR in 1929). It was during the years 1924-1928 that DeBroglie, Schrödinger, Heisenberg, Born, Jordan, Bohr, Dirac, and others formulated the 'new' quantum theory, which was more systematic than the old, much more successful in its predictions, and more revolutionary in its departures from

* This essay is published here with the permission of Cornell University Press (Ithaca, New York) and George Allen and Unwin, Ltd. (London). It appears in their volume *Philosophy in America,* edited by Max Black.

1 The abbreviations of the titles of Whitehead's works and the editions to which the page numbers refer are as follows:
SMW—*Science and the Modern World* (New York, 1925)
PR—*Process and Reality* (New York, 1929)
AI—*Adventures of Ideas* (New York, 1933)

classical physics. Whitehead never refers to the new quantum theory, and it would be unreasonable to expect that even so imaginative a philosopher and scientist as he could have anticipated it except in the most general terms. Nevertheless, it is important in evaluating the philosophy of organism to determine how well its physical implications agree with quantum theory and with contemporary microphysical theory in general. To do this is the primary purpose of the present essay. It will appear that the agreement is only partial and that there are several crucial discrepancies both in detailed predictions and in general spirit.

The second purpose of this essay is to suggest the possibility of a modified philosophy of organism, which would preserve Whitehead's essential ideas but would accord with the discoveries of modern physics. This is a very ambitious program, and only a few tentative speculations on how one might proceed will be presented. It is, furthermore, a somewhat hazardous program, for one can cite many examples in the history of thought of philosophical schemes which attempted to conform to current science but succeeded in amplifying scientific error. One can only hope that the comparison of quantum theory with a philosophical system of such great scope as Whitehead's will not only improve the philosophical system, but will throw light upon some of the conceptual difficulties in modern physics.

II. *Implications of the Philosophy of Organism for Microphysics*

In SMW Whitehead criticizes classical physics for supposing that there exist material entities without any intrinsic mental characteristics, and for supposing that these entities are simply located, i.e. "*Here* in space and *here* in time, or *here* in space-time, in a perfectly definite sense which does not require for its explanation any reference to other regions of space-time." (SMW 72). Both of these suppositions, he claims, are instances of the 'fallacy of misplaced concreteness,' which consists in regarding abstract characteristics of things as their complete and concrete natures. The philosophy of organism, which is an attempt to avoid misplaced concreteness, requires a radically different conception of physical reality from that of classical materialism. In this Section some of the propositions of what may be called "Whiteheadian physics" will be formulated, partly on the basis of Whitehead's explicit remarks and partly by inference. For this purpose, the following theses of the philosophy of organism are relevant.[3]

(1) The ultimate concrete objects in the universe[4] are the 'actual occasions,' each of which has proto-mental characteristics and can be characterized as a unit of experience. The word "experience" is obscure, and various of Whitehead's statements have the effect of postulating that the word may be applied to entities usually considered to be inorganic without entirely vitiating its ordinary intention.

(2) Every actual occasion is distinguishable from every other in virtue of its intrinsic character and not merely because of its external relations to the rest of the world.

(3) An actual occasion 'prehends' each occasion antecedent to it, i.e. it recognizes the experience of the antecedent occasions in qualitative detail though with loss of immediacy and shift of emphasis. It is the relation of prehension which prevents actual occasions from being simply located (cf. PR 208).

(4) The temporal duration of an actual occasion is finite, and even though phases of becoming of an occasion can be distinguished, each phase is only derivatively real and is incomplete without reference to the entire occasion.

(5) Each actual occasion occupies a definite spatial region and is indefinitely divisible, but the parts have only derivative reality relative to the whole occasion (PR 434-5).

(6) The total set of prehensions of antecedent occasions by an

[2] Hereafter the qualification "new" will be omitted.

[3] Except for thesis (5), for which a special reference is given, these theses are selected from 'categories of explanation' and 'categoreal obligations' (PR 33-41) with much condensation, rearrangement, and simplification.

[4] The one exception to this statement is the actual entity God, who shares some but not all of the characteristics of the actual occasions.

occasion in process of becoming does not suffice to determine that process in all its details. There is, thus, an element of freedom in the process, which is neglible in low-order occasions but which permits radical novelty in those of higher order.

The foregoing fragment of Whitehead's philosophy not only contains implicitly most of his views on physics, but also contains his explanation of the existence of the subject-matter for special sciences such as physics. Most occasions are almost entirely constituted by their prehensions of antecedent occasions, since spontaneity and originality are usually neglible. As a result, the world, or at least that part of it lying in our 'cosmic epoch', is populated largely by 'enduring objects.' An enduring object is a temporally ordered chain of actual occasions, all sharing a common defining characteristic and sharing it because it is the dominant element in the prehensions of each successive occasion in the chain. Thus the enduring object is, in a sense, self-sustaining. Even in a heterogeneous society of occasions there may be characteristics common to all or nearly all members, so that the prehensions of new occasions will be virtually uniform in certain respects and the common characteristics will tend to persist. Conformity to a basic law of physics is an outstanding example of a persistent set of characteristics in a heterogenous society. Special laws, such as the laws of biology, may hold in sub-societies of the vast society governed by the basic physical laws. The special laws may not be derivable from the more basic laws, and yet they may presuppose the order established by the latter. Similarly, the basic laws governing a heterogeneous society determine its tolerance for various kinds of simple enduring objects, with the result that the number of species of elementary particles may be small, though the number of exemplars of each may be enormous. (Cf. AI 257 and 264, PR 138-140.)

It follows from Whitehead's view of the nature of physical laws that in principle they cannot be derived from his philosophical first principles. In fact, since he supposes that the propagation of dominant characteristics in a society is subject to lapses, he infers that the type or order expressed by physical laws may decay or change. Nevertheless, Whitehead's philosophical principles do seem to have at least the following physical consequences, which would presumably be valid in any cosmic epoch.

(i) Even though physics abstracts from the detailed content of an actual occasion, it cannot overlook the spatial and temporal extendedness of occasions. *Hence, the basic microphysical consequence of Whitehead's scheme is atomicity.* Thus, the physical fact that energy is transferred in quanta follows from the supposition that "physical energy . . . must then be conceived as an abstraction from the complex energy, emotional and purposeful, inherent in the subjective form of the final synthesis in which each occasion completes itself" (AI 239), in conjunction with the thesis that the actual occasion is an extended atom of experience. Furthermore, there must be a temporal atomicity in physical processes, since the individual actual occasion is not divisible into concrete events, one of which is earlier than the other (PR 107). Whitehead seems to believe that from the fundamental atomicity of actual occasions there follow other types of physical atomicity—e.g. the integral character of elementary particles and the indivisibility of electric charge into units smaller than the charge of the electron—but he is vague on this point.[5] One would also expect him to claim that the elementary particles of physics exhibit spatial extendedness in spite of their integral character, since each occasion constituting a link in the career of a particle has an internal spatial structure. Whitehead does not make such a claim explicitly, but perhaps it is implicit in his discussion of 'vibratory organic deformations' of a proton (SMW 195).

(ii) *Elementary particles should be capable of creation and destruction.* According to Whitehead an elementary particle is an especially simple kind of enduring object, and the continuation of any enduring object depends upon the degree to which each new occasion in the appropriate neighborhood will re-enact the experience of earlier occasions. Since the experience of a new occasion is partially colored by prehensions of other occasions than those constituting the enduring object, as well as by exercise of its

[5] He is also careless, for example in saying that "Electrons and protons and *photons* are unit charges of electricity" (AI 238, italics not in the original text).

intrinsic spontaneity, the elementary particle will almost certainly end after a finite number of links. An inverse argument explains how an elementary particle can be initiated.

(iii) *A consequence of atomicity, according to Whitehead, is the association of some sort of vibratory motion with all elementary particles.* An elementary particle is a chain of occasions, all having nearly the same internal development, so that the particle has a definite periodic structure with a definite frequency. It is clear, however, that Whitehead conceives of waves as more generic than particles, for particles always exhibit some characteristics of waves but not conversely. Thus,

> the doctrine, here explained, conciliates Newton's corpuscular theory of light with the wave theory. For both a corpuscle, and an advancing element of wave front, are merely a permanent form propagated from atomic creature to atomic creature. A corpuscle is in fact an 'enduring object.' The notion of an 'enduring object' is, however, capable of more or less completeness of realization. Thus, in different stages of its career, a wave of light may be more or less corpuscular. A train of such waves at all stages of its career involves social order; but in the earlier stages this social order takes the more special form of loosely related strands of personal order. This dominant personal order gradually vanishes as the time advances. Its defining characteristics become less and less important, as their various features peter out. The waves then become a nexus with important social order, but with no strand of personal order. Thus the train of waves starts as a corpuscular society, and ends as a society which is not corpuscular. (PR 53-4)

(iv) *As a result of thesis (6) strict determinism cannot hold in physics.* However exhaustively the antecedents of a physical event are specified, the character of the event cannot, in principle, be predicted with certainty.

(v) *The specification of the state of a composite system containing several elementary particles is equivalent to the specifications of the states of the individual particles.* This follows from the natural identification of the state of an elementary particle at a given time as an actual occasion, together with Whitehead's theses that actual occasions are pre-eminently real and that all groupings of occasions have derivative status.

Several further physical consequences of Whitehead's philosophy appear plausible, except for doubts regarding the extent to which physics abstracts from concreteness.

(vi) If concrete reality is considered there is clear asymmetry between past and future, since the occasions of the past are fully determinate while those of the future are not. *It is reasonable, consequently, to expect the asymmetry between the past and the future to be exhibited in the laws of microphysics,* and not merely in macroscopic laws such as those of thermodynamics.

(vii) An occasion prehends, with suitable gradations of relevance, all previous actual occasions. *Consequently, the physical properties of an elementary particle (e.g. its charge or its magnetic moment) should be slightly modified by the inclusion of the particle in a highly organized society such as an animal body.*

(viii) If the occasions of an enduring object are considered in full concreteness, they will exhibit 'aging' from earlier to later parts of the chain, simply in virtue of the accumulation of prehensions. *Consequently, one expects a systematic development of physically observable characteristics of elementary particles*—perhaps a drift towards instability, indicative of primitive feelings of satiation, or perhaps a drift towards greater stability, indicative of the entrenchment of a habit.

III. *Comparison with Current Physics*

Propositions (i)-(viii) will now be examined in the light of current physics. Whenever possible, these propositions will be confronted with direct experimental evidence. In most cases, however, one can do no more than compare them with their counterparts in quantum theory, so that experimental evidence can be invoked only to the extent that quantum theory as a whole is experimentally confirmed.[6]

(i) Part of Whitehead's conception of atomicity is in excellent agreement with current physics. Specifically, energy is transferred in quanta, and matter has a granular structure in the small which prevents indefinite divisibility. There is also evidence, though it is not decisive, that the integral character of an elementary particle is compatible with spatial extendedness and internal structure. Thus, experiments in which protons are scattered by protons indicate that these particles have a definite charge and current distribution, even though it is not possible to subdivide their charge into discrete parts.[7] Furthermore, it is difficult to envisage how a particle which is localized at a mathematical point could exhibit spin angular momentum; but one should be cautious in advancing this argument, since in actual calculations theoretical physicists are able to treat the spin simply as a 'quantum number' without any commitments to the extendedness or non-extendedness of the particles. With regard to temporal atomicity, which is the most radical of Whitehead's assertions concerning atomism, the testimony of current physics is unfavorable, but not decisively so. One of Heisenberg's uncertainty relations is $\triangle E \triangle t \sim h$, i.e. the uncertainty of the duration of a physical process can be reduced only at the price of increasing the uncertainty of the energy of the system during the process. The limitation which Heisenberg's

[6] Individual references will not be given for each of the propositions of quantum theory mentioned in the following discussion. Although these propositions are explained in every standard exposition of quantum theory, several books are particularly worth noting for their treatment of topics which are philosophically significant: D. Bohm, *Quantum Theory* (New York, 1951), J. von Neumann, *Mathematical Foundations of Quantum Mechanics* (Princeton, 1955), F. London and E. Bauer, *La Théorie de l'observation en mécanique quantique* (Paris, 1939), P.A.M. Dirac, *The Principles of Quantum Mechanics*, 3rd edition (Oxford, 1947).

[7] Cf. R. Hofstadter, F. Bumiller, and M. R. Yearian, *Reviews of Modern Physics 30*, 482 (1958). The standard model for interpreting these experimental results pictures the observed proton as a 'cloud' of virtual particles, fluctuating in and out of existence. Although the cloud exhibits a statistically describable structure, the virtual particles are supposed to be 'bare' and perhaps without structure. Cf. E. M. Henley and W. Thirring, *Elementary Quantum Field Theory* (New York, 1962), pp. 77-8, 219-231. This model is so different from Whitehead's conception of a spatially extended actual occasion that one must hesitate to say that the scattering experiments really support his conception.

relation places upon the sharpness of temporal specification is reminiscent of Whitehead's proposition, yet it is not the same. Whitehead is not attributing an *indeterminate* stretch of time to the actual occasion, but rather a *determinate finite* stretch; he denies temporal definiteness only to the phases within the occasion, for "this genetic passage from phase to phase is not in physical time" (PR 434). Moreover, there is no hint in Whitehead's work that an occasion of short duration has a less definite energy than one of longer duration. On the contrary, he says that every actual occasion, when it has completed its process of becoming, is completely definite with respect to every family of attributes (PR 38, Category of Explanation xxv). It should be noted, incidentally, that although the duration of each actual occasion is indivisible, Whitehead does not assume a lower limit to the set of all durations (cf. SMW, 198). Nevertheless, his proposal of temporal atomicity would be supported if the postulation by March, Darling, *et al.* of a minimum length or minimum space-time region proved successful in removing the troublesome 'divergences' of quantum field theory.[8] So far their postulates have not led to outstanding successes, but in view of the great mathematical difficulties and the large number of variants to be examined their failure is not decisive.

(ii) The Whiteheadian proposition that elementary particles can be created and destroyed has been strikingly confirmed by experiments, with regard to the 'stable' as well as the 'unstable' particles. For example, physicists have not only observed ordinary beta-decay, in which the unstable neutron decays into three stable particles, the proton, the electron, and the anti-neutrino, but they have also detected inverse beta-decay, in which a proton and an anti-neutrino are annihilated and a neutron and positron are created. Thus, the integral character of elementary particles is

[8] B. T. Darling, *Physical Review 80,* 460 (1950). A. March, *Quantum Mechanics of Particles and Wave Fields* (New York, 1951). Their postulates are motivated by the desire to eliminate the 'divergences' of quantum field theory. Because of the quantum mechanical proportionality of energy to frequency and the fact that a minimum length would imply a maximum frequency, their postulates would cut off the high range of frequencies responsible for the theoretically computed infinite energies.

not associated with permanence, contrary to Democritean atomism but in accordance with Whiteheadian physics.

(iii) Whitehead's account of the association of waves with particles is only superficially in accord with quantum theory. The passage from pp. 53-4 of PR, quoted in (iii) of the preceding Section, shows that he conceives of a wave front as a set of actual occasions (a 'nexus') with each occasion occupying a small region of the front. The mutual relations among contiguous occasions, whereby they begin and end in unison, ensure that all parts of the wave front are in phase with each other. Whitehead's picture is reminiscent of Schrödinger's attempt to interpret the wave function of a particle as a description of an ordinary physical field, which manifests particle-like properties whenever the region of high field intensity is very small. This interpretation was abandoned, because Schrödinger's own equation for the time dependence of the wave function implies that in the absence of confining forces an initially concentrated wave packet will disperse.[9] In order to account for the experimental fact that particles do not disperse, Born proposed that the physical content of the wave function was to determine the probabilities of experimental outcomes (e.g. $|(\psi(x,y,z)|^2 dxdydz$ is the probability of finding the particle in a small volume dxdydz). In the Born interpretation the wave function characterizes the state of the particle in its entirety and does not describe the physically real parts of a field. There is a sharp discrepancy between this point of view, which at present is accepted by most physicists, and Whitehead's attribution of primary reality to the occasions of a wave front.

Deeper insight into this discrepancy can be achieved by considering the following fundamental principle of quantum theory, called "the superposition principle": if $u_1, u_2, u_3, \ldots$ represent physically possible states, then the combination $\sum c_i u_i$, where c_i are arbitrary complex numbers, represents a superposition of the original states which 'overlaps' each of them in a certain sense and which, moreover, is itself a physically possible state. The nature

[9] Cf. M. Born, *Atomic Physics* (Fifth edition, New York, 1951), p. 93 and pp. 142-4 for a discussion of the reasons for abandoning Schrödinger's interpretation in favor of Born's and for a general discussion of attempts to rationalize wave-particle dualism.

of the 'overlap' is most clearly exhibited if there is an observable property A of the system having definite values a_i in the states represented by the u_i, such that all the a_i are different. If the u_i are then specified as being of a standard length ('normalized' to 1), then according to quantum theory the probability of finding that $A=a_i$ when a measurement of A is performed is $|c_i|^2/|c_1|^2+|c_2|^2 \ldots$ It must be insisted that $\psi=\Sigma c_i u_i$ is a maximum specification of the system; it is not a statistical description of a system which really is in a state represented by a definite, but unknown, one of the u_i. The superposition therefore has the counter-intuitive characteristic of being a state in which the observable A is objectively indefinite and not merely unknown. It follows that the uncertainty principle of Heisenberg, which limits the simultaneous determination of complementary quantities such as position and momentum, refers to the objective properties of the particle and not simply to human knowledge about these properties. The wave function $\psi(x,y,z)$ can now be understood in terms of the superposition principle: roughly, $\psi(x,y,z)$ represents a superposition of states in each of which the particle has a position localized within a small region dxdydz, with the numbers $\psi(x,y,z)$dxdydz serving as the expansion coefficients in the superposition (i.e., they are the c_i in $\Sigma c_i u_i$). It should now be obvious that the quantum mechanical account of the waves associated with particles is entirely alien to the spirit of Whitehead's philosophy. In particular, the postulation of indefinite values of observable quantities, as required by the quantum mechanical account, would be repugnant to Whitehead. He might admit that indefiniteness is characteristic of the spatial or temporal parts of actual occasions, but surely not of the complete occasions: the twenty-fifth Category of Explanation, on p. 38 of PR, asserts,

> The final phase in the process of concrescence, constituting an actual entity, is one complex, fully determinate feeling. . . . It is fully determinate (a) as to its genesis, (b) as to its objective character for the transcendent creativity, and (c) as to its prehension—positive or negative—of every item in its universe.

An important technical disagreement follows from the fundamental conceptual difference between Whitehead's analysis of wave-particle dualism and that of quantum theory. According to Whitehead, a sharp frequency may be characteristic of a well-localized particle, since the periodicity is essentially due to the reiteration of a basic pattern in successive actual occasions. In quantum theory, on the other hand, a sharp frequency is characteristic of a particle which has a definite momentum, and therefore, by the uncertainty principle, a completely indefinite position; and conversely, the wave function of a well-localized particle has a very broad spectrum of frequencies.

A final note on Whitehead's account of waves and particles is to point out a striking disconfirmation by experiment. He explicitely states, in the passage quoted above from p. 53 of PR, that the corpuscular character of light is gradually lost as the light is propagated—a reasonable remark given his general analysis of particles. However, the phenomenon which most vividly illustrates the particle aspect of electromagnetic radiation, the photoelectric effect, is entirely independent of the age of the radiation. In fact, photoelectric cells are attached to telescopes in order to study starlight which has travelled for millions of years.

(iv) Quantum theory and Whiteheadian physics are both indeterministic, but in quite different ways. According to quantum theory, the state of a physical system evolves in a continuous and fully deterministic manner as long as the system is isolated. Probability enters only when a measurement is made of an observable which does not have a sharp value in the state of the system prior to the measurement. The interruption of the deterministic evolution of the state does not contradict the Schrödinger equation, since a measurement requires the interaction of the system with another system—a macroscopic piece of apparatus and perhaps a conscious observer. According to Whitehead, the evolution of an isolated system (the concept of which is an idealization, since an actual occasion prehends all occasions in its past) cannot be entirely deterministic, because of the element of freedom in each occasion. If one attempts to adjudicate between these two different accounts of indeterminism by considering the success of statistical predictions, one must certainly prefer quantum theory, for its statistical predictions are remarkably good, while Whitehead's proposals are too vague to yield any quantitative statistics. Nevertheless, there may be strong reasons for exploring the hypothesis that chance events occur in isolated systems, since, as will be discussed in Section IV, the quantum theoretical account of indeterministic transitions leads to conceptual difficulties.

(v) The Whiteheadian treatment of the state of a composite system is at odds with a quantum mechanical principle which has attracted little attention in spite of its revolutionary philosophical implications: *that a several-particle system may be in a definite state, i.e., may have as definite properties as quantum theory permits, without the individual particles being in definite states.* To illustrate this principle, consider two systems I and II and let ϕ_1 (I) and ϕ_2(I) represent states of I in which observable A has values a_1 and a_2 respectively ($a_1 \neq a_2$), and ψ_1(II) and ψ_2(II) represent states of II in which observable B has values b_1 and b_2 respectively ($b_1 \neq b_2$). Let $\phi_1(\mathrm{I}) \otimes \Psi_1(\mathrm{II})$ represent that state of the composite system I and II in which system I is described by ϕ_1(I) and system II by ψ_1(II); and let $\phi_2(\mathrm{I}) \otimes \Psi_2(\mathrm{II})$ be similar. The superposition principle can now be invoked to affirm the physical possibility of a remarkable state, namely, the one represented by

$$\Psi = \frac{1}{\sqrt{2}}(\phi_1(\mathrm{I}) \otimes \psi_1(\mathrm{II})) + \frac{1}{\sqrt{2}}(\phi_2(\mathrm{I}) \otimes \psi_2(\mathrm{II})).$$

Neither I nor II is in a definite state when Ψ represents the state of I plus II. A rigorous proof will be omitted, but the following rough argument indicates the essential reason: if one tries to claim that I is to some extent described by ϕ_1, and to some extent by ϕ_2, the claim is vague unless the expansion coefficients of ϕ_1 and ϕ_2 are specified, but in the above expression for Ψ it is clear that the expansion coefficients refer to states of II. One might try to reconcile the existence of such states of several-particle systems with Whitehead's consequence (v) by identifying the state of a several-particle system with a single actual occasion, and identifying the individual particles at a given moment with subdivisions of the occasion. Divisions of an occasion could be reasonably expected to lack the specificity implied in saying that an entity is in a definite state, for "in dividing the region we are ignoring the subjective unity which is inconsistent with such divisions" (PR 435). This attempt at reconciliation fails, however, because quantum mechanics permits the parts of a system described by Ψ to be indefinitely far apart spatially.[10] An actual occasion which not only is macroscopically extended, but even broken into non-contiguous parts, is evidently repugnant to Whitehead. To be sure, he is reticent about the exact extent of an occasion, but he seems to fear that the identification of a large scale process as a single occasion will remove the barriers to Spinozistic monism (cf. PR 10-11).

It is appropriate at this point to interpolate a discussion of another quantum mechanical principle concerning several-particle systems, even though it has no clear counterpart in Whiteheadian physics: *that if a system contains several elementary particles of the same species, they must all play the same role in the system as a whole.* Consider, for example, a system composed of two pi-mesons, and suppose ϕ_1 and ϕ_2 each represent a possible state of a single pi-meson. Then the principle forbids the composite system to be in the state represented by $\phi_1 \otimes \phi_2$, for in this state pi-mesons I and II have different roles; but the principle permits the state represented by

$$\frac{1}{\sqrt{2}}(\phi_1\,(\mathrm{I}) \otimes \phi_2\,(\mathrm{II})) + \frac{1}{\sqrt{2}}(\phi_2\,(\mathrm{I}) \otimes \phi_1\,(\mathrm{II})),$$

for this state is 'symmetrized' with respect to the two particles.[11]

To fit symmetrization into the framework of Whiteheadian physics is a delicate matter, since it implies a kind of loss of identity of the individual particles. In particular, the simple model of particle I as a chain of occasions $O_1, O_1', O_1'', \ldots$ and of particle II as a chain $O_2, O_2', O_2'', \ldots$ will not work, because individual occasions are always distinguishable in virtue of intrinsic characteristics (thesis 2 of Sect. II) and hence the two chains are distinguish-

[10] Cf. D. Bohm and Y. Aharonov, *Physical Review 108,* 1070 (1957).

[11] Symmetrization is actually a property of states of systems composed of particles with 'integral spin.' When the particles have 'half-integer spin,' as do electrons, neutrinos, neutrons, and protons, the state of the composite system must be 'anti-symmetrized.' However, the difference between symmetrization and anti-symmetrization is irrelevant for the present purpose, which is to insist upon the identity of roles of all particles of the same species.

able. A possible reply is that physics does not treat actual occasions *in concreto* and hence can fail to take cognizance of the respect in which the chains $O_1, O_1', O_1'', \ldots$ and $O_2, O_2', O_2'', \ldots$ are different; but the symmetrization obtained in this manner would appear to be a coincidence rather than a general law.

A more promising explanation is that the chains $O_1, O_1', O_1'', \ldots$ and $O_2, O_2', O_2'', \ldots$ intersect and the occasions which are shared by several chains have a quite different character—the physical manifestation of which is symmetrization—from the character of the occasions prior to the merger of the chains or after their separation. This explanation conforms to Whitehead's general view that the corpuscular nature of a society of occasions is highly special and easily dissipated. However, the explanation encounters the same difficulty that was noticed in the preceding paragraph. Composite systems containing particles of the same species can have indefinitely large spatial extents as in the case of the conduction electrons in a bar of metal. Since all the particles play the same role, it becomes necessary to identify the state of a macroscopically extended system with a single actual occasion, contrary to Whitehead's conception of the actual occasion as a microscopic entity.

(vi) In contemporary physics the only laws which involve a definite direction of time are macroscopic laws, the outstanding one being the thermodynamic law of entropy increase. No microphysical law has yet been discovered which is not invariant under reversal of the direction of time. If this peculiarity remains a permanent feature of physics, it would constitute evidence detrimental to Whitehead's scheme, in which the asymmetry of past and future is essential. The only defense of a Whiteheadian physics would then be the desperate one that the symmetry of past and future is one of the features of concrete reality neglected by physics. However, since the discovery of parity non-conservation by Yang and Lee *et al.*, the confidence of physicists in some of the physical symmetry principles has been shaken, and many suspect that a violation of time-reversal invariance will also be detected.[12]

[12] Cf. P. Morrison, *American Journal of Physics, 26* (1958). Note added in proof: An experiment by Christensen *et al.*, described in *Physical Review Letters 13,* 138, (1964) (and reported in *The New York Times,* Aug. 5, 1964, p. 56), indicates such a violation.

(vii) The Whiteheadian expectation that the physical properties of an elementary particle are slightly modified when the particle enters a structural society is counter to the reductionist spirit of physics, chemistry, and biophysics. Many of the predictions of these sciences rest upon the assumption that such properties of elementary particles as charge, mass, and magnetic moment are unchanged by the incorporation of particles into highly structured macroscopic objects. Of course, the same predictions would be made if the changes due to incorporation are extremely small. Since Whitehead gives no indication of the amount of modification to be expected there can be no crucial experiment.

(viii) The 'aging' of elementary particles implied by Whitehead's philosophy is contrary to current physical theory and is not confirmed by experiment. The intrinsic properties of newly created elementary particles seem to be no different, at least statistically, from those of particles which have endured a long while. Particularly significant is the decay rate in a population of unstable particles. The number of neutrons decaying per unit time is proportional to the number of neutrons in the population (with allowance, of course, for statistical fluctuations), and is independent of the 'age' of the population. It is reasonable to infer that the probability that a given neutron will decay during an interval of time is independent of the age of the neutron, thus suggesting that no physically significant changes occur in the neutron due to aging. Again, however, Whitehead's statements permit no quantitative estimate of the change to be anticipated in the decay rate, and therefore no crucial experiment is possible.

IV. *Proposals for Reconstruction*

The discrepancies noted in Sect. III between Whiteheadian physics and current microphysics constitute strong disconfirmation of Whitehead's philosophy as a whole. The possibility remains, however, of constructing a philosophical system which would be

Whiteheadian in its general conceptions though not in details, and which would accord with the fundamental discoveries of science. A few tentative suggestions will be given here concerning the initiation of such a large philosophical undertaking.

A useful first step is to distinguish both in the philosophy of organism and in quantum theory those elements which are radical by the standards of classical physics from those which are conservative. Most radical in Whitehead's philosophy are the attribution of proto-mental properties to entities normally considered to be physical, and the postulation of prehension as the fundamental relation between occasions. The assumption of the complete definiteness of the occasion in its final stage is conservative, though the correlative assumption of indefiniteness in the early stages of concrescence is not. Also conservative is his reductionist assumption that the characteristics of a nexus are entirely determined by the characteristics of its constituent occasions. Quantum theory is conservative in supposing that certain quantities initially introduced in the study of macroscopic physical objects—especially spatio-temporal position, energy, momentum, angular momentum, charge, and magnetic moment—can be used meaningfully in characterizing microscopic entities. On the other hand, the superposition principle is radical, for it has the consequence that a physical quantity can have an indefinite value in a maximally specific state of a microscopic system and has a sharp value only in exceptional states (the 'eigenstates' of that quantity). Quantum theory is also radical in its treatment of the relation between the state of a several-particle system and the states of its constituent particles.

The foregoing juxtaposition suggests a program of reconstruction: *to graft the radical elements of quantum theory onto the radical elements of the philosophy of organism, by assuming that elementary entities have proto-mental characteristics while treating the states of those entities in accordance with the combinatory principles of quantum theory.* The synthesis contemplated here does not seem forced from a Whiteheadian point of view. Whitehead often engages in a dialectical analysis which is reminiscent of the quantum mechanical treatment of complementary quantities,[13] but he never achieves what is most remarkable in quantum theory—a set of systematic rules for predicting statistically what will appear when a shift is made from one description to a complementary description. A modification of Whitehead's philosophy in accordance with the combinatory principles of quantum theory would perhaps make explicit and precise certain tendencies which are implicit and haphazard in his work.

Such a modification of Whitehead's system would surely change the conception of an actual occasion. For example, it would be impossible to maintain that an actual occasion, in its final phase, is definite with respect to every family of attributes (PR 38, Category of Explanation xxv). Instead, in accordance with the uncertainty principle, the specificity of any attribute is always attained at the price of indefiniteness of other attributes. In particular, an occasion may have a quite sharp location in time and an arbitrarily short duration, provided that properties complementary to duration were sufficiently indefinite.[14] It is also possible that the actual occasions may lose their status of being (along with God) the only ultimate real entities, and they may appear instead only as special cases of ultimate reality. Elementary particle theory and quantum electrodynamics may provide a hint as to the more general form of ultimate reality: i.e., it might be some kind of 'field' of diffused primitive feeling, of which the actual occasions are 'quanta' existing whenever there are individual loci of feeling. The hypothesis of diffused feelings is no more of an extrapolation from psychological data than is Whitehead's attribution of protomental characteristics to elementary particles, and indeed our everyday experience of sensitivity pervading the whole human body may possibly be construed as confirmation of the hypothesis.

[13] This is most striking in Whitehead's theology: "It is as true to say that God is permanent and the World fluent, as that the World is permanent and God is fluent . . ." (PR 528). Whitehead explains that "In each antithesis there is a shift of meaning which converts the opposition into a contrast" (*ibid.*). Dialectical analysis is also exhibited in the more mundane parts of his philosophy, for example in his statement that an actual occasion is prehended in its concreteness and yet with loss of immediacy.

[14] According to *Heisenberg's* $\triangle E \triangle t \sim h$, the property complementary to duration is energy. But if Whitehead's thesis is maintained that physical energy is an abstraction from emotional and purposive energy (AI 239), then Heisenberg's relation may require supplementation.

The physical evidence concerning composite systems, discussed in (v) of Sect. III, suggest a quantum theoretical refinement of Whitehead's treatment of the relation between the nexus and its constituent occasions. Whitehead conceives of the nexus in a reductionist manner, as the totality of its constituent occasions, and he supposes the internal relations exhibited in a nexus to be completely explicable in terms of the prehensions of earlier occasions of the nexus by each new occasion. Quantum theory, on the other hand, treats a composite system in a subtle manner, which at first seems paradoxical: it allows the state of the composite system to be described, in a certain sense, in terms of its components, and yet it permits the composite system to be in a definite state even when its components are not. This treatment, of course, is intimately bound up with the superposition principle. Thus, in the example cited in the previous Section, the composite system I + II is in the state represented by

$$\Psi = \frac{1}{\sqrt{2}}(\phi_1\,(\mathrm{I}) \otimes \psi_1\,(\mathrm{II})) + \frac{1}{\sqrt{2}}(\phi_2\,(\mathrm{I}) \otimes \psi_2\,(\mathrm{II})),$$

which is clearly describable in terms of the states ϕ_1 and ϕ_2 of component I and ψ_1 and ψ_2 of component II; yet, because of the character of the superposition, neither I nor II is in a definite state. If this quantum theoretical treatment of the whole-part relationship is introduced into the philosophy of organism, it opens a number of possible lines of exploration. For example, the 'field of feeling,' the existence of which was hypothesized in the preceding paragraph, might be characterized holistically as being in a definite state. This state could always be described as a superposition of field states, in each of which there is a definite set of actual occasions, just as in quantum electrodynamics the state of the entire electromagnetic field can be described as a superposition of field states in each of which there is a definite set of photons. Moreover, special states of the 'field of feeling' can exist in which the superposition is in part reduced, so that definite actual occasions exist. Consequently, the existence of many independent loci of feeling, which is an essential aspect of the experienced plurality of the world, is permitted—though not required—in this modified Whiteheadian scheme. A nexus of the type described by Whitehead is also permitted but in a rather special case: it occurs when there is a network of occasions sharing some common characteristic, each in a definite single-quantum state. What the modified scheme permits which Whitehead's does not is the existence of a composite system more complex than a nexus: an 'n-quanta system' in which each of the n occasions is so correlated with the others, *via* the superposition principle, that none is in a definite single-quantum state.

An evident advantage of the quantum theoretical modification of the philosophy of organism is that it removes some of the discrepancies with modern physics noted in Sect. III—a virtue which is not surprising, since the modification was inspired by these discrepancies. A further advantage is a possible improvement in treating the question of 'simple location.' Whitehead's rejection of simple location in SMW is a dramatic criticism of classical physics, but his sketch of an alternative in PR is somewhat disappointing. His alternative is essentially to postulate the relation of prehension, whereby an actual occasion is felt *in concreto* in the initial phase of each later occasion. Even if the ambiguities inherent in the conception of prehension can be dispelled, the relation of prehension can at best provide a kind of multiple location in time, i.e., the occasion as 'subject' and the same occasion as 'ingredient' in later subjects. By contrast, the quantum theoretical modification exhibits a breakdown in simple location in space in two respects: first the quantum state of an individual occasion may be such that its position is indefinite; and second, a composite system can have spatially separated components which are not in definite states, but which are so correlated with each other that the composite system is in a definite state. Finally, there is an advantage which was briefly mentioned earlier but which deserves amplification: the possibility that the modification of Whitehead's philosophy will greatly improve the account of high-order mental phenomena—which, after all, are the only mental phenomena we know about without resorting to radical hypotheses and extrapolations. Because Whitehead conceives actual occasions to be microscopic in size, and because the human personality at any moment has a unity which entitles it to the status of an actual occasion, he is led to the strange doctrine of a microscopic locus of high-order experience wandering through the society of occasions which compose the brain:

> Thus in an animal body the presiding occasion, if there be one, is the final node, or intersection, of a complex structure of many enduring objects. . . . There is also an enduring object formed by the inheritance from presiding occasion to presiding occasion. This endurance of the mind is only one more example of the general principle on which the body is constructed. This route of presiding occasions probably wanders from part to part of the brain, dissociated from the physical material atoms. (PR 166-7)

The question of the location of mentality is extremely complicated, but introspection seems to indicate that it is diffused throughout the body, and neuro-physiology has not yielded evidence of extreme localization. Various of the concepts of the modified Whiteheadian scheme seem relevant in describing high-order mentality: the concept of an indefinitely located actual occasion, the concept of a field of feeling which is generally diffused but occasionally quantized, and the concept of quantum correlations among the components of a composite system. Which one of these, or which combination of them, will be most fruitful is a matter of speculation, but all of them seem preferable to Whitehead's microscopic localization of high-order mentality.[15]

This list of advantages must be weighed against some strong reservations. The first reservation concerns a particular proposal made above rather than the general program of a quantum theoretical modification of the philosophy of organism. The proposal was to consider the entire 'field of feeling' as being in a definite state and to consider actual occasions as quanta in this field. One may wonder, in view of the notorious difficulties of quantum field theories, whether they are suitable models for a metaphysical scheme describing all of reality, and one may wonder whether a single field suffices for this purpose. Setting aside these doubts,

[15] Some speculations on the application of quantum theoretical concepts to mentality are given in D. Bohm, *Quantum Theory*, 168-72.

however, one may still be skeptical about the adequacy of this proposal to account for the experienced plurality of the world. It was noted that the existence of a definite state of the entire field is permissive of individual loci of feeling, but is permissiveness sufficient? Is it a mere contingency, characteristic perhaps of our scientific epoch, that there are definite actual occasions, or are there deep-lying reasons why this should always be so? Merely raising these questions exhibits the vagueness of the proposal and its need for supplementation.

A second reservation arises from the apparent absence of any clear psychological manifestations of the superposition principle, which one would expect if the combinatory principles of quantum theory apply to actual occasions. There are, to be sure, psychological phenomena which at first sight could be construed as evidence of the superimposing of mental states—e.g., perceptual vagueness, emotional ambiguity, conflict of loyalties, and the symbolism of dreams. Yet in all these cases the quantitative characteristics of the superposition principle, as it is exhibited in physics, are missing.[16] This reservation, however, is not decisive, for one can optimistically reply that the absence of confirming evidence is not equivalent to the existence of disconfirming evidence. It is possible that the psychological manifestations of the superposition principle are too delicate for the introspective, behavioristic, and physiological techniques in current use to detect.[17] It is also possible that these techniques are sufficient, but that no one has yet been sufficiently serious about using the superposition principle in psychology to design a good experiment.

A third reservation is that quantum theory, in spite of its striking successes in physics, is beset by a serious conceptual difficulty, which should perhaps be resolved before it is incorporated into a philosophical system. This conceptual difficulty concerns the 'reduction of a superposition,' which occurs if an observable A, having sharp and distinct values in the states represented by u_i, is measured when the state is represented by

[16] This point is discussed in A. Shimony, *American Journal of Physics, 31,* 755 (1963).

[17] Dr. Karl Kornacker pointed out in a private communication that the interference effects characteristic of superpositions are no more to be expected in gross emotional and perceptual phenomena than in macroscopic physical phenomena.

$\sum c_i u_i$. As a result of the measurement there is a non-deterministic transition from the initial state to a final state represented by a definite one of the u_i. Although many textbooks and popular accounts say that the transition occurs when a microscopic system interacts with the macroscopic measuring apparatus, this explanation is inadequate, since the Schrödinger equation, which is the equation governing the evolution of the state, implies that the final state of the apparatus plus microscopic system will also be a superposition with respect to the observable A. Some theoreticians have concluded that the reduction of the superposition does not take place until the result of the measurement is registered in the consciousness of an observer. This is a desperate conclusion and is unsatisfactory for several reasons.[18] It suffices to say here that if observation is a natural process, it is difficult to understand why a non-deterministic transition should occur when an observation is made, while all other natural processes are deterministic. One possible solution to the problem of the reduction of superpositions is to suppose that the evolution of the quantum state is to some extent stochastic and hence only approximately governed by the Schrödinger equation. The non-deterministic reduction of a superposition could then occur in a system remote from anything ordinarily called "an observer." In this way the superposition principle could be maintained in microphysics, but at the price of changing the dynamics of quantum theory. The success of a solution along this line would remove the third reservation about a quantum theoretical modification of Whitehead's philosophy, for the modification depends primarily upon extending the application of the superposition principle. The proposal of a chance element in the evolution of the state does not disrupt the program envisaged here, for in fact this proposal is closer to Whitehead's version of indeterminism than is the indeterminism of current quantum theory. One could even speculate that Whitehead's account of the concrescence of an actual occasion provides some insight into the way in which the reduction of the superposition occurs. Such speculations, however, are rather empty until stochas-

[18] E. Wigner, article in *The Scientist Speculates,* ed. I. J. Good (London, 1962); also the paper by A. Shimony cited in footnote 16.

tic generalizations of the Schrödinger equation are proposed and their physical consequences are studied.

A methodological remark is appropriate in conclusion. It has been tacitly assumed throughout this paper that the hypothetico-deductive method is an appropriate instrument in philosophical inquiry. This assumption is in the spirit of Whitehead's philosophy, for he deliberately formulated a categoreal scheme which could be confirmed or disconfirmed only by examining its remote consequences (PR 7-8). Regrettably, the difficulties of employing the hypothetico-deductive method in philosophy are illustrated only too clearly in this paper. The conclusions which can be drawn from philosophical first principles are generally qualitative, and therefore their confrontation with experience lacks sharpness. When they are confirmed, it is gross rather than fine confirmation, and when they are disconfirmed there is often a plausible way of saving appearances. The moral, however, is not that the hypothetico-deductive method should be abandoned. Rather, it is to seek refinements of philosophical first principles and liaisons of these principles with scientific hypotheses, in such a way that sharp predictions and fine confirmations may result.[19]

Dept. of Humanities
Massachusetts Institute of Technology
Cambridge, Mass.

[19] I am deeply indebted to Dr. Howard Stein for his criticism of an early draft of this paper and, more important, for stimulating and suggestive conversations over many years concerning the topics which it treats.

J. M. BURGERS

COMMENTS ON SHIMONY'S PAPER

In presenting some observations on Shimony's paper *"Quantum Physics and the Philosophy of Whitehead"* I would allow myself to start from the final paragraph of the paper, in which Dr. Shimony speaks about the "hypothetico-deductive method" in philosophy and asks whether this is an appropriate instrument in philosophical inquiry.

My reply is as follows: Every form of philosophy is an attempt to bring into an ordered picture the information we have collected about the phenomena around us and inside us. The data which are taken into consideration may cover a wider or a more restricted domain, but the picture must give an elucidation and an interpretation of the data. It is an aim of most philosophies to make the picture such that the relationships in which we are interested, will seem to follow from a few general principles. If this aim has been reached we can read the philosophical picture as a deductive system, and we can enjoy its clarity and the scope of its interpretations in a similar way as we enjoy a great artistic achievement. Sometimes we can extend the deductive scheme beyond its original limits in order to explore further forms of relationship, and this can furnish an important method of investigation. But we should never forget that the deductive scheme has been the result of analysis and of thinking based upon that which has been experienced.

Ultimately it is always experience, in so far as the experience is considered to be of interest and relevance, which has put us on our way. The final test of any philosophical scheme is its applicability as a means for the interpretation of the world of experience which we consider to be of importance.

Hence it is perfectly legitimate to attempt to change the scheme if we become aware of insufficient conformity. It follows that it is fully acceptable that we make an attempt to extend Whitehead's philosophical scheme, or change parts of it, or abandon parts, when its power of interpretation is deemed unsatisfactory. Judging by Whitehead's writings I would presume that he himself would gladly partake in discussions to this purpose.

Now when we speak of the possibility of changing some features of Whitehead's philosophy, this presupposes that there are certain basic ideas which we wish to retain. Let us consider therefore what are these basic ideas.

Dr. Shimony has given a brief picture of the principles of Whitehead's philosophy in section II of his paper, sub (1)– (6). In the first thesis Dr. Shimony mentions that the "actual occasions" of the universe according to Whitehead have "proto-mental characteristics," but he does not give a further explanation of this term. I believe that it is of great importance that we clearly see which step is made by Whitehead, and I would present the following statement, taken from my own work on Whitehead's philosophy (to be published in book form by the M.I.T. Press).

The functioning of the universe must be conceived as an immense complex of "events" ("actual occasions," in my terminology: "processes"), progressing and extending itself without end. In each step, that is, in each process, influences derived from the outcome of past events act together with forms of valuation directed towards a future. Each process has experience, obtained from what has gone before, which experience it subjects to valuations directed towards forms of future relationship. This means: the process is an anticipation of possibilities, and the actual outcome of the process is the result of a decision between these anticipated possibilities. The experience derived from the past does not itself completely direct the developing process; it is assumed that this incompleteness does not lead to mere statistical randomness, and there is room for a certain coordinative activity, that is, for conceptual activity which evaluates possibilities and decides the final outcome. This conceptual activity has some freedom and is creative, although it is influenced by a transmission of subjective form from an immediately preceding process. Each process is a unique occurrence, never repeated in precisely the same form.

To introduce such a thesis as the basis of the philosophical picture, means a definite step beyond causality in the physical sense. A new principle for the concatenation of events is postulated. Although this is not the subject of tonight's discussion, it will be evident that the assumption of this principle is of fundamental importance in the explanation we shall give to the phenomena of life.

It is with this fundamental thesis before us that we can discuss whether there are parts of the superstructure built upon it, which we wish to criticize and perhaps to change.

In approaching this problem we must be aware of the following point. Whitehead postulates that the constantly growing complex of processes, branching out from each other, joining again, and developing always further, is a feature of the universe more fundamental than the structure which we ascribe to the present framework of space and time. The relations of space and time, and their connections with the phenomena of matter, belong to the superstructure. We must be careful, therefore, when speaking of relations existing in the complex of processes, not to ascribe to them spatial and temporal characteristics, too closely modeled upon those which we use in present day physics. It is necessary to assume certain basic features of arrangement; they are stated in Whitehead's description, and I have collected them in my own presentation; they include statements which make possible a distinction between "past," "future," and "co-existent world," but (for instance) there is not a definite "light velocity," which makes a sharp distinction of such nature as is assumed in the theory of relativity.

I mention this point, because it is my feeling that in his discussion Dr. Shimony sometimes has introduced a concreteness which should not be brought into the fundamental assumptions.

There is a further difficulty: Whitehead's statements are not always easy to correlate and Whitehead made changes when he wrote a new book. Moreover, Whitehead gave illustrations of his ideas which may have been useful at the time he wrote them, but which cannot be retained in the view of present knowledge. Whitehead was not a physicist; some of his observations on vibrations and certain of his ideas on quanta cannot be considered as pertinent. I think that nowadays we should better leave them aside. I believe that Whitehead himself would do so, if we still could talk with him.

I will now attempt to follow Dr. Shimony's discussion in more detail.

In connection with Dr. Shimony's statements (1)–(6), section II, I would observe that the sentence (2) "Every actual occasion is distinguishable from every other in virtue of its intrinsic character" must be read: "Every actual occasion is a unique occurrence as regards its intrinsic, subjective development." This does not necessarily postulate that the outcome of each such actual occasion, which is the only result recognizable for an outside observer, is always distinguishable from the outcome of every other occasion.—Further: I do not think that the assumptions concerning "indefinite divisibility" and "derivative reality" in (5) are pertinent.

While I am, of course, fully convinced that atomicity and "quanticity" are results of the basic doctrine which describes the universe as a complex of processes, we must be careful when we speak of "particles," and we must not try to link them too closely to Whitehead's scheme. Whitehead's scheme makes it possible to consider "matter" as the historic route of complexes or of chains of processes, which for long stretches of time present a common characteristic. But the word "particle," as used in physics, is no more than a convenient abstract term, applied in the formulation of relations which ultimately refer to the outcome of certain experiments. We must not consider "particles" as existing real things; the term is only a guideword used to label certain observable manifestations of complexes of processes. The result of an actual occasion, occurring as the outcome of a physical experiment, can give us reason to frame a description in which we say "a particle was here," but another experiment carried out on the same complex of processes might give a result which better can be described by speaking of "waves."—The discussion of "vibratory organic deformation" of a proton, which occurred in Whitehead's "Science and the Modern World," should be discarded.

Elementary particles can originate and disappear, according to Whitehead's scheme, which considers them as manifestations of complexes or series of processes: it is possible that such a complex can develop in another direction. But Whitehead himself already observed that certain complexes of processes (certain "societies" of processes, to use the technical term introduced by Whitehead) can persist over enormous stretches of time: the societies featuring "atoms" can persist over several billions of years.

As to Whitehead's statement concerning the association of some sort of vibratory motion with all elementary particles, I again would advise caution. That "rhythmicity" is a primary feature, yes—but I am not inclined to consider all vibrations which have been introduced into the description of physics, as real effects of chains of processes, in the sense that each single oscillation would represent a process. The stress should be upon the words "some sort"—but beyond that one should be careful.

That Whitehead's philosophy is constructed so that there is no strict determinism, is true. We may say that this was the purpose of his scheme. But as was noted before, in Whitehead's picture the absence of determinism (meaning: that events are not completely determined by what has gone before) does not lead to mere statistical randomness, or to the statistical form of causality which is accepted in atomic physics: there is room for conceptual activity.

From that which we experience around us, we must conclude that on the whole there are two types of societies of processes: one type in which there is no evidence for sustained and coordinated conceptual activity, so that statistical rules are found to prevail; and another type where we find evidence for aim and purpose, and ultimately for consciousness. The first type is basic for the behavior of non-living matter; the second type carries the living world. To elucidate the origin of the dichotomy, and to understand more clearly what it means, forms an important problem for those who wish to work out Whitehead's philosophy. I presume that a feature in the origin of this dichotomy has been the occurrence of cyclic sets of processes, which gave an opportunity for collecting subjective experience in a form of memory.

As to Dr. Shimony's thesis (v), section II, "The specification of the state of a composite system containing several elementary particles is equivalent to the specification of the states of the individual particles," I do not see that this is a necessary consequence of Whitehead's philosophy. Whitehead's scheme refers to the relations between processes and complexes of processes; as mentioned above, the notion of a "particle" is a derivative notion, and the state of a particle, in so far as we can know anything about it through the outcome of an experiment (that is, from the outcome of a process which has subsided into a fact), cannot be identified with an actual occasion, which has a subjective development of which an outside observer is not aware. I believe that Dr. Shimony here introduces an assumption which is not warranted, and which would lead us to unrealistic conclusions.

That Whitehead's scheme introduces a fundamental asymmetry between past and future, is evident and is important. That nevertheless many phenomena in physics seem to run so as if there was symmetry, is a derivative effect, holding only within a limited domain. I contend that quantum physics embodies an essential asymmetry in its statistical rules.—The phenomena of life nowhere suggest symmetry between past and future.

I am also convinced that the properties of particles, for instance those of atoms, are different when they are embedded in a living body. That our experimental data so far do not show this difference clearly, does not constitute a proof, for we cannot observe the behavior of an atom closely enough, and if we could, we would kill the life. But general observation makes us aware of a wide difference in behavior between living matter and dead matter, and somehow this must be connected with different behavior of atoms. In my interpretation the difference of behavior is connected with the problem of the equality or inequality of the weights of configurations, such as are considered in the statistical interpretation of entropy.

Coming now to Dr. Shimony's "Comparisons with Current Physics" (section III), I believe that his sub (i) suffers from misplaced concreteness. I do not believe that Whitehead attributes a definite and finite stretch of time to an actual occasion, and if per chance he should have suggested this somewhere, we do well to discard the idea. Nor is there in Whitehead's books a statement according to which an actual occasion, when it has completed its process of becoming, is completely definite with respect to every family of attributes. The statement referred to in "Process and Reality" (Category of Explanation xxv) does not say this in so many words; I think that Dr. Shimony's interpretation puts too much into it. From other places in "Process and Reality" it clearly comes forward that the fact which is the outcome of an actual occasion or a process, does not give expression to many features of the experience. As the outside observer (and also later processes which do not immediately follow upon the process under consideration) can only have experience from the fact, a great deal of information concerning what has gone on subjectively in the process, is lost.

I accept what is said in (ii), but (iii) again brings ideas which we should better leave aside. In particular, we must not attribute reality to "actual occasions connected with a wave front."

All assertions of quantum theory are predictions concerning the outcome of experiments. The elaborate mathematical scheme, powerful as it is, must not be interpreted as meaning that definite states exist as objectively recognizable entities.

The superposition principle of quantum theory is one of its powerful tools. Since it seems to be so fundamental in physics, it might be expected that it should be related to basic aspects of the philosophical picture. Still, this principle, as is also the case with the applications of the wave function, has a much more definite character than is present in the fundamental postulates of Whitehead's scheme. We must therefore presume that these definite, quantitative relations are the result of a particular development. This is a point that requires further investigation.

I do not believe that "the postulation of indefinite values of observable quantities, as required by the quantum mechanical account, would be repugnant to Whitehead" [Shimony, sub (iii)]. As I mentioned earlier, I believe that it is a misconception to interpret the twenty-fifth category of explanation in the sense that Whitehead would require complete definiteness. To say the least, the term "complete" seems ambiguous in Whitehead's text here, and it cannot be applied to the fact in which a process results.—As a consequence I do not think the discussion in the rest of (iii) to be relevant.

As regards (iv), I presume that the philosophical scheme is compatible with the existence of statistical rules in the whole domain of phenomena which we classify as "non-living"—we use this qualification, because we do not see anything else than statistical rules. The precision of the observed statistical rules is the result of a long development. It must also be this development which has ultimately led to a definite numerical value for Planck's constant. We may not exclude that this "constant" could still change, and perhaps might slowly decrease. I would not even exclude the possibility that what we call "constants of nature" have their origin in fluctuating relations, and may perhaps still be somewhat fluctuating.

In connection with (v), I do think it possible that an actual occasion or process can refer to the state of a complicated extended system, which upon the observer makes the impression of a system consisting of disconnected particles. The ultimate relations in the complex of processes are more basic than the present geometry of space-time, and connections between regions of space-time which seem far apart to us, may be possible in the basic scheme. Therefore I do not believe that the suggestion that an actual occasion can extend over non-contiguous parts, as judged in our space-time, should be repugnant to Whitehead. Perhaps it is too difficult to decide what Whitehead would have thought; let me say that I consider it possible to find an interpretation of the fundamental scheme in which such forms of extension can be included. I mentioned earlier that the definite value of the light velocity, which at present sets a clear distinction between past, future, and co-existent world in the domain of physical happenings, may be a derivate from a more vague relationship, in which contacts extending over space and time are possible, which to us may suggest the transmission of influences with speeds exceeding the speed of light.

Coming now to section IV, "Proposals for Reconstruction," of Dr. Shimony's paper, I fully agree that it will be rewarding to look for an elaboration of Whitehead's philosophical scheme. However, I cannot accept the argument that "the discrepancies noted in Section III between Whiteheadian physics and current microphysics constitute strong disconfirmation of Whitehead's philosophy as a whole." I deny that we can speak of Whiteheadian physics. We can say that there are marked discrepancies between micro-physics and some of the ideas with which Whitehead illustrated various possibilities of his scheme, but these illustrations do not involve the fundamental character of the philosophy and can (and must) be discarded. Their rejection does not disrupt the picture in itself.

To follow Dr. Shimony's argumentation in these pages, some repetition may be desirable of what I have said before. I do not believe that Whitehead made an assumption concerning the complete definiteness of an occasion in its final phases. In each process there is definiteness in the results of its subjective interior development with regard to the perspective under which the process accepted its experience; however, in the outcome of the process there is no completeness with regard to all topics which an outsider might bring forward. Whitehead's picture of the subjective development of a process can be best compared with that which happens in the mind (or in the soul) of an artist: when a composer makes a sonata, his work has become definite as a piece of music when he has arrived at his final decision, but it does not tell us anything about physics or about history. To maintain that an actual occasion in its final phase should be definite with respect to every family of attributes, in my opinion constitutes an unwarranted conclusion.—Neither do I think that we should assume that the characteristics of a "nexus" are entirely determined by the characteristics of its constituent occasions. A nexus develops by the development of new processes; so long as these have not subsided into facts, they are not determined.

The idea that an occasion can have a sharp location in time and an arbitrarily short duration, provided that properties in a sense "complementary" to duration were sufficiently indefinite, is acceptable. Again one must keep in mind that the statement refers to the fact which is the result of the occasion, and which is the only experience which an outside observer can have of it: the subjective, interior development of the occasion or process is a phenomenon which cannot be described in physical time. A description in physical time requires that some process is brought to conclusion, which combines experiences derived from the process under discussion (or derived from a chain of processes) with experience derived from a mechanical system such as we use for "time-keeping"; every statement based upon an observation can give only completed, exterior data.

The most important part of Dr. Shimony's considerations are contained in the question whether the superposition principle of quantum theory can be grafted on the philosophy of organism. It is difficult at this moment to say in how far this may be possible. It is certainly compatible with Whitehead's scheme to assume that the state of an individual occasion may be such that its location in space is indefinite. A composite system can be featured by a society of processes which are not each in a definite state; nevertheless, there can arise a process which gives a definite interior development to the system, ultimately leading to a definite state with regard to the perspective relevant for its subjective form. But when we ask about spatial location or separateness, these terms refer to the observations of an outside observer: it is the fact, or the facts, which he sees that may have locations in space and may be separated. His results cannot be construed in the sense that the actual occasion should exist in separate parts of physical space-time; the actual occasion develops within spatio-temporal relations that cannot be mapped unambiguously onto physical space and time.—I would warn, by the way, against the idea that we should consider actual occasions as "microscopic" in size, and that Whitehead believed this to be relevant for the explanation of life, as Dr. Shimony seems to suggest.

I doubt whether the combinatorial principles of quantum theory, as these are exhibited in the mathematical formulation of the superposition principle, can be connected with the fundamental ideas of Whitehead's philosophy. My feeling is that a chain of intermediate assumptions is needed, before this point can be settled.

Characteristic for physics and quantum theory is the division of the observables into two groups, say coordinates, and canonically conjugate momenta. The formulation of the principle of complementarity and of the uncertainty principle are closely bound up with this duality; it is at the basis of the scheme of calculations connected with the Schrödinger equation and with the application of a Hilbert space for the representation of physical relationship. Whitehead's scheme in its present form is too wide for this, and does not have the dualistic form of complementarity.

Perhaps it will become possible to introduce supplementary assumptions which will make understandable why in those societies of processes in which initiative and free play of conceptual activity has been lost, this duality has arisen as a necessity. When this point could be cleared, we will be in a better position to discuss the superposition principle. This is one of the urgent problems connected with Whitehead's philosophy; it may well be a test case. (The apparent absence of clear psychological manifestations of the superposition principle, which Dr. Shimony mentions, might be in agreement with my idea.)

Whitehead, in some final chapters of "Process and Reality," seems to have been struggling for a more elaborate theory of relationship in space and time (see in Part IV the chapters on "Extensive Connection," "Flat Loci," "Strains," and "Measurement"). I have never been greatly impressed by the discussion in these chapters, and it does not seem to me that they are organically well connected with the basic ideas of the philosophical scheme. I find it very interesting that Dr. Shimony does not refer to these chapters in his paper; apparently he did not see much help in them either. It looks as if we both have the feeling that a far reaching reconstruction is needed here.

I would believe that for an elaboration of the fundamental scheme we will need more insight into relations between what we call mental processes, or states of mind—in particular what relations they may have in space and time, and whether they may be able to bridge distances, which to us look as gaps. The set of forms of relationship which we use in our investigations, is too heavily leaning upon the language and the methods of physical theory, and may be insufficient. The heavy stress upon physical relationship, which often is presented as the uniquely valid relationship in nature, may have narrowed our faculties for penetration beyond it. Let us not forget that physical relationship is a set of rules for calculation, for prediction—but never does it tell us what is, or what acts. We are often taught that one should not ask for anything beyond its limits.

With all the great power of physical knowledge, difficulties of interpretation are continually coming up. It may be that in order to have a better understanding of the meaning of observation and measurement, we must develop deeper insights into metaphysical forms of relationship, and a different attitude with regard to the wonders of the universe may be a prerequisite for this. The importance of Whitehead's philosophy is that it is an attempt to make an opening beyond present limits and to widen our outlook. Much meditation will be needed before we shall be able to grasp all that which may become visible.

I hope that with these considerations I may have given a helpful reply to Dr. Shimony's observations on Whitehead's philosophy.

Institute for Fluid Dynamics and Applied Mathematics
University of Maryland
College Park, Maryland

11

NORMAN RUDICH

THE DIALECTICS OF POESIS: LITERATURE AS A MODE OF COGNITION

Presented March 19, 1964

Tragedy, by means of legends and emotions, creates a deception in which the deceiver is more honest than the non-deceiver; and the deceived is wiser than the non-deceived.

GORGIAS

Admiration and love, to which all knowledge truly vital must tend, are felt by men of real genius in proportion as their discoveries in natural philosophies are enlarged; and the beauty of form in plant or animal is not made less but more apparent as a whole by more accurate insight into its constituent properties and powers.

WORDSWORTH

For constantly I felt I was moving among two groups—comparable in intelligence, identical in race, not grossly different in social origin, earning about the same incomes, who had almost ceased to communicate at all.

C. P. SNOW, *The Two Cultures and The Scientific Revolution*

The moral aim of this essay is to contribute positively to the hopeful dialogue now in progress between the two cultures, the scientific and the literary.

The mutual incomprehension of scientists and humanists is morally and socially pernicious. No matter what reservations may be justifiably held concerning the provocative essay of C. P. Snow, it has obliged scientists and humanists to re-examine in the vital context of political and social reality the purposes and the import of intellectual endeavor. Although the issues discussed in this paper are philosophical, and do not correspond directly to the actualities treated by Snow, I hope that its political implication will be clear.

His criticism of the snobbery, self-centeredness, irresponsible pessimism and political indifference of many modern authors and critics must be taken seriously by everyone concerned with the healthy development of our cultural life.[1] And Snow is definitely so concerned. It is time to recognize that there is truth enough in his charges to warrant a serious reappraisal of the relations between literature and science in the realistic and humanitarian context on which he insists. The scene of his dialogue is a world dangerously divided against itself, in which hundreds of millions of suffering human beings are beginning to understand the tremendous promise for economic and cultural betterment offered by the "scientific revolution". They are also realizing that science and technology do not develop in a vacuum, but interact with all the other branches of human activity. Revolution in science, society, education, culture, and religion defines our age; universal change has never been so palpable a fact of life as it is now.

How about art? How about literature? Just as no age can match ours in rate and violence of change, neither has any witnessed so dizzying a multiplicity of revolutions in the practice and theory of the arts. In the nineteenth century, when an artist was identified as believing in or practicing, say, naturalism, it was still possible to make a reasonable guess at some of the features of his works; the notion of an artistic school, although never a very precise concept, still had useful meaning. Today the "isms" of art blur, fuse, disintegrate, and reform faster than their manifestoes can be printed. In the last analysis, each artist is a school unto himself. In no other domain do we see so purely reflected the cataclysmic individuation of modern man as in art. The colors, sounds, and words of our century have been shaped by artists to express alienation, estrangement between man and his world, between man and man. Some are resigned as to a perennial condition which the life of the times has brought to the surface. Some welcome the crisis as a final confrontation of man and his potentialities in which the only free choice possible is self-realization. But no matter what the despair or what the hope the diagnosis is quite generally the same. Human psyche, ideology, personal and social institutions, moral development have been jeopardized by the unparalleled pressures created by the industrial, economic, and political revolutions of the past two centuries.

The arts were not alone in making this diagnosis. Psychoanalysis, philosophy, history, etc., have participated in depicting the phenomenon and exploring its causes. But so decisively has the artistic, especially the literary, mode of consciousness impregnated these disciplines that some leading modern thinkers—Nietzsche, Jung, Spengler, Sartre, Camus, Marcuse, to name a few—have turned to myth, the novel, theatre, etc., in order to give adequate expression to their philosophical insights.[2] Literature is still perhaps the foremost of the arts in shaping awareness, purpose, and moral conviction. Contrary to all the prognosticators of doom for humanistic studies "by-passed" by the scientific revolution, I believe that we are in the process of discovering new values in the humanistic tradition. My reasons for so believing will appear in the course of this essay. Let it be clear from the outset, however, that the relation of literature to science is not an academic question. The universal reverberations produced by Snow's essay are proof of that.

The present essay explores the epistemological status of literature. It is my belief that the poet (any writer of fiction or imaginative literature in verse or prose) is in some specific, definable, unmystical, rational sense engaged in a search for truths about life. In this he resembles the scientist, and it is this resemblance which makes me believe that the polemics which artists and scientists have so frequently directed against one another are based on philosophical and social misunderstandings which it is in the interest of both parties to clear up. I am here suggesting a line of thought which may be helpful in achieving that goal.

Philosophically, I wish to present in outline, though with some elaboration, my reasons for believing that poetry has, among other virtues, genuinely cognitive content and that criteria of truth and falsehood are pertinent in understanding and judging it. *Poesis* ("making something out of words" in Greek) here refers to the creative process which generates poetic truths and falsehoods, and to the ways in which this process is reflected linguistically in the poem (Greek *poema:* "the word-thing achieved"). In poetry creation entails two notions apparently irreconcilable with the accepted usage of "cognitive." The fact that poetic language eminently reflects the psychological and ideological sources which inform its structure is one of the major differences between literary and scientific languages. The poet's language expresses his personality, his attitudes, wishes, moods. His utterance is thereby deflected from propositional objectivity by subjective content and by emotional interferences. In contrast to these qualities not much prized in our metallic civilization, the language of science displays the virtues of semantic purity, referential precision and perfect communicability without which the world as we know it would cease to function. The nondescript quarrel between the hard-headed scientists and analytic philosophers who look upon poetry as a kind of epistemological self-indulgence to be more or less tolerated, and the poets, seconded by their would-be supporters in literary quarterlies and English departments, who depict science as the arch-enemy of the human spirit, provide excellent material for a satire which doubtless will be written someday soon and to the confusion of both parties. That is not the present task. The body of this essay is an attempt to reply seriously to the scientists and philosophers concerned with the semantics of art. The present point is to make clear that a defense of poetry does not necessarily imply a contempt for science. Thus from the outset I wish to discard the myth of the unfeeling, unintuitive, unimaginative scientist, the monstrous intellectual machine impassively presiding over our collective dehumanization. The various modern schools of irrationalism, and their literary hangers-on who have promulgated this image of the scientist in books and journals, are confusing the *language* of science, its effort to state exactly and efficiently a given state of affairs in nature, with the *total activity* of science in which imagination, intuition, feeling and hunch are as much at play as they are in artistic creation. The impersonal and objective languages of science are among the greatest achievements of the human spirit. To attack scientific language as abstract or mechanical is to misunderstand the historical process by which the various sciences elaborated languages most suited to express their contents. It is to attack the entire scientific enterprise on the incorrect tacit assumption that poetic style is a universal model of excellence. Yet if fitness to its object, accuracy, and efficiency of statement are desirable, indeed indispensable aesthetic norms, then few poems can rival the perfect elegance of Newton's law of universal gravitation. Nor is scientific language arbitrary:

> "The mathematical style of science is not to be explained as mere human contrivance or convention. Nature has style as well and the major developments of mathematics are major because they imitate a pervasive style of nature."[3]

Writing as a literary scholar, I shall depend on philosophic and scientific authorities for the descriptive analysis of scientific language, insofar as such discussion may throw light upon the cognitive value of literature. My point of departure is the general comparison of the main properties of scientific and literary language. The distinction between them, by now a commonplace, leads to the question of the reasons for two such divergent languages, and to the major hypothesis of this essay: that they are necessary because their "object-matters" differ. (I prefer this term to "subject-matter" because I wish to emphasize that the root difference is an objective one, that it can only be explained with reference to the kinds of realities each chooses as objects of study.) Now both the linguistic distinction and speculative inquiry into its causes are to be found in Aristotle's *Poetics,* still in my opinion the single most profound treatise on the arts. Through analysis of key passages in the *Poetics,* some major consequences of the Aristotelian theory of literature and of its further development by Kenneth Burke will be examined. The next and major exhibit will be a minute analysis of the first quatrain of Gray's *"Elegy in a Country Churchyard"* in which the theory will be illustrated by concrete application to a specimen of poetic language. The conclusion will expose and examine several outstanding propositions arising out of that analysis.

The breadth of my subject naturally excludes exhaustive treatment of each topic. I am counting on the *explication de texte* to answer some of the questions raised in the theoretical sections. As for the others, I shall be satisfied to have contributed to their clarification.

I

> In Shakespeare's day, there was little distinction between styles, often none between prose and poetry, certainly only the rudiments of any self-conscious separation between the language of the poet and that of the scientist. Kepler, as we shall see, often spoke like Dante on the one hand, Marlowe on the other. The language of poetry and of science was one when the world was one.
>
> Marjorie Hope Nicolson,
> *The Breaking of the Circle*
> (Evanston, 1960, p. 106)

> In the language of everyday life it very often happens that the same word signifies in two different ways—and therefore belongs to two different symbols—or that two words, which signify in different ways, are apparently applied in the same way in the proposition. (3.323)
>
> In order to avoid these errors, we must employ a symbolism which excludes them, by not applying the same sign in different symbols and by not applying signs in the same way which signify in different ways. A symbolism, that is to say, which obeys the rules of *logical* grammar—or logical syntax. (3.325)
>
> Ludwig Wittgenstein,
> *Tractatus Logico-Philosophicus*

> Many linguistic utterances are analogous to laughing in that they have only an expressive function, no representative function. Examples of this are cries like 'oh, oh', or, on a higher level, lyrical verses. The aim of a lyrical poem, in which occur the words 'sunshine' and 'clouds' is not to inform us of certain meteorological facts but to express certain feelings of the poet and to excite similar feelings in us.
>
> RUDOLF CARNAP,
> *Philosophy and Logical Syntax*
> (London, 1935, p. 28)

"The language of poetry and of science was one when the world was one." But as the sciences developed, ramified, specialized in modern times, they discovered the necessity of inventing precise, technical languages, because everyday language was inadequate to give a scientific account of nature's properties and operations. This broke the old linguistic fraternity of poetry and science, for poetry, dealing as it does with matters of everyday life, had no reason to abandon ordinary language. T. S. Eliot has written, "The poetry of a people takes its life from the people's speech and in turn gives life to it; and represents its highest point of conciousness, its greatest power and its most delicate sensibility."[4]

While the languages of science were evolving toward a formal (and usually mathematical) ideal, poetry, especially with the romantic revolution, was casting aside aristocratic conventions of form which had arbitrarily limited its scope and depth. The entire range of actual and possible human experience from the most routine to the most esoteric and eccentric is laid open to poetic exploration. Concomitantly the poet appropriates to his use all the expressive potentialities of his national language. The modern individualism of which he has become the strident paradigm is reflected in the multiplicity of styles, in the violent liberties taken with grammar, syntax, punctuation, etc., in the obscurity and difficulty of many works which have contributed to the growing isolation of the literary elite, not only from society in general, but from intellectuals in all fields. The point is, however, that Joyce's

Finnegan's Wake, the poems built on nonsense syllables by Lewis Carroll[5] and the enigmatic prose of Gertrude Stein are written in English, that they initially depend for their meanings as well as for their effects on our knowledge and everyday use of the English language. Poetry is a special case of ordinary language, but it is not a technical language like that of the chemist. Poetry is an inventive re-shaping of everyday speech, a mobilization of its diffuse material and ideational substances for the purpose of displaying the movement and structure of significant experience.

The 19th century saw an acute crisis of the literary consciousness in the face of scientific and industrial revolutions. Many in France, England, and Russia felt called upon to define literature and art in general in relation to science. The age-old assumption that art was engaged in a quest for truth, just as surely as "the natural philosophies", was being called into question by scientists themselves, and by philosophers, e.g., Claude Bernard and Auguste Comte. The French realists felt the challenge most acutely. Balzac thought of himself as a modern historian applying to the problems of literature techniques developed by Geoffroy Saint-Hilaire. Flaubert in his correspondence frequently reflects on the problem of objectivity and sees the impersonality of the scientist as a model of rigor which the artist should strive to attain. Zola climaxes this trend of thought in his *Roman experimental,* a defense of naturalism which makes of the artist a kind of universal social scientist, competing with psychology, sociology, economics, etc. The artist writing a novel based on exact observation, on the careful accumulation of evidence, on unprejudiced interpretation, is analogous, and then identical with, the chemist or physiologist working in his laboratory.

But it is only in the twentieth century, with the development of analytic philosophy and its probing of the structure and nature of meaning, that the problem is placed in a linguistic perspective. The contrast between the logical syntax of scientific discourse and the emotional, metaphoric, and ambiguous statements of poetry is by now an intellectual commonplace. Consider the following.[6]

	Poetic Language	*Scientific Language*
Susanne K. Langer	presentational symbols	discursive symbols
Charles W. Morris	iconic signs	statemental signs

Ernst Cassirer	mythic thought	discursive thought
Morris R. Cohen	metaphorical hypothesis	scientific confirmation
Philip Wheelwright	expressive language	steno-language
I. A. Richards	pseudo-statement	statement
F. S. C. Northrop	concept by intuition	concept by postulation
A. J. Ayer	emotive	scientific

Not all these writers follow Carnap and Wittgenstein in asserting that the language of science is alone cognitive and thereby consigning all other forms of discourse from everyday language to pure metaphysics to the realms of wish-fulfillment, moral exhortation, psycho-biological gratification, etc., poetry being merely the purest case, the furthest removed from the empirical, rational ways of cognition. Langer and Wheelwright in particular have undertaken the case for the defense of poetry as meaningful discourse. All, however, agree that the linguistic cleavage is a fact of prime significance.

In *The Burning Fountain,*[7] Philip Wheelwright has worked out in a systematic way the contrast between the steno-language of plain sense which in its purest form becomes scientific statement and expressive language, a blend of emotive and referential elements proper to poetry. Some of his points are highly pertinent to the present argument.[8]

Steno-language is characterized by:

1. semantic discreteness: a symbol is always distinct from its referent.
2. univocation: a symbol has one meaning, refers only to one thing or one set of things.
3. definiteness: non-vagueness.
4. semantic invariance: a given symbol must keep the same meaning throughout the course of a given argument.
5. bidimensional significance: the symbol has either logical universality and extensional particularity, or both, i.e., common names defining general classes (horse) and/or particularized and proper names, distinguishing a member from its class (this horse, Man-o-War).

6. contradictories: a proposition and its contradiction cannot both be true.

Expressive language, by contrast, has these properties:

1. iconic signification: symbols with self-intentive, as well as external reference; they mean by resembling.
2. plurisignation: symbols which have more than one legitimate reference.
3. soft focus: symbols whose meaning depends on connotation as much as denotation: thus subjective, vague, sometimes obscure.
4. contextualism: the shifting of symbol-meaning on different levels, within a given context.
5. paralogical dimensionality: symbol having other semantic dimensions than logical universality and extensional particularity.
6. paradox: contradictory assertions can co-exist meaningfully.

Professor Wheelright, throughout his discussion of these properties of steno-language and expressive language, and throughout his book, cogently points out the arbitrariness and narrowness of the positivists' restriction of meaning and cognition to the canons of scientific discourse. Reality has breadths and depths which the sciences alone do not and cannot explore. The psychological processes involved in knowing (and our diverse manners of translating them into adequate symbolic systems) cannot be understood from a positivist point of view. Such positivism banishes all rational content from poetry on the grounds that it also expresses emotion, makes value judgments and vague philosophical claims. To all schools and schoolmen engaged in this type of analysis, Professor Wheelwright is saying—there are more things in heaven and earth than are dreamed of in your philosophy. In this respect my agreement with him is complete.

What is needed, however, is a more precise explanation than is furnished in *The Burning Fountain,* or in any single work I have discovered to date for the breakdown of the unity of language, and a careful consideration of its philosophic significance. The nineteenth century practitioners of the realistic novel mentioned above were in my opinion on the right track when they posed the question of artistic truth in terms of science. Unlike many contemporary artists, critics and literary scholars, they harbored no hostile feelings toward scientists; on the contrary, they saw them as co-workers in the search for truth and exulted in their discoveries. They were conversant enough with the sciences of their day to apply them in their novels. In the *Rougon-Macquart* a vast series of novels depicting French life in the second half of the nineteenth century, Zola consciously based his characterization on the laws of heredity and environment as he understood them, an approach common to all his followers in the naturalistic school.

But there's the rub. The underlying assumption which I think led them all into error and which made it impossible for them adequately to think out the relation of art to science was the assumption that the object-matter of their research was identical with that of the natural and social sciences. They, and many after them, have been satisfied with the idea that both art and science explore the same "reality". The naturalistic novel teems with examples of physiological, psychological and social determinism; it becomes an imaginative and crude illustration of general laws or principles enunciated by science at a given level of its development. The artistic quest for truth is thereby reduced to that of popularizing scientific discoveries. It would follow from this that as soon as the sciences have once more progressed, and reformulated the laws of physiology, psychology, etc., the old novels would cease to be of any value, except perhaps as mementos of interest to the historians of science. Clearly, this has not occurred, and for good reason, as I will try to show later.

Furthest removed from the naturalist view of literature as illustration of scientific truths is the school of "art for art's sake" as it was called a century ago, or the New Critics, as we say today. John Crowe Ransom, R. P. Blackmur, Robert Penn Warren and Cleanth Brooks have proclaimed the autonomy of literature from history, sociology, psychology, politics, etc. They accuse the literary historians, whose methods derive from the social sciences, of undermining the autonomy of art, or reducing it to a source of information for social and intellectual historians. At their worst, it is true, these methods reduce the work of art to piecemeal reportage. The New Critics identify the outlook of this scholarship with the deterministic naturalism of Taine.

However, whether or not they realize it, their rejection of naturalism leads them to a formalism incompatible with a serious theory of art as a mode of cognition. They are driven to grave ambiguities regarding poetry's referential power. Since a poem is a world in itself, thing in itself, artifact, since in Wheelwright's words, its language has "self-intentive reference" and its meaning is "radically implicit", its apparent capacity to refer to dimensions of reality outside of itself is hard to explain. Many of these critics have philosophized about poetry, science, and knowledge, but I believe that their real and unique contribution to the subject is their persistent and sensitive focusing of attention on qualities of language exclusively poetic. Their awareness that part of the problem of cognition is situated in language provides us with a perspective which the nineteenth century lacked.

Neither naturalism nor formalism, despite their contributions, can be of moment in the present study. Zola's theory of the *Roman experimental* achieves a semblance of consistency by sacrificing the subjective component in a work of art; for in order to reproduce with fidelity the laws of nature and society which the artist, like the scientist, explores with experimental rigor, the novel must be liberated from the arbitrary influences of individual prejudice, feeling, point of view. Zola's way of pressing art's claim to be a mode of seeking truth is to eliminate the artist from his work. On the other hand, the defenders of *l'Art pour l'Art,* in the main, abandon the search for truth-content in favor of the cultivation and analysis of refined techniques. They justify themselves in terms of a purely aesthetic experience, postulated as an ethical value, a metaphysical principle or as a separate psychological category. In their view, science is utility. They try to save literature by exalting its pristine uselessness; they eliminate the outer world from the work of art.

These contrasts between scientific and poetic language and between the naturalistic and formalistic approaches to the relationship of science and literature make it possible to situate the point of view of this essay. I shall do so in four points:

1. The subjective component in expressive language is a fact which any cognitive theory of art must account for. The poet, be he Zola or Valéry, injects willy-nilly his manner of perceiving, feeling, or judging whatever he describes into his language. To remove his personality from his style is to change this meaning completely, to transform expressive into steno-language.

2. It does not follow therefrom that the statements made in expressive language are false or meaningless. The particular bias of a style may, on the contrary, turn out to be the best one for revealing the most important content of a situation. Expressive language reveals a genuine objective content of which the subject is a necessary constituent. It is not the question of subjective content but the problem of verification of poetic statements which presents the greatest hurdle to theory.

3. Thus poetry must have both subjective and objective content. This point of view joins a long tradition (Aristotle, Diderot, Kenneth Burke) which believes that poetry, and art in general, can only describe a reality in which subject and object are seen as one.

4. The same tradition holds the belief that the arts are the various languages we use to explore and describe such a reality.

This essay deals only with one of the arts—literature—whose language is words. Its major illustration is drawn from verse for the sake of clarity, but I think that the general principles are also valid for literary prose. At a number of points they appear to be valid for one or more of the other arts, but no attempt can here be made to justify that opinion.

II

> But is it true that men acting, and only they, pretty much always develop well-defined characters? Yes, the fundamental principle of Aristotle's theory of character-development is that *we become what we do,* that our acts harden into character.
>
> GERALD F. ELSE, *Aristotle's Poetics: The Argument*[9]

The major hypothesis of this essay is that the steno-language of science and the expressive language of poetry refer to two different aspects of reality.

Steno-language deals with reality as it exists or is supposed to exist in itself. It is necessary, but not sufficient, to say that this reality, which I shall call nature, is independent of the observer's consciousness, and that the language which describes it aims to neutralize or eliminate the presence of the observer. The exclusion from steno-language of subjectivity—the variations in perceptual acuity, span of attention, personal biases of different observers—goes hand in hand with a progressive delimitation of the aspects of nature it can report upon. Each of the sciences necessarily restricts its field to specific properties of nature, and eliminates from its language all notations for the characteristics of objects which are not needed to explain the kinds of behavior with which it is concerned.

The object of steno-language is nature, the properties of nature, and the laws of motion and change which explain its operations. The names of the various sciences specify which properties each investigates, e.g., physics, biology, chemistry, psychology. As the sciences progress regularities are discovered in the interrelations of these properties, and new sciences emerge whose names define the kind of laws to be found—physical chemistry, biophysics, psychosomatic medicine. In all cases the object-matter is isolable properties of nature which a steno-language aims to describe objectively, quantitatively and universally.

What David Hawkins says about mathematics as a mode of imitation of the style of nature holds true, in differing degrees, for scientific language in general, for it is by and large agreed that all the sciences see mathematics as their ideal language. The characteristics of steno-language analyzed by Professor Wheelwright are, then, dictated by the nature of nature—are necessary to communicate truths about the laws that govern its operations.

That does not mean that the sciences attend to nature to the exclusion of man. The social sciences, sometimes called human sciences, take man as their specific object of study, and these sciences too are named according to the particular properties of the object to be investigated: psychology, sociology, political sciences, economics, history. Insofar as one investigates man scientifically as a natural phenomenon, he is an ensemble of distinct properties like any other natural object. As such he is indistinguishable from the rest of nature except in the sense that he may exhibit a number of properties, e.g. speech and changing culture, which cannot be found elsewhere in nature. The unit of study is always an objective and isolable property or set of properties of nature, in this case human nature. The objective of the social sciences is to state the general laws governing these properties. Therefore, their languages become more technical and abstract with time and their linguistic ideal more mathematical.

The object-matter of literature is *praxis,* Greek for "action", the imitation of which, according to Aristotle, is the procedure common to all the arts. I prefer the Greek term because "action" too strongly connotes man's initiating or innovating activity, his power to bring about controlled changes in his environment. Praxis is here meant to refer to the total process of human interaction with his personal, social, and natural environment. Thus praxis means the resistance man encounters in the world as well as his action upon it. Indeed the comprehensiveness of the term is so great that Aristotle practically equates praxis with "life", a term which I could use were it not so heavily laden with emotional connotations and mystical accretions. Praxis refers to the dialectic of men with the world (society and nature), which is not the same as the relationship of subject and object in epistemology, for that denotes only the mental activities involved in cognition. Praxis refers to the entire process of man's transactions with the world among which the cognitive is one of the most important and highly prized. In praxis man is present with all his physical and spiritual faculties, the kinesthesia, sensations, mental images, perceptions and thoughts involved in his particular activity. For praxis must not be understood as a simple act performed by the body, as rudimentary muscular motion; that capacity is more highly developed in other animals. Praxis is a peculiarly human transaction in which individual personality achieves expression, reveals itself as a unity of mind and body.

> "O body swayed to music, O brightening glance,
> How can we know the dancer from the dance?"[10]

Just as the characteristic impersonality of steno-language reflects its objective and the purpose behind it, so do the properties of expressive language spring from the nature of the reality it must symbolize. Expressive language investigates praxis and can do so only by modifying the kind of objectivity indispensable to the sciences. The observer and interpreter of praxis, the poet, must be present in his language in order to describe praxis accurately.[11] Given the dialectical structure of praxis, his words must simultaneously designate objects, situations, and relationships and express possible human attitudes toward them—must refer to the continuum of things and to our experience of them. I believe that imitation is the only method for achieving that end. However, it is necessary first to explore briefly some of the salient features of praxis in order to distinguish it unmistakably from the domain of nature.

III

> Still, there is a difference, and a radical difference, between building a house and writing a poem about building a house —and a poem about having children by marriage is not the same thing as having children by marriage. There are practical acts, and there are symbolic acts (nor is the distinction, clear enough in its extremes, to be dropped simply because there is a borderline area wherein many practical acts take on a symbolic ingredient, as one may buy a certain commodity not merely to use it, but also because its possession testifies to his enrollment in a certain stratum of society).
>
> KENNETH BURKE, *The Philosophy of Literary Form*

Praxis refers to human presence in nature, a compound *unit* of man and world, which therefore cannot be reduced to simple properties of nature. Praxis is composed of human events and no matter how finely one subdivides any one of these events, it must always retain its two-fold essence in order to maintain its identity as praxis. Failure to do so will result in a change of object-matter. The separation of the unit, man/world, into its constituent parts will lead to the investigation of nature and its isolable properties, including those of man. The fact of human presence constitutes a qualitative change in the reality under investigation and requires a specialized cognitive activity to ascertain its properties. *Poesis* is that activity.[12]

Aristotle speaks in the *Poetics* about praxis mainly in connection with drama. That is to say he discusses predominantly imitated praxis, which has already been transformed by the poet into literature, rather than praxis as the raw-material, the object-matter of literature. Nevertheless there are a number of statements which clearly indicate that he understood the implications of that distinction. He is forced to use it in order to obviate what he apparently considers to be the most tempting and therefore the most dangerous confusion in which a philosopher dealing with the arts could become involved, i.e., the confusion of character and praxis. Thus he asserts:

> But most important of all is the structure of the incidents. For tragedy is an imitation not of men, but of an action and of life, and life consists in action, and its end is a mode of action, not of quality. Now character determines men's qualities, but it is by their action that they are happy or the reverse. Dramatic action therefore is not with a view to the representation of character: character comes in as subsidiary to the actions. Hence the incidents and the plot are the end of a tragedy and the end is the chief thing of all. *Poetics* VI (Butcher tr.)

If I read the implications of the passage correctly, Aristotle is differentiating between poetry on one hand, and psychology or moral philosophy on the other. The distinction is based simply on the different object-matter treated by each. The poet is interested in praxis, the psychologist and moral philosopher in "character" which is a "quality", i.e., a property of human nature, only one of the components of praxis.[13]

The key difference would therefore seem to be that praxis is a form of motion, a specifically human form of motion, whereas character is conceived as a static property. As we have seen, Gerald Else concludes in his exhaustive analysis of the *Poetics,* that Aristotle saw character as a result of praxis—"we become what we do". Praxis is the total dynamic process of human life, as it is lived by real individuals in their concrete day-to-day transactions with the world. The vocabulary, grammar and syntax used to describe praxis are those of ordinary language which itself is an aspect of praxis, for it is one of our most important means for dealing with life. Drama is an imitation "of persons acting and doing", says Aristotle. In drama, it is most frequently persons engaged in the act of speaking who are imitated. The imitation, of course, is not an echo copy, but a poetic version designed to suggest the original, and bring out its most important features. We can and do talk about praxis in ordinary language, but *poesis,* the creative remaking of praxis for our consciousness, is a mimetic transformation, a representation, not a copy. Sometimes ordinary speech becomes art, as in wit, or when we mime the mannerisms of a fop or pedant. Ordinarily, however, these isolated moments are drowned in the swift and complex currents of practical affairs.

Theater is the most direct, but not the only way of imitating action. In his *Poetry of Ezra Pound,* Mr. Hugh Kenner[14] cites in a footnote a different translation of the very text from the *Poetics* which I chose as my point of departure. He comments as follows:

> "When it is recalled that action (praxis) includes deeds, fortunes, and mental and emotional happenings, 'plot' will be seen to be applicable to a lyric as well as a tragedy. A sonnet, a *hokku,* depend on 'peripeteia' and 'discovery' as much as the 'Oedipus'."

In other words, praxis is intrinsically dramatic in its structure. Drama is a universal condition of human experience, the name for a kind of motion which comes into being only with the emergence of man. When we call natural events like storms or avalanches "dramatic", we are indulging in anthropomorphic projections, attributing praxis to nature. The expressive language of poetry imitates and reflects the varieties of dynamic structure we discover as actors in the drama which is the human condition.

That is, nature and praxis, motion and drama, are two modes of reality not reducible to one another.

Ontologically they are equivalent. Nature is neither more nor less real than praxis. Both are objective facts of the universe and open to scrutiny as objects of possible knowledge. These statements are not infirmed by the fact that the natural sciences have on hand a body of indisputable knowledge about nature, whereas the arts apparently cannot point to any such achievement. I am simply pointing to the objective existence of both nature and praxis. That does not imply anything about characteristics in either which facilitate or hinder cognition.

Again, no one up to now has actually questioned the history of literature for its specific cognitive content. Most histories of art tell us about the arts, but do not tell us very much about their relation to the reality they depict. Such analysis should theoretically be done by the aestheticians, but almost to a man they eschew it in favor of speculations and researches into the formal qualities of art works and the problems of perception. Until the emerging field of philosophy of art, and more particularly philosophy of literature, gains a foothold as an independent branch of aesthetics, the question of the cognitive content of the arts will not receive the systematic attention I think it deserves.[15]

The preceding analysis of praxis is the basis of Kenneth Burke's[16] "dramatistic" philosophy which constitutes, in my view, a major advance in our understanding of poetry and language in general. Burke breaks down action into five components deduced from Aristotle's exposition of the elements of tragedy. These are: agent, scene, purpose, agency and act. In other words, in any action, someone, somewhere, wishing to achieve a goal and having certain instruments (skill or tool) does something, moves, acts. Burke finds these categories present in all language, not only poetry. He sees reflected in the depersonalization of the purely semantic language of science an act of ascetic self-abnegation in which, so to speak, the scientist's heroic deed is made notable by the absence of any trace of personality. The silence of the scientist allows nature to speak through him, and in this sense, Burke's position joins that of David Hawkins for whom mathematics imitates nature.

I cannot here go into my disagreements with Kenneth Burke. Suffice it to say that he is more interested in the analysis of motive than in the problem of poetic cognition. Language is for him "symbolic action" in which men reveal their strategies for attaining their goals. He is more of a sociologist than a historian[17] and has recently shown more interest in psychoanalysis than in general social questions.

For my purposes, however, Burke's anatomizing of action clarifies the distinction between praxis and nature, between action and motion. To isolate the agent from the other four elements of the "pentad" locates the object-matter of the physiologist or psychologist or, in a sense, the philosophical anthropologist. To separate out the scene as the object of study is to delimit the areas of the natural and social sciences.[18] Burke, after Fenellosa, sees all language as a species of imitation and even the simplest sentence as containing a "plot". In my opinion he does not sufficiently distinguish poesis, conscious imitation of an action by means of words, from the kind of imitation which is an automatic result of utterance. I am particularly concerned here with poetic imitation of praxis because I believe that the consciousness involved in the creation of a poem is the key to understanding its cognitive content.

Just as Burke appropriates Aristotle's insights for very un-Aristotelian purposes, I now intend to apply the pentad to an examination of the properties of praxis in a manner most un-Burkeian. Burke's approach is insufficiently dialectical probably because his concept of action is largely dominated by verbal action, i.e., communication as the basic power formative of human attitudes. Men speaking is his model for men acting. In my view work is a more satisfactory because a more complete model of human action. It displays more prominently than communication the multi-dimensional contours of the human *engagement* in the world. To know what a man "does in life" is one of the first spontaneous requirements of the mind seeking to locate the major coordinates of a personality. To say he digs ditches, he is a doctor, she is a secretary, is to name three ways of life whose patterns are common knowledge in a culture.

For my purposes physical labor provides the most efficient model for discussing the dialectics of praxis. In the practical action of transforming nature to satisfy human needs, human agency, purpose, and instrumentalities differentiate themselves from known being. Communication is essential to the labor process and probably arose as a necessary product of its evolution. At this level communication is inseparable from praxis; it is the most fundamental sense of the term "ordinary language." From this matrix are born the varieties of scientific, technical, and literary languages. In order to grasp the theoretical nature of poetry, the symbolizing function of symbolic action, it is indispensable to keep in mind the objectivity of praxis which is being theorized and symbolized through art. That is why an effort must be made at this point to characterize praxis as such. This description, no matter how incomplete, will facilitate the matter in hand—the elucidation of poetic language as a mode of cognition.

1. Praxis is historically conditioned. It is a process involving a reciprocal transfer of energies between men and the world as in physical labor. The individual and collective *agent* applies physical and mental powers to the transformation of the *scene* into *agencies* for the achievement of *purposes.* Bacon's "knowledge is power" should keep us from confusing action with act, from conceiving praxis as mere physical effort. Man makes nature work for him by discovering its properties and laws and using them to create the conditions for his own survival and the preconditions for new conquests. His social *agencies* are to a large extent determined by the necessities imposed at any given stage of his development by his struggle with nature. This means that the content of praxis is historically determined, that its properties are not the same in different climates, geographical conditions, stages of civilization or historical epochs. War is practiced by the chief of staff of the modern state, and by the aboriginal head-hunter. Their praxis differs completely. Picture the state of mind of the head-hunter on a bombing mission or of the chief of staff, naked and tattooed, blowgun in hand, stalking through a snake-ridden swamp, and the shift of *agents, scenes,* and *agencies,* not to speak of *purposes,* take on its full poetic meaning.[19]

Poetic language is integrally historical in its depiction of praxis, not only because language itself has a history, but because that history is inseparable from the historical reality which it reflects.

2. Praxis is the mode of being of human individuals, and that mode is always *social*. That is to say that heredity and environment are facts of life, that we are all born of parents at a specific time and place into a given society. The political and social structures, economic and educational, legal and religious institutions, cultural and moral heritage, prejudices, practical and theoretical wisdom and unwisdom of society are instilled in the individual with language, with school, military training and job. Of course, "society" is an elastic notion, almost as elastic as "individual". The "individual" born in a village of 12th century France or of Central Africa today and the "individual" contemporary American, are no more alike than the "societies" from which they stem. Again, American society, like all modern societies, is no longer isolated from foreign influence. Our social reality is cosmopolitan, our education more and more pan-cultural. No matter what the framework, however, the individual is *socialized* from birth.

This does not mean that the category of individuality is exhausted by social determinism. After all, the content, movement and direction of social change can be no more than the ensemble of the chaotic cross-patterns of individual praxis, including the wielding of power and the assent, dissent, or indifference of those governed. Society cannot preserve or change itself; both of these are done by individuals who are led to use their energies politically. The dialectic of individual in society is, of course, historically and socially determined, and that is not the same as saying "society determines the individual".

The individual retains his social character even when he is separated from society. Robinson Crusoe remains an 18th century Englishman on his desert island. He survives because the attitudes and skills inculcated in him by a middle-class family and a long maritime experience enabled him to face his new situation with resolution and intelligence; ten years earlier might have been too soon.

But language is the best example of a social product without which individuality loses almost all meaning. To speak English, French, Spanish, is to participate in a cultural tradition. This is of particular importance to the poet, for whom language is an end as well as a means. When he inherits it, it is already laden with the living traditions of his people, with emotional and literary connotations which are specific to their historical experience. No matter how personal his poetry the very medium of his craft obliges him to exploit these social elements on pain of not being understood.

Poetry displays the dialectic of individual praxis and social praxis in the social medium of language.

3. Praxis is always concrete, complex, and strictly speaking, unique, unrepeatable.

Concreteness means that each element of the pentad is absolutely specific. Change any one and you change the structure and character of the action. For example, "In ringing words, President X called upon the nation to repel the aggressor in the name of every man's God-given right to 'life, liberty and the pursuit of happiness.'" Let X be Wilson, then Roosevelt, then Truman, and nothing remains the same except the pure *act:* "Called upon". The shift in time, the different kinds of war involved, the political purposes underlying the rhetoric in each case, the agencies for mass-diffusion, the national ethos at three distinct moments of its history radically transform the dramatic content of the action.[20]

Complexity refers to the infinity of facts and factors involved in an action. Take President X delivering his climactic call to the nation. Tone of voice, physical appearance, gesture, expressiveness, the listening Congress, nation, world—all the myriad attributes of agent and scene can never be exhausted by verbal description. The "strategy" of the hypothetical newspaper description cited above is to center attention on voice and rhetoric—the determination of the "ringing", the pathos of the Declaration of Independence. Another approach might stress the set jaw, the indignation barely controlled behind calm mask and knit brows, the unanimity of national assent, etc. In any case, the description must necessarily remain incomplete, selective.

The uniqueness of any given action means that the conditions existing at the time of the performance of action A and A′ are not identical. Suppose for some reason President X had to deliver his speech a second time. Not only would his state of mind reflect itself in a different style of delivery, but as a result of A, the scene of A′ would be radically affected. On the second occasion, for example, the scene might be dominated by pacifist pickets, alerted by A. This change could, in turn, influence the style of the speaker.

The expressive language of poetry must communicate the concreteness, complexity, and uniqueness of praxis. Because it is selective, its names for the process of praxis must somehow say more than it says.

4. The concreteness and uniqueness of praxis do not exclude generality. No two actions are strictly identical, but they may bear to one another relations of similarity and difference of form and content, helpful for discerning their meanings. Conceptually we relate actions to one another by analogy, by metaphor. We "cross the Rubicon", "climb the ladder of success",[21] "put the cart before the horse", "cut the Gordian knot", etc. Suppose instead that we were to take decisive action, endangering our future career, or to get rich and powerful boldly, or to reverse the proper order of procedure in thought or practical affairs, or to solve a complicated problem with a single brilliant insight.

Something has been lost. What, precisely? The aesthetic savor of concreteness. The pleasure of affirming our identification with a cultural heritage by drawing on its stock of handy phrases, incisive brevity of formulation, richness of association. Those are formal and emotional values we are learning to appreciate, thanks in large part to our literary critics, at a time when propaganda, advertising, and pseudo-scientific and other jargons from all quarters are debasing the language.

More to the point here, however, is the loss of content. "To cross the Rubicon" is not only to risk one's entire fortune in a single gamble, but to identify oneself with Caesar, to cast one's act in an heroic mold. When we "climb the ladder of success", there is effort and exaltation as well as the bold risk of a fall from high places. The image, "the cart before the horse", displays clumsiness and comic perplexity, not only an error of judgment. And even "the Gordian knot", for all its triteness, is still useful in an age of increasingly tangled red tape. The suppression of metaphor (Burke calls it "anesthesia") has added neither clarity nor objectivity. Their translation into steno-language simply changes the object-matter from praxis to poorly achieved general description. The metaphors generalize about a dramatic situation, accounting for agent, scene, purpose, etc. The steno-sentences simply point to general facts and gloss over the relationships and attitudes they entail.

Those expressions have become proverbial because of their paradigmatic value, their efficiency in summarizing the essential dramatic elements involved in acting in a particular way, because they display patterns to which many actions are analogous. Their concreteness is the key to their conceptual reach.

The fact that proverbs contradict one another (example: "look before you leap," "for he who hesitates is lost") simply means that life is contradictory, that our contradictory experiences are habitual enough to warrant opposing formulations. Poetic language must generalize the concrete.

5. Praxis has its own mode of consciousness, its own language, its own kind of knowledge.

This will become clear by recalling that praxis is a synonym for everyday life. The language, consciousness and knowledge of praxis are those with which we get from day to day, and this is as true for the Arctic explorer, the telephone operator and the bushman as it is for the poet and the scientist.

In brief I am talking about the practical side of life, not identical with work, although work is one of its most important features. Practical life also includes getting food, shelter and rest, and carrying on our relationships with family, friends, strangers and enemies.

The language of practical life is more or less directly linked to matters at hand and reflects a low level of articulated consciousness, consciousness of life's foreground. The paradigm of practicality is the starving or drowning man, or the man pursued by deadly enemies. Survival requires from each the most stringent practicality. One distracted, abstract or noble thought may be his last, and any larger vocabulary than "help!" is a disastrous waste of energy. In these examples there can be no doubt about the content of consciousness. It is obsessed with one object.

The foreground of practical life, however, is usually broader. Where the issue is not survival but managing to get things done or even improving them a little, practical consciousness can stretch out, notice this or that "new" aspect in the familiar field of its concerns. Sometimes this can turn out to be a useful, a practical discovery, made "accidentally" or by "trying something out" in the normal course of the daily round.

And that goes as well for writing an essay about cognition in literature as for gardening, washing dishes or pleasing the boss. That kind of discovery–*praktische Menschenkenntnis*–is a kind of knowledge and a very important kind, but it does not transcend praxis enough to give us a view of its forms and processes as a whole.

The acts by which we acquire *praktische Menschenkenntnis* absorb us, dominating consciousness. Science and art are two ways of gaining perspective on reality, of moving from the pre-reflective consciousness of praxis to the higher ground of conscious reflection on nature and praxis. The goal in both cases is to discover the general properties of the object and the how and why of its behavior.

The consciousness, language, and knowledge involved in *poesis* are reflective in two senses, that they transcend *praktische Menschenkenntnis* making it part of their theoretical concern, and, that they "hold the mirror up to" praxis, reflecting, in ways to be discussed hereafter, its diverse shapes and movements.

This brief characterization of praxis is, of course, not exhaustive, but sufficient, I hope, for locating its universal property: the inescapable *hic et nunc* of human experience, analyzed by Hegel in the *Phenomenology of Mind;* or it may be more accurate to say that Hegel's analysis of perceptual acts is an irreducible model of praxis, perception being the minimal transaction between man and world, constituting them a dialectical *unit* having properties over and above those of each term taken singly. I prefer the more elaborate model of labor because it displays the interpenetration of man and world more completely, and typifies better the human condition than does bare perception, which seems onesidedly contemplative.

Still, especially when talking about poets who make symbolic *acts* out of words, it should be kept in mind that man perceiving means man apperceiving, thinking, i.e. conceiving images, words, ideas in terms of motives and purposes, which imply action in the world, a change of some kind in the natural and human status quo. All the premises of praxis are present; only the *act* is lacking. Man writing is man acting.

IV

> The Curfew tolls the knell of parting day,
> The lowing herd winds slowly o'er the lea,
> The ploughman homeward plods his weary way
> And leaves the world to darkness and to me.
>
> THOMAS GRAY
> *Elegy Written in a Country Churchyard.*

Poetry, then, is a special use of ordinary language, which, by deliberate imitation of praxis, seeks to present to consciousness symbolic, theoretical versions of the multifarious forms and processes of our daily (*hic et nunc*) transactions with the world.

What is special about poetic language? How can words *imitate* actions? In what sense is poetry *theoretical?* What can it alone present to consciousness? I shall attempt to answer these questions through a dialetical analysis of the first quatrain of Gray's *"Elegy"* (1748), a thirty-two stanza reflection on the unsung lives and deaths of the humble, virtuous, and hard-working folk of rural England, "the rude forefathers of the hamlet". His ambition is to raise a monument of words in pious commemoration of those simple, stalwart, and productive lives which are so swiftly and unjustly forgotten, or mocked by the proud of the earth. We shall never know how many potentially great poets, statesmen, or defenders of liberty lie buried beneath the churchyard turf,—"Chill Penury" nipped their talents in the bud. Gray wishes to be buried with them, and not with the famous and the powerful. Because of the simplicity of the language and Gray's skill in mobilizing all its forces, it is an excellent illustration of what I mean by the dialectic of *poesis.* The study of the first quatrain should suffice to display my method.

In order to show that literature gives knowledge of praxis, I shall try to establish that:

1. Literature is mimesis, the imitation of praxis.
2. Mimesis necessarily entails interpretation.
3. Interpretation implies evaluation.
4. To interpret and to evaluate in expressive language is to theorize about the meaning and value of praxis.

In other words, I am convinced that the only way in which to account for the referential or propositional aspects of expressive language is through a dialectical reinterpretation of the Aristotelian view of art as the imitation of praxis, which for convenience I shall call mimesis.

The mimetic act, *poesis,* initiates a dialectical process whose nodal points are interpretation, evaluation, and theoretical generalization. Expressive language presents these to us simultaneously, in a state of fusion, so that they can only be distinguished by an effort of analysis. They are inherent in mimesis.

In the following pages, I shall try to show how Gray's language constitutes a mimetic interpretation, evaluation, and generalization of a simple action. Each of the four points will be taken up in turn.

The poet is surveying from the vantage point of an English country churchyard the sights and sounds of a rural landscape as evening falls.

1. *Mimesis.*

> But the value of examining objective characteristics [of specific poems] carefully before talking at large about the imaginative constructs reared in the foundation of words is that this results, at least, in a recognition of the part played by the corporeality of the words and by the structures which connect them, not only in determining lesser poetic effects but also in directing the larger mental and

> imaginative processes activated by the poem; it may well lead, further, to a recognition of the fact that the various elements of poetic language interpenetrate one another with an intimacy which is of first importance in any consideration of how poetry "works".
>
> WINIFRED NOVOTNY
>
> *The Language Poets Use* (London, 1962), pp. 1-2.

The application of the methods of dialectical analysis to Gray's Elegy is not arbitrary. As soon as we try to answer the question, "By what principle may we explain the structure of the quatrain?", we are confronted with the problem of the unity of opposites. An examination of the first two lines will bring out this duality and tension.

The most immediate and perceptible fact about these verses is that we are dealing with English words which have, within a certain range of possible variation, recognizable denotative and connotative meanings. But the simple act of reading them aloud discloses a second dimension. Various properties of sound which we articulate in the very process of stating those meanings in speech assume a measure of independent existence and require separate analytical treatment. The propositions needed for the description of sonic properties are different from those which refer to meanings. Corporeality and ideation are co-present in Gray's language. We shall discuss each separately and then as a dynamic unit.

The first two lines display several interesting differences of sonic organization. In the first, the rhythm and the meter are identical, and give the perfect regularity of iambic pentameter. This regularity is re-enforced by means of several devices. The iambic rhythm leads us naturally to emphasize the consonants which introduce the stressed syllables. Four of these—*c, t, p, d*—are plosives. The *n* at the center of the line stands out from the preceding fricative and the following liquid and sibilant by virtue both of its position and nasal quality. Thus it plays a double role: rhythmically as a plosive and tonally as a continuant consonant. It finds its echo in the unstressed *ing* of "parting". When we consider that the consonants of all the unstressed syllables in the line (the *t* of "parting" being the only exception) are fricative, liquid, or sibilant, or absorbed like the *w* in "curfew" or the *y* in "day" into diphthongs, we may conclude that this verse is composed of a regular alternation of staccato and legato phonemes and the *n* by its position and quality partakes of both components of sonic movement. This analysis is confirmed by another important fact: all the vowels of the stressed syllables (except in "curfew") are open and those of the unstressed closed. The combination of plosive consonants and open vowels concentrated in the stressed syllables of the regular iambic meter achieves by contrast to the continuant consonants and the closed vowels of the unstressed syllables, the effect of a strong beat within a continuous flow of sound.

The second verse differs from the first both tonally and rhythmically. Its only two plosives are the *d* of "herd" and the *d* of "winds", both in weak positions. Vowels dominate consonants, the rest of which are continuants. Three of the six stressed syllables contain *o*. The *n* occupies a subordinate position in both "lowing" and "winds". All the techniques which were used in the first line to emphasize the staccato beat of the rhythm are here subdued or eliminated. Continuity is the keynote.

The two most notable sonic differences between the first and the second verses are: A.) Read naturally, the second is slower than the first. B.) In the second, rhythm and meter are not identical.

A. Three long *o*'s, the long *i* of "winds", and the long *e* of "lea" go far to explain why it takes more time to read the second verse. There is only one *o* in the first line and none of its other stressed vowels have to be held as long as those of the second in order to be pronounced correctly. Add to this that most of the consonants of the first line occur singly whereas in the second we have such sequences as–*ng h, d w, nds sl, r th,*–and it becomes apparent that to give full value to each sound imposes a slackening of the reading pace. Six stresses instead of five also contribute to this end.

B. The second line is not strictly iambic. "Herd winds" and the *slow* of "slowly" span the middle of the verse and are all stressed. "Winds" breaks the meter. "Winds" occupies a central position. It is the single longest vocable and yet falls precisely where we would expect to find the unstressed beat of an iamb. In brief, the second verse is legato, metrically irregular, and slower to articulate.

As we pointed out at the beginning of this analysis of sound patterns, the discovery of their relative independence rests on a previous recognition of their role in composing vehicles of meaning.

What then is Gray imitating by moulding the plastic material of language? Presumably, the things and relationships which are referred to in the two verses under examination. The first verse, regularly punctuated by strong plosive consonants and by the centrality of the nasal, reproduces in conjunction with, and by contrast to, the continuant consonants and open vowels and their regular distribution, the ringing of the curfew bell, whose continuous resonance is sustained and interrupted by the strokes of its swinging tongue. The sonic structure of the second verse imitates the single, slow-moving curvature of the herd and the grave monotone of their lowing. In brief, if we assume that mimesis is the organizing principle of these verses, we can explain the relationship of the material and the ideational content of their language.

But is this not a self-contradiction? Mimesis means imitation of praxis and the analysis up to now seems to imply that Gray is imitating nature, an idea already specifically rejected as an impossibility. In order to answer this objection, I shall have to discuss ideation. Its function up to now has been limited to that of assisting in the analysis of corporeality and for that purpose an intuitive or common sense definition should have sufficed. The elements of ideation which are pertinent to this analysis of the dialectic of mimesis are denotation, connotation, syntax, context, texture, and voice-directives.

2. *Interpretation.*

> Poetry is nothing but a way of perceiving exterior objects, a special organ which filters matter and which, without changing it, transfigures it. Well, if you see the world exclusively through that lens, the world will be colored as it is, and the words for expressing your feeling will turn out to be in necessary relation with the reality that caused it. In order to do a thing well, that thing

> must become part of your constitution . . . a botanist should have neither the hands nor the eyes nor the head of an astronomer, but should see the stars only in their relationship to herbs.
>
> GUSTAVE FLAUBERT
> *Correspondance.*

An examination of the strictly literal reference of some of the words in the first verse will indicate that there is more involved in mimesis than a simple sonic reproduction of an objectively ascertainable state of affairs. In reality, a curfew does not sound like the tolling of a knell. It is more lively and rapid. "Tolling" and "knell" taken together, denote slow and regular ringing for a death or a funeral. From the point of view of strict denotation, the first line is open to criticism.

When we consider, however, that it is the "day" which is dying, (parting is a shortened form of departing, and, like "tolls" and "knell", suggests death) it becomes apparent that Gray is counting on the connotations of these words rather than their denotations in order to communicate his meaning. The verse is a metaphor constructed with words commonly associated with death, and it rests on an analogy between the approach of evening and the slow decline of a person dying peacefully. The factual occurrences reported in this line may be stated denotatively as follows: "the ringing of the curfew bell indicates that the day is ending". The use of "tolls", "knell" and "parting" can only be explained in terms of connotation. They indicate the presence in the language of a fact for which the purely denotative paraphrase does not account. This fact is the presence of some *agent* who is *interpreting*, by means of metaphor, the mingled sounds and sights of the scene. This agent is just as real a part of the landscape as the curfew or the darkening sky. Their presence is explicit, and his, implicit. In order to make the agent's presence explicit as part of the objective situation being described, at least a second analytical, denotative statement is obligatory, i.e., "The speaker or writer of this line finds the scene before him reminiscent of death". Thus it is not the bell alone which the verse embodies and reproduces by means of mimesis; it is the bell experienced by the agent as a dirge which is being imitated. In a word, this is not imitation of nature, but mimesis, imitation of an action.

(If we have chosen to demonstrate the presence in poetic language of an agent through an analysis of the role played by connotation, this does not mean that connotation is the sole or even the main determinant of poetic language.)

Like results are reached if we examine the verse for its syntax. The simple word order does not deviate from that of normal prose utterance. Yet, there is one peculiarity. The agent who is interpreting the scene before him uses the present tense of the verb, rather than the progressive form. For him the bell *is tolling.* His use of the word "tolls" (and this holds true for "winds" and "plods" in the next verses) is an interpretation as well as a simple report. The present tense informs us that the agent is aware that these events form a pattern of traditional routine. He is introducing by this device one of the most important themes of the poem, i.e., the immemorial cycles of days, seasons and generations which have ritually guided the laborious lives of the pastoral and agricultural communities of England. The curfew is the nightly ceremonial of assuagement. He communicates this to us through *syntax.*

But this syntax is not in itself poetic. It is plain English. The use of the present tense for description of occurrences which repeat themselves (as when we explain the workings of the solar system) is a trait of our everyday speech. A close examination of the second verse should enable us to reconfirm the inadequacy of either connotation or syntax to supply a useful criterion for distinguishing poetic from ordinary language.

The discussion of the sonic patterns of the second verse could do no more than point to a strong analogy between its phonetic structure and its denotations. The long vowels, the consonantal clusters and the crucial position of "winds" seem to reproduce the slow progress of the lowing cattle. Where, then, is the agent? An examination of connotations is not as fruitful in the second as in the first verse. The rural antiquity of "lea" calls up associations which harmonize with the themes suggested by the analysis of the syntax of the first verse. But there is no sharp contrast here between the denotative and connotative aspects of the words. The syntax of the present tense functions differently for "winds" than for "tolls"; for whereas the curfew in its own right denotes an occurrence repeated daily, the significance of the movement of the cattle requires the previous context (the curfew and the evening) as well as the present tense of the verb "winds" in order to achieve the precise ideological nuance of idyllic security which "tolls" achieves unsupported. The curfew is the main event, driving in the herd a subsidiary consequence, the day's last task.

Of course we have introduced a new and undefined term: *context.* And there seems to be an implication that the agent manifests himself therein. We shall return to the question of context later. The present task is to show that it is possible within the limits of the defined terminology to identify the agent's presence in the second verse.

I have emphasized the importance of "winds" in the second verse; its length, sonic qualities and central position make it the pivot of the sonic structure. The fact that "winds" is also the verb adds to this emphasis. But it is the denotation of the word which offers a plausible explanation of its preponderance. "Winds" denotes two facts which determine the character of the total image. It includes both shape and motion. It indicates that the agent is at a distance from the herd which permits him to perceive its movement as a total arc, or sinuous procession. There are two other coordinates of the visual perspective which locate the agent-observer in relation to the cattle. He can hear them lowing and they are moving across a grassy plain. We may presume that he is seeing them from the higher ground of a churchyard and that the plain is not at a great distance. Life is carried on within earshot of the bell. The proximity of the horizon is in keeping with the communal ethos.

The analysis of sound pattern could only locate the problem. It takes the analysis of meaning to propose a solution. The slowest point in the slow verse falls on "winds" and on the *sl* of "slowly", and it is precisely at this point that the visual elements reveal the active presence of the agent. It is also here that the image receives its distinctive shape. Replace "winds" with "moves"; almost all the essential sonic properties have been preserved, but the aesthetic distance, scenic particularly, and the precise location of the observer have been seriously compromised.

The transposition of the second verse into denotational language would therefore also require at least two statements: A.) "The lowing cattle are moving slowly in an arc across a green plain", and B.) "The distance separating the observer from the events described permits him to observe them simultaneously". If we juxtapose on one hand Gray's two verses with the denotational transcriptions proposed above, two important differences may be noted.

A. The two verses require at least four denotational statements in order to account for the situation described to date.
B. A quality makes its appearance in the verses we have not analysed up to now and which does not exist in the denotational transcriptions, i.e. similarities and contrasts which are aspects of context.

The denotational translation of the two verses loses more than the "beauty" of the language. The loss includes a most important aspect of meaning, namely context. The two denotational couples have no necessary relation to one another. Only a comma intimates weakly that the tolling of the bell and the movement of the herd are going on at the same time and place. This could be made fully explicit by prefixing the word "simultaneously" to the objective statement about the cattle. But this addition would differ in an important respect from the other words in the sentence, all of which were derived from an analysis of the words appearing in Gray's text. The word "simultaneously" has a different source which becomes perceptible only after reading the two verses in their entirety. In other words the two verses, taken as a unit, contain a fact, simultaneity, for which neither their denotation, connotation, nor syntax can account, and which disappears in denotational transposition unless it is expressly added by the critical interpreter.

The context created by the two verses is manifest in their differences and similiarities. Their purely ideational content cannot account for either.

In the discussion of mimesis I emphasized the contrasts of sonic pattern between the verses; and this emphasis permitted me to state my reasons for rejecting the theory of literature as imitation of nature. Beside this sonic contrast, the only other relationship between the two verses I had occasion to suggest was the dependency of the meaning of the second verse on the reference of "tolls". A third relationship, the interpenetration of similarities and contrasts, becomes evident when we examine both verses from the point of view of their sonic resemblances.

The sounds *o, l, s* and *n* reoccur throughout both verses. This technique keeps the bell ringing while the herd is being described. *S, l* and *w* which appear at the end of words in the first line assume a dominant position of initial letters to words and syllables in the second. The *o* which occurs only once in the first is repeated three times in the second. The ringing continues but it is the continuous quality of its resonance rather than its individual strokes which receives emphasis. Thus it is that, taken together, the two verses realistically reproduce the ringing of the bell; for when it begins to ring we hear the strokes most distinctly, whereas its resonance grows louder and more sonorous as it gains momentum. In this respect the *n* of "winds" performs an intermediate function analogous to the *n* in "knell". Its position in the line obliges us to stress it and to hold it longer than its counterpart. Its onomatopoetic quality contributes to the continued resonance of the ringing curfew. But its weak position within the word–outdone by the long *i* and cut off by the *ds*–acts as a check on its consonantal value, too much of which would disrupt the desired continuity. The same may be said of the two *d*'s. All these sonic patterns suit the second verse for its double function, for they are the very ones required in order to reproduce the movement and the lowing of the herd. Moreover it is this simultaneity, inherent in the context created by the two verses, that the agent is revealed to be a necessary part of the objective situation which they imitate. This is equivalent to saying that poetic context is a result of mimesis.

This interpenetration, interplay of sonic materials is part of the semantic value of the language. It is the specifically poetic way of preserving the continuity of praxis, of suggesting the unity of manifold experience. At the same time, however, Gray creates an atmosphere, mood, tone, feeling which the notion of "context" does not cover. In real life we acknowledge the importance of these qualities of experience as well as their elusiveness when we use them as a last recourse in explaining to others why we liked or did not like this or that city, hotel, food, rug, etc.—"it made me feel uncomfortable"; "it left a bad taste in my mouth", etc. The trouble, I think, is due to the infinity of variables, both subjective and objective, which unite to create the atmosphere and mood of a situation, always "unique," which defies the abstracting genius of language. It is like the task of describing the precise taste of a peach or the touch of a fine silk garment. This uncompromising specificity of experience is recreated in poetry as *texture;* i.e., it is atmosphere and mood disclosed as qualities of language, essentialized and made available to consciousness. The texture of Gray's quatrain is woven of all the thematic threads previously discussed. It is the empathy of the poet with the scene about him.

The analysis of corporeality seemed to show that it is through the manipulation of sound that the poet incorporates objective nature into his work. This is part truth and part illusion. The truth lies in the fact that the mimetic use of sound reproduces by analogy certain qualities of nature. The illusory part arises from our tendency to take reference for granted so that we forget that meaning or ideation tells us immediately what it is that the sounds are imitating. In order for sounds to reproduce nature, they must be an interpretation of nature; they must include an active consciousness. Mimesis includes the object by interpreting it, not by photographing it.

An examination of the third and fourth verses from the point of view of their material and ideational structures may now be undertaken in terms of the categories developed previously. The rhythm of the third verse is a replica of the first; it coincides with the meter. The *o* of "tolls" and the *o* of "homeward" both fall on the stressed beat of the second iamb, while "parting day" and "weary way" are brought into correspondence both by rhythm and rhyme. In the first three feet, plosives, nasals and the diphthong reinforce the resemblance by picking up again the staccato strokes of the ringing bell. But this sound no longer has the sharp clarity which the unvoiced single consonants of the first verse suggest. The *pl* of "ploughman" and of "plods", the three nasals concentrated in the second foot, the shift from the closed *o* of "homeward" to the open *o* of "plods" and the distance separating the two *p*'s indicate that the bell is slowing down. This is confirmed in the last two feet in which the only sound reminiscent of the curfew bell is the final rhyming diphthong. The alliterative *w* carries forward the techniques of the second verse, as do the *ea* of "weary" corresponding to "lea", the continuant *s* and *r* and the sharp sonic contrast of the last two with the first three feet. In brief, the third verse materially synthesizes the first two while introducing something new, the fading of the curfew.

These sonic echoes are used to reproduce the tired walk of the ploughman and then his slow withdrawal into the distance. The verse connotes the hard labor and simple joys of the idyllic life. "Homeward" is the only word in the quatrain which deviates from the order of normal prose syntax. Its position lends value to its rich ideological connotation. The observer watches the solitary ploughman disappear into the darkening horizon. With the fading away of the bell, the herd, and the ploughman, the poet will be free to reflect upon the significance of the twilight scene. The solitary ploughman introduces these new elements into the context of the quatrain and provides a transition to the solitude of the poet.

This transition is marked by the conjunction "and" which introduces the fourth verse and links it syntactically to the third. Muffled sonic reminiscences of the preceding three verses are so scattered throughout the line that they can no longer summon up distinct sensations. The real concern of this line is to introduce the person of the observer. The iambic movement of the first three feet is followed by the unstressed *ness* of "darkness", then by a slight pause and finally by an anapest. In effect this change of rhythm marks a halt to the characteristic sounds and movements which have prevailed up to now. "Leaves the world" restates in a metaphoric circumlocution with theological overtones the theme of death which the tolling of the knell earlier embodied. Abstraction and religious elevation are necessary at this point. The rhetorical euphemism, reminiscent of funeral orations, gives a voice to the poet, a tone and a context to his thought and an intimation of his personality. "Darkness," by now the pictorial equivalent of death, is to be the subject matter of the ensuing disquisition. The poet makes his appearance, grammatically speaking, in the modest guise of the monosyllabic, indirect object of the verb, but now for the first time, manifest at the very end of the quatrain. The task of explaining the significance of the rural life surrounding him falls to his lot because he alone was present to witness the "privileged moment" of that poetic dusk.

Each of the first three verses describes an important aspect of the external scene. The three images taken together provide a context and texture which unite the individual facts in a single picture. The fourth verse is not scenic in its intent. Darkness abolishes the scene, and the observer, in a flash, effectively focuses our attention on him: "and to me". The implicit is made explicit. The rhythmical variation of the last two feet in the quatrain therefore plays a double role:

a.) By breaking the rhythmical iambic regularity which is by this time dominantly identified in the reader's experience with the motion and sound of the bell, etc., it definitely marks an end to the ringing and, in conjunction with "darkness", establishes silence and solitude. Mimesis reproduces not only ringing, but non-ringing.

b.) We said that "and to me" is an anapest. We must mention its strong rhyme with "lea" as another of its specifically poetic qualities. Looked upon as an element of verse both these facts define its formal poetic nature. But looked upon as elements of ideation in the light of its rhythmical distinctness from the rest of the quatrain, of its purely denotative character, of its lack of imagistic value and of its direct explicitness, the final anapest is strongly reminiscent of steno-language, i.e. straightforward referential discourse.

The poet designates his presence as an explicit and distinct fact requiring its own separate denotational sign. In the previous three verses his presence is implicit, synthetic, and recondite. What happens then to the idea of mimesis, the imitation of praxis, when applied to a phrase like "and to me", the literalness of which suggests that the phrase refers to nature rather than imitating praxis?

Even without recourse to earlier arguments by which I tried to establish that the phrase designates sonically the cessation of the curfew, we can still maintain that the line has throughout a mimetic structure. Its explicitness contains a contrast of considerable importance with the implicitness of the previous three verses. "Leaves the world" is a direct statement of the poet-observer which is neither a concrete image nor a total abstraction. Its figurativeness inclines towards metaphor; its theological connotations lend to it an ethereal dimension. It has neither the imagistic concreteness of the third verse nor the simple factual character of the concluding anapest. It marks a transition from the state of contemplative absorption, during which the poet emphatically identifies himself with the surrounding scene, to the gradual awakening of his self-awareness which parallels the fading of the landscape. The fourth verse imitates the emergence of this self-awareness from a state of concentrated absorption, and the simple statement of the fact which terminates the verse is the mimetic rendering of this final emergence. Achievement of self-awareness, the pre-condition for didactic exposition which dominates the poem, is also a form of praxis and is thus available to mimetic formulation. This interpretation would also explain the handling of the temporal exposition. The darkness which falls so suddenly catches the poet-observer by surprise. The quatrain imitates the fleeting subjective time of his concentrated absorption in the scene. The explicitness of the phrase "and to me" is mimetic because it corresponds to the moment when the poet rediscovers his individuality *vis-à-vis* the external world.

The analysis of mimetic structure has revealed that it involves interpretation. In his encounter with this pastoral dusk the poet finds a number of dominant qualities: melancholy, tranquility, simplicity, orderliness, tradition, familiarity, security, and spiritual elevation. These are all qualities of the action as interpreted by Gray. But these are not only qualities which can be identified objectively like the properties of nature as the real content of praxis; they are value judgements as well. The severe music of the verses, the poet's sympathetic attention to the scene, and the final elevation of tone all affirm the poet's belief in the worth of these qualities. His express intention, to stay on after darkness, is itself an implicit value judgement; for it means that the significance of the action extends beyond immediate sense experience. The rest of the poem is a justification of this judgement.

3. *Evaluation*

> If the work of art can be regarded as a sign, the central question is as to the nature of the designating sign and the nature of what is designated (in technical terms, the nature of the sign-vehicle and the designatum). The view proposed is that the aesthetic sign designates the value-properties of actual or possible situations, and that it is an iconic sign (an image) in that it embodies these values in some medium where they can be directly inspected (in short, the aesthetic sign is an iconic sign whose designatum is a value).
>
> CHARLES MORRIS

"Three Primary Forms of Discourse." *Kenyon Review 1* (Autumn, 1939), pp. 410-420

The difference between interpretation and evaluation becomes clear as soon as we explicitly introduce another necessary element of meaning into the analysis. This is voice—its pitch, timbre, intensity, modulation, and volume. It would, however, be fallacious to identify the interpretive and evaluative contents of the mimetic act. If we preserve all the material and ideational relations which we have discussed, and read the quatrain several times, louder each time, or more tremulously, the meaning changes. The result is grotesque or ridiculous. So long as such a reading is consistent, the interpretation of the scene does not change. The qualities of melancholy, tranquility, etc., which are deeply involved in the denotations, connotations, context and texture of the language are still discernible. The change of voice expresses a change in the evaluation of these qualities. This experiment should make it clear that interpretation and evaluation are not identical.

Why then do they appear to coincide? We think that in good poetry, interpretation and evaluation coincide because the reader adapts his voice to the voice directives which the poet blends into his language.

Gray wishes his quatrain to be read quietly, gravely, reflectively. The word "elegy" in the title is actually an explicit voice directive, and in another way, so is "churchyard." A sympathetic reading of the quatrain fuses interpretation and evaluation, or in other words expresses the value which the poet found inherent in the qualities of the action which he interpreted mimetically. Thus a reader could agree with Gray about the existence of these qualities and disagree about their value.

Of course it is possible to read these lines so as to change both the interpretation and evaluation, or better yet, to indicate one's dissent from Gray's evaluation by changing the interpretation. Imagine a dead-pan satirist of urban persuasion listlessly droning the quatrain. He can make the nightly curfew sound like "that damn bell again", "the lowing herd" like very bovine cows, the ploughman like a stupid peasant; and, by yawning the last verse, he might suggest that when night falls upon the hamlet, there is nothing to do but hit the sack. Gray's sense of immemorial tradition is reinterpreted as anesthetizing humdrum. His sense of quiet security becomes the ruminating satisfaction of the milked cow. Even here, however, we can discuss the difference between interpretation and evaluation; for the satiric actor is himself suggesting that certain qualities inhere in the action. The change of voice is simply adding the notion "show me the shortest way back to Piccadilly".

In other words, the poet's language has a kind of built-in musical notation which must be read in and between the lines, a complex system of cues, emanating simultaneously from all levels of meaning previously discussed. "The poet's song" is less metaphorical than it appears: for by this notation he tells us how best to modulate our voices so as to harmonize with his meaning. It is a direct appeal for assent to his interpretation, an appeal so fully integrated in the interpretation as to defy complete extrapolation, yet distinct from it. Mimesis and interpretation taken alone would constitute only a relationship between poet and scene. Add the subtle music of vocal cues and a new dimension immediately appears.

The poet is addressing an audience, appealing to English-speaking men and women to glorify the time-honored ways of England's rural life. His elegy makes available to them the quintessence of a personal experience he has found to be of great value. Implicitly he is claiming that those who consent to relive it through his song will, each in his own way, be vicariously enriched. By reliving that solemn dusk, they will experience a unity with the English earth; and sympathy for its noble poor will deepen their understanding and appreciation of their national heritage.

> Let not Ambition mock their useful toil,
> Their homely joys, and destiny obscure;
> Nor Grandeur hear, with a disdainful smile,
> The short and simple annals of the poor.

In the last two centuries, industrial and political revolutions have profoundly transformed our attitudes towards city and country life. In *The Two Cultures and the Scientific Revolution,* C. P. Snow waxes very unsentimental about the true conditions of rural misery and squalor of pre-industrial times. And Mark Blitzstein reminds us in his translation of Brecht's *"Threepenny Opera"* that "the noble poor are nobly underfed". Resignation to hard labor is no longer the dominant attitude of the dispossessed, and in many quarters, apprehension is rapidly displacing pity for their lamentable lot. Ideologically, Gray's *"Elegy"* can hardly guide our attitudes and actions in the modern world, for in praising English peasantry as he does, he accepts, and wants us to accept with a resignation equal to theirs, the maintenance of the *status quo* which keeps them poor. Fortunately, or unfortunately, that choice is no longer ours to make. I shall return a little later to the question of why Gray's *"Elegy"* is still a living part of our literary culture.

To sum up, Gray's mimetic interpretation of an action contains voice directives for its evaluation. We must, therefore, add evaluation to the ideational coordinates which are inseparable from the mimetic act and from the meaning it expresses. This is a decisive turning point in our argument, for the voice directives latent in the language of the *"Elegy"* are Gray's most personal and intimate appeal to the intuitive sympathy of his public, and there is nothing in the language that can guarantee that the appeal will be heeded. The analysis of poetic structure itself, in brief, poses the problem of communication between poet and society. The presence within mimetic structure of the necessity to envisage that structure within a larger historical framework defines our stand *vis-à-vis* the formalists (e.g. New Critics) and toward traditional literary scholarship.

The formalists are inadequate theoreticians of poetic language, for they fail to account for communication, which is one of its dimensions, namely its tendency to become image, feeling, attitude, action on others. Language reproduces and negates itself in communication. The formalists isolate language from life, i.e., from praxis. Historical critics and theoreticians go to the other extreme; for in their desire to explore and demonstrate the continuity of life with literature, they frequently lose sight of the specific mimetic structures which distinguish poetic from other forms of language. They see nothing but communication in literature, and this violates its significant subjective and emotional content. Bad as well as good poetry reflects history, after all.

The dialectic of linguistic corporeality and ideation exemplified above pays tribute to the formalist contribution to criticism by employing its techniques in so far as they shed light on the internal dynamics of poetry. This same dialectical analysis, however, constitutes a rejection, I hope a refutation, of formalism because it leads to the conclusion that purely formal analysis contradicts and negates itself by disclosing the categories within poetic language e.g., evaluation and communication, which force us to overreach formal limits. My remaining task is to inquire into the extra-poetic dimension of poetry, into the relationships among poet, poem, and public. I can only justify my critique of traditional historical criticism by defining in general the sense in which the aesthetic dimension of literary research is indispensable for an understanding of literary history.

4. *Generalization.*

"La forme scientifique de la vie".

Gustave Flaubert, *Correspondance*

To say that all language generalizes, abstracts, is by now a truism. That language does not cease to generalize when its mimetic potentialities are utilized in poetry seems to me self-evident. If it did not, poetry would communicate nothing.

Romantic, mystical and expressionistic glorifications of the personality of "The Poet" have obscured for many some basic humdrum facts essential for understanding the nature of poetic truth. The poet (every poet) is born and brought up in a society which he does not choose. The economic, social, political and cultural institutions which he finds at a particular historical stage of their growth or decay constitute a framework, sometimes rigid, sometimes elastic, which defines his possible range and intensity of experience. The social situation of his family, for example, may determine to a large extent the patterns of his interaction with society: his religious beliefs and political outlook, his friendships and loves, the extent and variety of his education, his possibilities for travel, leisure, etc. Such circumstances may effect his entire emotional and intellectual life. Most important, however, is that the poet's interaction with his society has a direct influence upon his knowledge of and feeling for language which serves as the raw materials, tools and products of his craft.

Language too has a history. Along with the institutions of the society in which the poet participates, he inherits a language at a certain stage of development. This language, his mastering obsession and yet obedient slave, belongs to an entire people. It registers the historical experience and activities of that people. It is heavy, therefore, when the poet receives it, with the complex connotations of good and evil, tabu, superstition, illusion, ideology, knowledge and practical wisdom which that history involves both as cause and effect. The poetic sensibility is a linguistic sensibility, i.e., of all the products which reveal human presence, language is most revealing to the poet. His task is to reveal that human presence to others and to explain its significance. How does he achieve this?

Ordinary language, the language of praxis, is a confused mixture of signals, simple statements of facts, exclamations, opinions, exhortations, etc., which most frequently refer us to some immediate, practical situation with a view to action. We use it purely as means to an end, like a tool; for handled at maximum efficiency, it disappears, becomes transparent. We try to direct attention away from language "for its own sake" and toward the things, situations and actions which words designate. Ordinary language includes as part of its context the realities to which it refers and depends upon them for its meaning.

But just as nature sometimes resembles a painting and makes us feel that the sunset is a work of art, so also ordinary language gives birth to poetic moments which draw our attention away from the immediate import of words and toward their inherent qualities. Witty remarks, suggestive rhythmical or onomatopoetic combinations of sounds, striking metaphors, comparisons and innuendos, expressive tones of voice, significant anecdotes, jokes, etc., are examples of protoliterary outgrowths of everyday language. Ordinary language achieves the status of literature by excluding nonlinguistic elements of context and replacing them by linguistic elements. Each work of literature creates its own context; ordinary language needs the immediate support of external realities to achieve communication. This does not mean that poetry cannot be a call to action, but that it has to be something more. The attempt to make a clear distinction between propaganda and literature by classifying the one as action and the other as contemplation is futile. In terms of the present theory, poetry becomes propaganda when its signaling function stands in contradiction to its function as symbolic action. Thus, Shakespeare's *"Richard III"* is propaganda, whereas T. S. Eliot's *"Hippopotamus"* and Mayakovsky's political verses are poetry.

In analyzing Gray's quatrain we pointed out the importance of *context* as a distinguishing feature of poetic language in opposition to steno-language. *Texture* is the criterion for differentiating poetic and ordinary language. In the first opposition, the role played in poetry by sonic contrasts and analogies was the decisive point. In the second, it is the inter-penetration of all sonic and semantic elements, the weave of the language, which marks off poetic from ordinary language. The texture of the poetic work draws our attention to its specific qualities, to its existence as an artifact. Its *Dinglichkeit,* as John Crowe Ransom calls it, has so engaged the attention and interest of some critics that they have been deluded into believing it autonomous, self-sufficient, self-contained, self-referential, etc. My analysis has tried to show that these properties of poetry do not destroy the referential dimension which relates it to reality, but rather directs its reference to praxis.

A poem, I suppose, may be written using nothing but scientific signs to imitate an action. It might even remain true to scientifically established facts, but it would necessarily mean more than their simple references. The organization of their sonic qualities according to a mimetic pattern (*poesis*) would bring into play all the consequences I have been exemplifying through Gray. A context would be created in which the simple denotations of the formulae would take on a new set of qualifying attributes, that is, a texture such as voice-directives, connotations, plurality of reference, etc. The formulae would not lose their meaning; but their meaning, qualified by their subordination to a mimetic form, would change. Their purely scientific function would have to be isolated by an effort of analysis, similar to my isolation of the purely denotative aspect of Gray's quatrain. Their new meaning could only be gauged by taking the agent into account.

Poesis is the means of stating hypotheses about praxis, the means of typifying praxis and socializing the individual experience of praxis. *Poesis* results in a fusion of form and content, which bears an analogy to praxis, and which embodies the qualities of praxis while designating it. In the course of literary history it has been discovered that the qualities inherent in different forms of praxis can be grouped. Thus we get epic, tragedy, comedy, ode, satire, etc. A genre is a name for a recurrent pattern of praxis, as well as for a type of literary work. The names of genres therefore connote the interpretation of praxis which can be expressed by those mimetic patterns of language proper to each. Clearly also they have a history; they must be seen in terms of the society in which they take shape. But it is equally clear that their meaningfulness extends beyond the society in which they first appeared. Why do we still react to Sophocles and Shakespeare? It is my belief that the insights into the nature of praxis which those great men achieved still have validity, i.e., still throw light on our own actions. It is for good reason that the words tragic, comic, epic, etc., are in everyday use and applied to experience independently of literature.

Gray's *"Elegy"* is a genre in the sense that I have discussed the term. It imitates, interprets and celebrates an action. It selects those qualities of the country scene which bespeak the beauties of a way of life, rich with the tranquil harmonies of a humanized nature. Gray's experience reveals to him the universal meaningfulness, i.e., human value, of the rural life of England. He writes during the period when advancing industrialization is beginning to undermine the old village economy and ethos. Soon the exodus to the great urban centers will become general. An elegy is a poem of mourning. Gray is delivering a prophetic funeral oration for a condemned way of life.

Literature generalizes through typification. Each of Gray's details stands for a body of related activities to be readily completed by our imaginations. Of course his choice is guided and informed by attitudes, prejudices, tastes previously acquired and without which, perhaps, he could not have written so fine a poem. Thus literary typification turns out to express an ideology, i.e., an outlook on life relative to specific experience. Gray, the learned English gentleman of letters, cannot possibly see country life through the eyes of the ploughman. The distance that separates the poet from the plow engenders idyllic illusions, a city-dweller's arcadian idealization of the simple, healthful and artless pastoral existence. Gray has been by-passed as a spokesman for the poor. What the elegy now expresses, I think, is a yearning discernible in modern culture to overcome the dichotomy between civilization and nature, and between individual and community. Gray recorded a moment in the early development of the modern sense of this double alienation. We needed this insight in order to formulate our own.

Literature was for Flaubert *"la forme scientifique de la vie"*. His radical formulation has this advantage over my more reticent talk about cognition in literature and knowledge of praxis: that to defend or refute it requires a comparison of the poet and scientist according to vigorous scientific standards of knowledge. In the last section of this essay, I shall attempt to suggest the broad lines that such a comparison would follow. The immediate task is to spell out Flaubert's idea in the light of the foregoing analysis of Gray's quatrain.

Flaubert would not have liked the *"Elegy."* The first sense of *scientifique* was objective, he said "impersonal." His entire theory of literature is an attack on the Romantics for their first-person singular emotive exhibitionism and egocentrism. Their poetry distorted life, nature, man, God into modes of personal suffering or satisfaction. Flaubert would have felt, I think, that Gray was more interested in his own sympathy for the poor than he was in depicting their lives accurately. *Scientifique* means descriptive as opposed to lyrical.

It also means *true*. Flaubert remarks a number of times that he encountered in real life scenes and characters which he had created imaginatively. "Emma Bovary is suffering and weeping in twenty villages of France." The methods of realism give confirmable results. For reasons not clear to him the objective descriptive approach to *la vie,* integrated by the imagination, could discover qualities and values of experience verifiable in practice.

Finally *scientifique* meant that the poet should keep his personal predilections, sympathies, tastes to himself, that the poem should present life as it is. The foregoing analysis of Gray could be applied to Flaubert's novels with analogous results. It is, in my view, impossible for the artist not to express his personality in his work. The Olympian aloofness of Flaubert's realistic methods protected him against Romantic self-display; but, as the entire corpus of Flaubert scholarship has shown, he did not escape the fate of all artists. Psychologically, biographically, stylistically, Flaubert is in his works.

La forme distinguishes poetry from *la vie*. It is the final product of the creative process. If the artist has been faithful to the reality he is depicting, *la forme* will not only correspond to his model, but attain an internal aesthetic coherence of its own, a structure of language embodying, and yet distinct from, real life. *La forme* is a simplification of the manifold complexities of life for the purpose of displaying its most *significant* patterns to consciousness. Poetry is therefore always a hypothesis about the significance and value of an action. I take *la vie* to mean praxis.

According to this corrected version of Flaubert's cryptic formula, we arrive at the following definition: a poem is a hypothesis constructed mimetically and designed to display those objective features of an action which best reveal its properties and values. *Poetry is theory of praxis.*

V.

> For the difference between the historian and the poet is not in their presenting accounts that are versified or not versified, since it would be possible for Herodotus' work

> to be put into verses and it would be no less a kind of history with verse than it is without verses; rather the difference is this: the one tells what has happened, the other the kind of things that can happen. And in fact that is why the writing of poetry is a more philosophical activity, and one to be taken more seriously, than the writing of history; for poetry tells us rather the universals, history the particulars.
>
> ARISTOTLE—*Poetics*
> (Wheelwright ed. N.Y. 1935, p. 170)

> But in all this we must bear constantly in mind our previous warning: not to expect the same degree of precision in all fields, but only so much as belongs to a given subject matter and is appropriate to a particular type of inquiry (*methodos*).
>
> ARISTOTLE—*Nicomachean Ethics*
> (G. Else, *ibid.*, p. 301)

It is no longer possible to accept Aristotle's characterization of history as a recounting of particulars. I quote the passage from the *Poetics* for two reasons. By raising the question of the relationship between history and poetry, the passage introduces a corrective to the method followed in this essay. The technique of defining poetry by contrasting it to natural and social science tends to obscure the fact that between these two poles there is a shaded spectrum of language-types which fit neither category and participate in both. Does history imitate action? Is it an art or a science? Both views have partisans, but all must agree that the object of history is to discover recurrent patterns of change in societies, cultures and civilizations. Parts of Gibbon's *Decline and Fall of the Roman Empire* may read like a novel, but it is none the less the work of a great historian because the events it recounts are almost unfailingly of historical significance, i.e., throw light on the causes of Roman decline.

For present purposes, however, the most important contribution of the passage is Aristotle's recognition that the poetic evocation of immediate experience is a means to an end in poetry. To say that the poet is a kind of philosopher, interested in "universals," is to accept the notion, no matter how vague, that the poet is a seeker after truth, that *poesis* is a method of inquiry. If, in addition, the reader keeps in mind what Aristotle says in the *Ethics* about method and the relative degrees of exactitude and certainty in various domains of inquiry, my concluding analysis of the extent and limitation of poetic cognition will be placed in its correct frame of reference.

Poesis, the creative molding of the body and mind of words, so as to represent by hypothetical analogy the essential features of praxis, has, in the first place, an aesthetic aim. By *aesthetic,* I mean the experience of vicarious *participation* in an action, imagining we are living in person the life depicted in words. We speak of "losing ourselves in a book," "being transported, fascinated, charmed, bewitched, ecstatic, etc." The vocabulary of magic and religion reminds us of the origins of art in primitive ritual, wherein the mimetic rain-dance, fertility-chant, and symbolic enactment of hunt or harvest were experienced as necessary parts of the activities they represented, and not as mere symbolic surrogates. Art was as much real life as hunt and harvest.

With the emergence of science and philosophy, the content of aesthetic experience changes. "Symbolic action" separates itself from praxis. Art sets itself over against life, gains perspective on its object, and assumes a measure of autonomy, which makes room for a revolutionary internal elaboration of forms and styles. Only after these differentiations are well underway is it possible for art criticism and scholarship to arise as independent fields of study. In ancient Greece, these developments appear in purest form. Homer, Aeschylus, Sophocles, Euripides, Aristophanes, Plato and Aristotle write the declaration of independence of the arts. Ever since, the relationship between art and life has been a subject for heated and frequently bitter debate.

The sublimation of the arts and the proliferation of new forms did not signify a complete break with magic and ritual participation in communal life: *Mutatis mutandi* the arts have preserved the marks of their origin in the iconic symbols of mimesis, the aesthetic versions of reality. Poetry is a theory of praxis which must embody its object in order to interpret, evaluate and generalize its content. *Poesis* thus presupposes the unity of mind and body. Poetry reconciles aesthetic and cognitive experience. We have to "hear" the curfew ringing in order to "know" what the action in the churchyard contains, and to judge the import and value of Gray's message to us.

Mimetic statement, then, is seriously limited in its powers of generalization. It cannot, like steno-language, simply give the final product of cognitive activity, as science textbooks do. Poetry is obliged to reproduce the process, which made significant insight possible. Its concreteness, historicity, individuality, etc., limit its conceptual scope. The variables in a poem are more numerous than its words. That is why any poet of genuine talent always outstrips his conscious intention, uncovering subtle patterns in action, or committing errors in consistency, which he is surprised to read about when the reviews come out. All any poet can do is identify patterns of action which typify best the various aspects of the life of his times. He must grasp their most salient features, display them in his language, so as to provoke in us a desire to re-enact them imaginatively, and make us feel, believe, think that he has discovered something of importance for us. Inevitably, by-products are created in the process, and it is impossible for the poet to know them all. It sometimes turns out that the quasi-accidental overgrowth is as important or even more important than the patterns of action mimed by the poet in the first place. That is at least one way to become a lesser or greater "precursor", e.g., Rabelais, Cyrano, Richardson, Rousseau. If it were not so, if the poet's language were exhausted in fulfilling his intent, critics would scarcely be needed.

The aesthetic experience is an inward reliving of an action in mimetic response to the movement of a poem. Real life is full of accident and irrelevancies which obscure the meaning of events. Poetry presents experience purified of such dross so that we can relive it efficiently and then judge for ourselves of its sense and worthwhileness. But because re-enactment is essential to the understanding of poetic statement, it is frequently very difficult in the long run to distinguish the knowledge we gained from poetry from that of everyday praxis. The cognitive content of poetry tends to be absorbed into *praktische Menschenkenntnis.* No wonder, then, that we cannot point to an accumulated body of specifically poetic knowledge as distinct from what is roughly called "the lessons of history," or "the wisdom of the ages."

The characters and situations of great literature provide an ever-broadening repertoire of paradigmatic metaphors by which we may efficiently identify, categorize, and evaluate actions of everyday life. Moses, Prometheus, Oedipus, Tristan, Robin Hood, Gargantua, Don Quixote, Hamlet, Tartuffe, Pangloss, Faust, Julien Sorel, Vautrin, Becky Sharp, Fagin, Raskolnikov, Anna Karenina, Huckleberry Finn, Swann, Hans Castorp, Mother Courage, Willie Loman, Dr. Strangelove evoke faces, gestures, actions which epitomize the ways of life of Western man. They cannot be abstracted from the context of action and suffering which gave them life without inseparable loss of meaning. The phrases "Faustian man," "quixotic ideas," cannot have the same meaning for those who know and do not know the works from which they spring. Summaries, critical articles, reviews cannot substitute for the original works, for, as I pointed out earlier, aesthetic re-enactment is a prerequisite to poetic understanding.

But what can these characters and stories tell us that we could not learn from historians, anthropologists, sociologists, etc.? *How is it that these names have the conceptual sweep of ideas?* It is necessary, but insufficient, to say that they typify people and situations. They achieved their status as types through portrayal of the concrete, complex, unique, and historically situated praxis in which they were inextricably engaged. What they give us that nothing else can give us is a sense of *how life felt* in the different times and places which we commemorate as culturally significant. Art is the record of the way men in different historical and social conditions discovered meaningful patterns in day-to-day praxis. These quintessential symbols are as close as we can ever get to the uniqueness of their experience.

The data of art and the data of history may or may not contradict one another. Tolstoy's description of the Battle of Borodino in *War and Peace* is quite accurate according to most authorities. Victor Hugo's chapters on Waterloo in *Les Miserables* are not. But strict fidelity to historical fact can never be the sole criterion of artistic truth. After all, even with the historical conscience of a Tolstoy, the poet cannot guarantee any more accuracy than can be furnished at any particular time by the witnesses, documents and research at his disposal. Hugo's Waterloo is probably useless to a military historian; yet it reveals one dimension of the experience of those who fought that day. Hugo captures with extraordinary power the sense of desperate heroism of the French army, conscious of the enormous stakes and of the enemy's superior firing power. In a military report, this consciousness would be commented upon separately under the heading "Troop Morale." As long as his historical errors are not necessary for the depiction of the heroic content of the action, his poetry retains a measure of validity. A more serious criticism of Hugo's interpretation of the battle is that the epic scale of the heroic style absorbs or excludes almost all other qualities of experience inseparable from war. A good antidote to Victor Hugo is Stendhal. *The Charterhouse of Parma* describes Waterloo as experienced by a young romantic hero who has voluntarily joined the French army in the field in search of glory. We get no bird's-eye view of the battlefield, no descriptions of flashing military panoply, no eloquent evocations of imperial sway. The young, romantic Fabricio gets violently hungry, then nauseated at the sight of a corpse; his horse is stolen by French officers; there is some running and shooting; then, suddenly, everyone is fleeing in panic. From his point of view, Waterloo is a flop; nothing of interest has happened. Hugo's truth and Stendhal's truth do not, however, exclude one another. In real life, epic, comic and pathetic ingredients coexist and intermingle in defiance of both logical and critical categories.

In other words, the information that poetry gives us about praxis necessarily includes agents. It may be wrong about specific facts and still portray significantly the ways in which particular persons experienced a given life-situation. But this is to say that such a portrayal must necessarily remain hypothetical, that poetry presents hypotheses about praxis. It would follow that the only possible confirmation of poetic hypotheses is praxis, that no experiment can be imagined that would tell us whether Gray, Stendhal or Tolstoy were right. Experiments can be and have been performed to ascertain the reactions of groups of readers to various poems. We may perhaps learn from these some valuable things about the different ways in which people interpret the hypothetical statements of poetry. But the only kind of report that might provide confirmation would have to deal with the discovery in praxis of qualities hypothesized in poetry.

Gray's hypothesis, for example, may be re-stated in the following metalanguage: "If the educated urban reader of my poem finds himself someday in a typical, quiet, English country hamlet, with its church and graveyard, pasturage, plowed fields, and cottages, he will find that my way of experiencing and describing scenes has revealed its most important qualities and their most abiding values." The hypothesis covers the categories of "Interpretation" and "Evaluation" discussed earlier. The kind of scene described is known to most people and its typicality transcends local time and place much more than Gray seems to have realized. I have had occasion to recall Gray's poem in upper New York State, Ohio, and central France. Such recollection accentuates the analogies which facilitate one kind of integrated imagery of agricultural civilization which we project. At the same time it helps us, by contrast, to become aware of the peculiar qualities of this or that rural scene for which Gray's typification does not account. Those aspects, in turn, might evoke scenes from Wordsworth, Vigny, Turgenev, Hardy, or Frost. And if the country traveler were a poet he might spot qualities of things and experiences never before set down in language.

That does not mean that Gray's *"Elegy"* can never be superseded. His diction, tone, stylistic texture are inseparable from their historical moment. He has transmitted to us the particular nuance of sensibility and eloquence of the English eighteenth century. A modern writer, made melancholy by the breakdown in America of the rural life and ethos, who tries to sing its praises in Gray's terms, will produce an interesting pastiche of a style gone by, and probably a satire. Part of the change in praxis since the eighteenth century is the change in the language and its uses. One of the reasons that the *"Elegy"* is lasting is because it is forever identified with the past.

A poem's power of generalization can only be known through praxis. The works of literature which have stood the test of time, which have outlived civilizations, historical epochs, and even the language in which they were written, have proved that the patterns of living they embody can still provide insight into modes of praxis of which the old poets never dreamed. It is frequently forgotten that the body of great works we treasure as classics actually represent a very small portion of the total literary production of our culture. Great art reveals itself as men discover that it helps them to live. In the medium of words, poetry is the only possible science of praxis.

Dept. of Romance Languages
Wesleyan University
Middletown, Connecticut

1 Moreover, to answer him by pointing up the failure of scientists to assume their measure of moral and political responsibility for the social uses made of their discoveries and inventions may be a needed corrective to his exaggerations, but does not refute his charges against literature.

2 See my essay "Individual as myth," *Chicago Review 12,* no. 2, 94-119, Summer 1958.

3 David Hawkins, *The Language of Nature* (San Francisco and London, 1964), p 2.

4 "On Poetry and Poets", (N.Y., 1961)

5 Lewis Carroll's *Jabberwocky* is English poetry.

6 Langer, *Philosophy in a New Key.*
Morris, *Technology, Science and Art.*
Cassirer, *Language and Myth.*
Cohen, *A Preface to Logic.*
Wheelwright, *The Burning Fountain*
Richards, *Science and Poetry.*
Northrop, *The Meeting of East and West.*
Ayer, *Language, Truth and Logic.*

7 *The Burning Fountain,* (Bloomington, Indiana, 1954).

8 Professor Wheelwright's defense of poetry against the various schools of positivism is among the best. However, the case he makes for the cognitive value of poetry involves mystical overtones which in my opinion goes far to explain why his argument never quite jells.

9 (Cambridge, Mass., 1957), p. 70.

10 W. B. Yeats, "Among School Children."

11 Does this mean that the observer is part of that which is observed? The answer is, yes. As for the Heisenberg uncertainty principle, the stand taken in this essay is adequately described by Mario Bunge in *Causality* (New York, 1963), especially pp. 346 ff.

12 The concentration of epistemologists on the language and logic of science sometimes obscures the fact that knowledge-getting is a concrete human activity involving the full exercise of all the physical and mental potentialities of real people. We learn by doing and our acts express what we have learned. Cf. Robert Guihéneuf, *Le Problème de la théorie Marxiste de la valeur*:

"Engels met l'accent sur le monde matériel sans insister sur le processus de la connaissance. Marx, au contraire, insiste sur le rapport dialectique fondamental qui unit l'homme et la nature. C'est ce rapport qu'on peut appeler connaissance, si on veut bien garder présent à l esprit la notion de "praxis." La connaissance est action, elle s'insère donc dans le monde materiel, et en même temps l'action est connaissance, c'est-à-dire qu'elle met en oeuvre les facultés intellectuelles de l'homme. L'important dans ce double mouvement est le rapport dialectique "homme-nature," rapport dont le travail est l'expression la plus haute." (Paris, 1952) p. 30.

13 Only literature can deal with moral *situations*. There is theoretical ethics, but practical ethics is drama.

14 (Norfolk, Connecticut) p. 57.

15 A number of thinkers in various fields and of various philosophical persuasions, both here and abroad, have been at work on these problems for some time. I can only name some of the most influential:

In the United States: Kenneth Burke, Hugh D. Duncan, Leo Lowenthal, John Dewey.

In England: Herbert Read, Arnold Hauser, E. H. Gombrich, Christopher Caudwell, Arnold Kettle, Ian Watt.

In France: Lucien Goldmann, Jean-Paul Sartre, Roland Barthes, Arnold Ghisselbrecht.

In Germany: Walter Benjamin, Theodor Adorno, Bertholt Brecht, Werner Kraus, Max Raphael.

In Hungary: Georg Lukács.

In Austria: Ernst Fischer.

In Soviet Union: Mikhail Lifschitz.

16 The major works of Kenneth Burke are: *Attitudes Toward History, Permanence and Change, A Grammar of Motives, A Rhetoric of Motives, The Philosophy of Literary Form.* All of them have contributed to my enlightenment but the detailed analysis of the idea of "action," most important to the present essay, is to be found in *A Grammar of Motives.* Anyone acquainted with Burke's thinking will realize the extent of my debt to him.

17 Burke is unhistorical in a sense that he is more interested in categorizing the forms of symbolic action, in other words in their similarities, than in their specific textures which derive from their historic character. Cf. Georg Lukács: ". . . Even beyond this, however, the objective historicity of being and its peculiarly emphatic appearance in human society have important consequences for grasping in principle the specific quality of the aesthetic," and "It is precisely in works of art that the historicity of objective reality receives its subjective as well as its objective shape." —The New Hungarian Quarterly, vol. V, No. 14. (1964)

18 *Purpose,* being an attribute of agent, and *agency,* of scene or agent, we may deduce that the pentad is based on the dialectical unit man/world discussed above.

19 The poetic way of making a point may be even better illustrated if we imagine his tattoo as five-star epaulettes.

20 In 1917 Wilson was asking the nation to send an expeditionary force to Europe for the first time in American history against a Germany which was in no sense our traditional enemy. In 1941 the President of the New Deal declared war in Asia and Europe against imperial and fascist powers already feared and hated by large sections of the American public. In 1953 how many Americans could have found Korea on a map? Add to these factors the differences of character of the three Presidents, the changes in methods of warfare, and the context of A-bomb diplomacy of the fifties and the content of the same "ringing words" changes color and texture before our unastonished eyes.

[21] Metaphors lend themselves to a kind of expansion or elaboration difficult to characterize. For example, Bertolt Brecht speaks somewhere of the "bloody shoe." Since many people are scrambling up the ladder of success (and most ladders are rather narrow) it is to be expected that some ill-bred climbers may use the face of the man below for extra leverage or safer balance. This might be expanded again by another metaphor; for once on top one is sure to find an obliging fellow who will lick them clean.

12

MICHAEL STOCK, O.P.

METHODOLOGICAL REFLECTIONS ON AQUINAS AND FREUD

Presented April 30, 1964

An Analysis of the Concept of the Unconscious Mind.

I. *Introduction*

The concept of an unconscious mind, which Freud was not the first to postulate, but which he developed far beyond what any other philosopher or psychologist had dreamed of, is not only a key concept in the total structure of psychoanalytic theory, but one of Freud's major contributions to general explanatory psychology, and an eminently fruitful theory in psychological research. It is not, nevertheless, a concept which has completely established itself in psychological circles, for where it is not simply denied, it is often depreciated, and where it is neither depreciated nor denied, it is still sometimes misunderstood. For all these reasons it seems worthwhile to submit the concept of the unconscious mind, and the data and argumentation which lead up to it, to critical scrutiny.

My approach to this scrutiny is conditioned by my own background, which is not psychoanalytic but philosophical along Thomistic lines. This raises a problem, since the validity of a cross-system critique of concepts is open to doubt, and perhaps cannot be sustained. The argument is that you can hardly avoid distorting ideas.

If you are schooled in the methods and ideas of one system, and are interested in a concept from a disparate line of research and speculation, it seems that there are two possible approaches. You could, on the one hand, try to take a detached and uncommitted point of view, outside your own categories and presuppositions, and pass a judgment from that standpoint. Or you could frankly submit the alien concept to your own categories of thought and see how it fares. It is here that the danger of distortion comes in. It seems to me however that the latter course is the only one psychologically possible, since the former course reduces to nothing more than a *de facto* acceptance of principles of operation broader than those which define the presumed commitment; that, in fact, your point of view includes categories and presuppositions which you choose to call detached in addition to the ones you choose to term committed.

Therefore the approach in this paper will be one which takes the unconscious mind as Freud proposed it and passes a judgment on it in the light of the methodological canons of the Aristotelian-Thomistic line of thought.

In fact, I doubt if there will be much straining of categories. While Thomas Aquinas had no concept of unconscious mind as Freud proposed it, he had a number of insights into the mind's workings which prevented him from assuming the one position which *a priori* rules out unconscious mind, the position, namely, which Freud went to such great efforts to dispel, that mental and conscious are synonomous terms. It will not distort Thomistic categories to develop a thesis of unconscious mind within their framework.

Moreover, it does no great injustice to Freud to assess his ideas along these lines. Freud obtained some acquaintance with Thomistic psychology and logic through his studies under Franz Brentano, and these ideas influenced his latter work, and can presumably provide a basis for some kind of valid comparison. (See: Jones, Vol. I, p. 37, Rappaport, pp. 61, 62).

The Materials and Their Logical Form.

The materials on which we will base our study are mainly the metapsychological paper: *The Unconscious* (Standard Edition, Vol. XIV, pp. 166-204) written in 1915, and the short paper: *A note on the unconscious in psychoanalysis* (Standard Edition,

Vol. XII, pp. 260-266) written three years earlier. The former is the more important work—in Freud's judgment, a major step in his own theoretical development (*Jones,* Vol. II, p. 21), but it seems somewhat incomplete without the latter. We will also make references to the paper on *Repression* (Standard Edition, Vol. XIV, pp. 146-158)

The question I will ask is whether Freud proved the existence of an unconscious mind and if he did, whether he defined it accurately.

I do not intend a detailed exegesis of the paper *The Unconscious,* since it would take far more time than we have at our disposal. Neither do I plan to make a critical study of the text as it is written. The reason for this is that I do want to examine the demonstrative force of Freud's arguments, but the paper *The Unconscious* as written, is not a piece of demonstration. In its first section, "The Justification for the Concept," Freud proceeds in a purely dialectical manner, starting not from facts of evidence but from common opinions which he supposes his listeners or readers will accept. "We have demonstration when the premises from which reasoning starts are true and primary, . . . reasoning on the other hand is dialectical if it reasons from opinions that are generally accepted." That is, generally accepted either by everyone, or by the majority, or by the experts. (*Post. Anal.* I, iv, #41; Topics, I, 100a) It would be inappropriate to take arguments formulated on dialectical rules and expect them to satisfy the requirements of demonstrative logic. And at the moment, we are not interested in Freud's skill as a dialectician. We will therefore draw our materials from the sources mentioned, but not limit ourselves to analysing their formal expression; what we will attempt is a conversion of the dialectics into demonstrations.

There is another good reason, I think, for not analysing the text of the paper *The Unconscious* as it stands. Most scientific and philosophical papers are written, at least in their final version, from a reflective and didactic point of view—they have been reformed and polished with the idea of bringing out the force of the argument in the best possible light and of making them most easily accessible to the reader. Freud however wrote this paper (as he wrote many other works) inventively. It is remarkable how much of Freud's writings are published accounts of his "thinking out loud." It is not unusual to find him apparently seriously proposing hypotheses on one page which he will proceed to demolish a few pages later. Most writers prefer to suppress their mental probings; to present, when they are ready to present anything, a polished rendering of their thoughts, with the gropings and false leads eliminated. Freud does not seem to have been concerned with this, and as a result has left us a fascinating record of a first class mind at work searching out problems and their solutions, an invaluable record if you reflect that even the false leads of a deep mind may be instructive.

In any event, there is no doubt that Freud was in the process of working his way through to the concept of the unconscious mind while he was in the process of writing this paper. He sets up two alternative hypotheses early in the paper, and finds, as he goes along, support for one side or the other, until at the end of the last section, he suddenly overthrows both in favor of a third. If we can accept Ernest Jones account of the process, Freud was not thinking of settling the question of the precise difference between conscious and unconscious mental operations when he began the final section of the paper (he was only trying to relate what he had worked out to other facts already known to his readers). But, having undertaken to review some of the facts (namely, those associated with schizophrenia) he bumped into his definitive statement of the nature of the unconscious—re-capturing, according to Jones again, an idea he had previously discovered and forgotten several times, and finally perceiving its significance! (Jones, Vol. II, p. 325)

Freud, then, is inventing as he goes along, leaping from hypothesis to hypothesis and juggling his ideas, and perhaps half conscious of where he is going even in the last section. It would be worthwhile trying to figure out the psychology of this paper, but that would be another study.* At present, we want to analyse its

* Everyone is aware of the role of exactness, of rigorous observation and logic, in science. Freud too admired these qualities intensely, in one of his early teachers, for instance, Ernst Brücke, whom he always regarded as a model of the scientific mind. But he himself was afflicted with a kind of 'sloppiness' of whose

demonstrative force, and since it is not written from a reflective point of view, with the purpose in mind of observing the canons of demonstration, we cannot hold Freud to the formalities of methodology. What we will try to do, however, is to show to what extent there is latent demonstrative strength in his arguments.

II. *Justification for the Concept of the Unconscious.*

The first of the seven sections of the metapsychological paper, *The Unconscious,* is entitled: "Justification for the Concept of the Unconscious." Jones calls it a "brilliant exposition of the necessity and legitimacy of the conception that there is an unconscious mind." (Jones, Vol. II, p. 322.) It is indeed a brilliant display of dialectical skill, undermining the prejudices which obstruct the acceptance of the concept, and pinning the reader down to an admission of unconscious psychic realities on the basis of facts and inferences he would be otherwise ready to allow. Nevertheless, Freud does not explicate in detail the fundamental facts and arguments on which he bases his assertions; he makes reference to them, but does not elaborate. Perhaps the explanation for this is that he had previously, in the paper *(A Note on the Unconscious in Psycho-Analysis)* delivered in 1912, spelled these details out at length. Now, apparently, having failed to convince all of his fellow psychologists of the legitimacy of his conception by direct demonstration, he is prepared to try flanking attacks and ambushes.

value in science he was not unaware. (See: Jones, Vol. I, p. 34) There is a kind of sloppiness which is more aptly described perhaps as a calculated resistance to the extremes of exactness and rigor, especially in the process of invention or discovery of new ideas. It is an unwillingness to refine a category too quickly, i.e., before an uneasiness about the situation is dispelled by the only thing that can legitimately dispel it, namely, an insight or vision of the true solution. This uneasiness arises when we see something and are not sure what it is, and suspect our initial formulas have not grasped it, adequately. It is an uneasiness we would like to dispel and are tempted to dispel, and the temptation has to be resisted, or we risk closing off the thinking process prematurely. Nor can the uneasiness be dispelled by rigorous application of method, procedural rules, known formulas, etc., for the solution might lie outside presently realized categories. So the invention of ideas depends to some extent on a tolerance for sloppiness, and this kind of tolerance Freud developed, or at least did not suppress.

The contents of the first section, "Justification for the Concept of the Unconscious," comprise four distinct lines of arguments. In the first, Freud lays down an outline of a direct demonstration, but instead of a step-by-step accounting of the facts and a close analysis of their significance for an unconscious mind, he simply indicates the facts in a general way, and appeals to the gain in meaning and the pragmatic successes which the assumption of an unconscious mind provides. He states that the data of consciousness show gaps which can be explained only by presupposing other acts of which consciousness affords no evidence. Parapraxes and dreams, psychical symptoms in the sick, and our ordinary experience of arriving at intellectual conclusions without knowing what our train of thought has been, are cited as examples of this. These otherwise inexplicable data, he continues, fall into demonstrable connection if we interpolate between them unconscious acts which we have inferred. And this, he states, is valid, because, if only we assume the unconscious, we first of all achieve a gain in meaning, and secondly, we are able to construct a successful procedure by which we can influence conscious processes.

Therefore, Freud says, the assumption of the unconscious is necessary and legitimate. By necessary, I believe Freud means in the nature of things, i.e., that he means that things could not be otherwise, (See: Freud as a realist, last paragraphs in *The Future of an Illusion.*) for he connects this necessity directly with the data he is bringing forth. By legitimate, I believe he means according to normal rules of thinking, since he connects the justifiability of his assumption with the gain in meaning it affords.

In Thomistic terms, the argument would run this way. Every effect requires a sufficient cause. But every psychic act is an effect. Therefore every psychic act has a sufficient cause. But of some psychic acts the cause is not apparent to consciousness. Therefore there are unconscious psychic acts, or an unconscious mind.

In general the argument is a simple statement of the necessity of inferring a cause. If you are presented with an event which has come into being, since it could not have come from nothing, it must have come from a cause. Therefore you have to posit causes if they are not directly knowable in themselves. To be necessary is the same as to be. Moreover, since a cause is a principle from which something comes with dependence in being, a cause makes a thing to be what it is, or determines it. To know a cause, therefore, is to know why a thing is what it is, and this is full knowledge of things. Therefore by positing an unconscious mind, we make it possible to understand what is otherwise unintelligible, and therefore it is justified. Moreover, it is possible to devise successful methods of influencing the mind, by positing unconscious causes of mental acts, and this further establishes its validity.

These arguments are valid as far as they go, but they do not make explicit the precise point needed to demonstrate the conclusion. What Freud must show is that the assumption of the unconscious mind is the only possible assumption which will explain the facts to be explained, and the only possible assumption on which you can construct a successful method of influencing the conscious. To do this he would have to show that the unconscious causes he discovers are actually themselves psychical acts; that a mental event, x, which is conscious, is the cause of an event, x-i, which is unconscious and psychic in character, and that event x-i is the cause of another event, x-ii, which is psychical and conscious. There could, of course, be several unconscious events, all psychical and all related in a cause-effect series, and this would strengthen the case, but the process described above would be sufficient to establish unconscious mind.

Now in fact the evidences which Freud brings forth are all of this nature, but he does not describe them explicitly at this point. Hence he is leaving himself open to the objection that the gaps between conscious mental events could be explained by postulating neurological (somatic) events rather than unconscious psychical events.

Therefore in the second line of argument, Freud takes up this challenge. He notes the fact that the vastly larger parts of our conscious knowledge are in a state of latency for long periods of time, and from this state of latency they pass into a conscious state, and this, he says, necessarily proves the existence of unconscious mind. If someone says that latent contents are present in the mind as residues of somatic processes, the answer is that they are manifestly residues of psychical processes, for example, ordinary memories, which obviously arise from previous psychic experiences. It is hard to know whether or not Freud had his tongue in his cheek when he said this. Obviously, the objection is that the memories latently present as residues are present *in the form of* somatic processes, and not that they are latently present as residues left from processes which were originally somatic. So, for the moment at least, Freud does not establish his demonstration convincingly.

He turns rather to the problem of undermining the prejudice that exists against unconscious mind—the assumption that mental and conscious are equivalent terms. He points out that this assertion is either a begging of the question or a convention, and if it is a convention, it is a bad one for four reasons. The reasons are closely connected and they all have to do with the relationship of pre-scientific knowledge to scientific knowledge, that is, the question of how one sets off or delimits the area of investigation of a single science. In sum, they amount to the statement that a convention which prevents a science from finding all the answers to the questions it investigates is a bad convention. And the assumption that conscious and mental are equivalents does this. First because it disrupts psychical continuities, that is, it rules out the possibility of finding psychical explanations for some psychical facts by accepting in principle the position that some psychical events can arise without pre-determining psychical causes. Secondly, it leads us into insoluble difficulties, namely, into the so far frustrating task of trying to locate the neurological events which adequately explain psychological events, the problems of psycho-somatic parallelism. Thirdly, it attributes the property of the part to the whole, without good reason, that is, it assumes without proof that consciousness plays a part in all mental activity. Fourthly, it obliges us to abandon research in psychology without giving us any compensation from another field, i.e., it rules out the chance of explaining psychical acts psychologically without showing that these facts can be explained from physiology or some other line of investigation. Taking these all together, Freud is simply asserting that if psychology is to be a science, it has to find psychological explanations for psychological facts, or, if this is impossible, it has to have some evidence that another science can explain its facts—any convention that rules out the first alternative and shows no hope of fulfilling the second, is invalid.

At this point, Freud sums up what he judges has been accomplished by his arguments, overstating his case perhaps in asserting that the only way mental contents in the latent state differ from mental contents in the conscious state is in the absence of consciousness. Freud's opponents would add: and in being somatic rather than psychic. As a kind of rejoinder, Freud again makes reference to the data of psychoanalytic investigations, which abundantly prove the existence of unconscious psychic processes, and he adds the piece of evidence from the experimental field, the evidence of post-hypnotic suggestion. But again, he does not draw out the full significance of these data.

In order to present the core of Freud's proof of the unconscious mind, we will have to examine these data more closely, but before doing that, we might describe his final line of argument. We are directly aware, he points out, only of our own consciousness. Yet we customarily infer the existence of consciousness in other people because we can explain their external actions only on this assumption. He admits that in times past this assumption was made uncritically, when consciousness was attributed to trees, rocks, animals and natural forces, but asserts that while critical reflection has caused us to withdraw our assumption in these cases, it has not been able to dispel our belief in the consciousness of other people. If therefore this belief is valid, then the belief in an unconscious mind within ourselves is equally valid.

Freud, of course, immediately notes that his analogy has proved too much. It logically leads to the assumption, not of an unconscious mind in us, but of a second and unknown consciousness in us, and Freud is quick to point out that this is a very shaky position to adopt. If an unconscious mind is hard to accept, how much more difficult it is to accept an unconscious consciousness. Moreover, as he mentions, the operations of the unconscious mind are so different from those of the conscious mind, that it should be set off as a different kind of thing, and not just another consciousness.

But having shown that he proved too much, Freud does not abandon the whole line of approach; he suggests that having proved more than he needed to prove, he will be content if the reader accepts a diminished version of his conclusion—an unconscious mind if not an unconscious consciousness! This, I suspect, Freud takes as an argument from *a fortiori,* and seems valid enough if you can show that unconscious mind is a stage half way between conscious mind and a somatic process. But that is still the crux of the process. What Freud has succeeded in doing, with readers who followed this line of argument, is to focus attention on modes of operation in virtue of which we customarily, i.e., spontaneously, attribute the properties of 'mind' in cases in which we have no direct awareness of mind. When the modes of operation are unintelligible without the interposition of psychical acts, we postulate the existence of a mind. We do this to other people; by the same token, we should do it to ourselves. The reason we do not spontaneously attribute a second mind to ourselves, one operating outside the range of consciousness, is because it seems bizarre to have a mind of which we are unaware, but this, Freud maintains, is an unjustifiable prejudice.

Freud's real or demonstrative assault on this prejudice is expressed more fully in the earlier paper, "A Note on the Unconscious in Psychoanalysis," to which we have already made reference. At this point, then, we want to examine what he says there, in the light of Thomistic rules for demonstration.

In essence, Freud must prove, as we noted above, that a conscious psychic event, x, causes an unconscious psychic event, x-i, which in turn causes a conscious psychic event, x-ii, either directly or through one or more other unconscious psychic events. X-i, and the other unconscious psychic events caused by it, if any, would constitute the unconscious mind.

As prerequisites for this proof, we must define the difference between a psychic event and a somatic event, and what we mean by a cause, and finally, precisely what we mean by a psychic cause. A process we will take to mean a series of events connected causally.

One type of psychic event is defined by the property of intentionality, i.e., it represents or stands for some object other than itself, presenting this object somehow to the subject of the event. It provides secondarily the basis for the attachment of significance to the object. By significance, I mean that the object as represented

—not necessarily as it exists physically—becomes an object of a directed response on the part of the subject, i.e., of a response which will employ the object for the purposes of the subject.

One characteristic of such psychic events is that, as psychic, they are not physically detectable in the subject, although it is reasonable to suppose that every psychic event is correlated with a determined somatic event, and in some cases correlations can be shown, at least roughly. Many conscious events are psychic in this sense, i.e., intentional or representative and/or significative.

Somatic events, on the other hand, are detectable by physiological investigation, are not intentional as such, and do not appear in consciousness as such.

Another kind of psychic event is called affectivity, which is characterized as a state of feeling with at least two general poles, pleasure and pain, each of which has many modalities—satisfaction, relief, enjoyment as opposed to dissatisfaction, frustration, anxiety, fear, anger, and so on. These states can be generated by physical or physiological causes operating on or in the body, and also by cognitive causes, ideas and images which represent physical or physiological causes, and arouse affects by acts of recognition and anticipation. In turn, affects can influence ideas and images, as love and fear stimulate ideas and images of actions appropriate to obtain or avoid the object loved or feared. While the influences of affects on physiological processes and physiological processes on affects do not belong purely to the psychic sphere, the influence of ideas on affects and affects on ideas do; these are part of psychology's domain.

A psychic process, then, is a series of psychic events one of which causes the other.

The question then is: when can we validly assert a true causal connection between events? One contemporary philosopher of science (Ernest Nagel, *The Structure of Science,* pp. 73-78) lays down four conditions for the valid assumption of causality: A can be said to cause B if the sequence of B from A is invariable, i.e., if A is the necessary and sufficient condition for B, if A and B are contiguous in time, and also in space, and if the relationship is asymmetircal, i.e., B follows from A but A does not follow from B. I would agree that these conditions when fulfilled, provide a sufficient reason for asserting a causal connection, of at least one kind of causality. But I would maintain also that there are other conditions which might be added, which, if fulfilled, give an even more convincing picture of the operation of causal influences.

In the thought of Thomas Aquinas, causality is a complex notion, fundamentally metaphysical, and knowledge of causes is the key to scientific knowledge. The causes of a thing are its complete determinants; therefore to know the causes of a thing is to know the thing fully, which leads toward scientific certitude. Not all the causes of everything are equally knowable; where the knowledge of causes is obscure or uncertain, the knowledge of the thing is correspondingly deficient.

The ideal of perfect knowledge can only be attained when all the causes are known, and these fall into four classes: the final cause, or purpose of the thing, the efficient cause, or agent that produces the thing, the formal cause, or intrinsic determinant of the thing, and the material cause, or what the thing was made out of (*Posterior Analytics,* Book II, Lesson 9). The four causes are interconnected. The final cause causes the efficient cause, since it establishes the purpose for which the agent acts, the efficient cause causes the formal cause, since the latter is the precise effect it produces, and the formal cause determines the material cause, since it is related to it as determinant to determinable.

Obviously, there are many events for which the four causes cannot be established. They are most manifest in human activity,—in the arts, in logic, in moral actions,—in which the purpose and the agent's activity are most knowable, often from within. To give a simple example, a table can, in theory, be known about as completely as is possible. The purpose or purposes which motivated the carpenter give its final cause, the skill in carpentry which he used can be analysed into its elementary parts, the shape that emerged is the direct result of the carpenter's actions, and the wood, glue, nails and other fasteners and paint and so forth comprise the material from which it was made. The final cause might be complex—the carpenter might be motivated by a desire for money, for reputation, for recreation, etc., but his purpose had to crystallize at some point into a resolve to make this table. His efficiency was modified and enhanced by his tools, and his tools might have been shaped for his immediate purposes by other tools, and his health and energy affected his working habits—all the aspects of the case that influence the actual production belong to the order of efficient causality. The model he had in mind, which is eventually realized in the actual shape of the table, the modifications which were introduced into the final form—aspects of this nature pertain to formal causality. And all the preparations which bring the materials to the final point at which they can be included in the table belong to the genus of material causality. Taking them all together, once they are known, the table is known as far as it is possible for man's mind to grasp a thing.

When we try to apply a paradigm like this to natural science, we immediately run into a variety of obstacles. First of all, the final cause cannot be directly known, nor can we always get inside the agent, to see any part of his efficacy at work. Nor are forms as simple as table-shapes, nor is any natural matter completely known, at least at present. All along the line there is a diminution of intelligibility.

Nevertheless, there are analogues of the four causes, more or less closely assimilable to the causes as they function in their clearest forms of intelligibility, and in virtue of these analogues, we can get more insight into the causal aspects of natural events than the four rules outlined above indicate. For instance, in natural objects we cannot ascertain the intention, if any, which motivated or motivates the agent, except in some human operations. We can, however, draw an analogy between the effect of purposive action as opposed to chance action, and postulate a form of final causality in those terms. Chance effect, for example, occurs according to determinable patterns of distribution. Where these patterns are not found, where the effect is, in Aquinas' words, something which occurs all the time or in the majority of cases, we can suppose that some influence is working which is not chance, but allied rather to purposive action, which achieves its end all the time or for the most part. If in addition we can determine that a given effect is a necessary condition for a further effect, and both of these are steps in a developmental process, and this within a closed system, we have an even closer analogy with true causality. A concept of causality of this sort can play a considerable role in biological and psychological science; even an anti-teleological psychologist like Freud uses terms such as purpose and goal and end to help explain mental processes.

Similarly, in efficient causality, we cannot get a look at the inner workings of agents, again apart from human agents. Nevertheless, it is also possible to determine something more than a necessary condition of an effect, contiguous in time and space and asymmetrical. We may not be able to discern the model in the agent according to which he operates,—the model which determines the form he is educing in the materials at his disposal—but we can often detect the pre-existence in the agent of the form which comes to be in the matter, and this supplies an additional basis for the assertion of a causal connection. The fact that the offspring of organisms are like the parent organisms, not only in specific form but often in individual characteristics, the fact that a printed or stamped product, from a book to a fossilized footprint, bears the specific and even individual peculiarities of its cause, gives added certification of the causal connection. In Aquinas' phrase: Every agent acts (to produce something) similar to itself.

We cannot go into all the ramifications of the theory of causality and its significance for demonstrative theory here, but what has been said will perhaps be enough for our present purposes. Fortunately, the materials studied in psychology often lend themselves readily to analysis in terms of all the genera of causes, and enable us to attain a deeper insight into their intelligibility than the materials of most other sciences allow.

With these definitions of 'psychic' and 'causal' in hand, we can propose a definition of a psychic process, i.e., a series of causally connected psychic events. A psychic process is a series of psychic events each of which predetermines the subsequent one and all operate at the level of intentionality or affect. In the cognitive order, examples of this causality are numerous. Classification by which a category is formed mentally to embrace two or more objects or experiences on the basis of a perceived likeness, and subsequently the inclusion or exclusion of newly perceived objects, is a phenomenon caused at the psychic level. Such categories or classes can subsequently become the subjects or predicates of propositions and be asserted or denied one of the other on the basis of experience or logic. Propositions can cause new propositions by simple or complex deductions, and propositions of all kinds can be put into modal forms by comparing them with their own conditions of mental existence, e.g., doubt, possibility, probability, certitude, etc. By division a class is broken up on the basis of differences represented within it; by analogy, a class is extended to dissimilar objects on the strength of causal connection, proportionalities, etc. At the sub-intellectual level, imagery and fantasy form combinations by association of similarity and contrast, and combinations of images and ideas operate to produce symbolic representations, and so on. All these are samples of psychic causes operating by psychic laws.

In the affective order, objects incite emotions, and the memories or anticipations of objects can arouse the same emotions; emotions arouse emotions (we fear fear) and inhibit emotions (anxiety inhibits desire), while the discharge of energies in action results in pleasure, and inhibition of discharge in frustration, rage, etc.

In terms of final causality, to explain or understand or express or organize operate as goals in the intentional order; to achieve pleasure and avoid pain operate as affective goals. In terms of efficient causality, the rules of logic, for example, govern the permissible sequence of ideas which are ordered to valid explanation or organization, the laws of association express the pre-determined sequences of movements from image to image, the exciting or inhibiting influences of different emotions exert appropriate influence on the sequence of ideas and of actions, which eventually terminate in pleasure or pain. Looking at the formal aspects of psychic phenomena, we can often trace the connections by formal similarities which amount almost to formal identities in some cases. Thus a memory is obviously a memory of a previously experienced event to which it points as a present re-presentation of the past, and this memory can operate as a purpose to be accomplished in the future to serve the same ends it obtained in the past.

In these terms, then, is there an unconscious mind? Are there psychic processes which operate outside the area of that psychic operation we call self-awareness or consciousness? Freud claims that this is manifest.

In his *Note on the Unconscious in Psychoanalysis* he spells out cases from the analysis of which it is impossible to avoid the conclusion that unconscious mind exists. He first presents the case of post-hypnotic suggestion. A person is put into a hypnotic state, and while he is in this state is ordered to execute a certain action at a certain fixed time after his awakening. Then he is aroused from the hypnotic state. He appears to be fully conscious and in his ordinary condition and has no recollection of his hypnotic state. At the prescribed moment, he feels an impulse to execute the order which was given, and does so consciously, although not knowing precisely why.

If we analyze this situation, we find first that the person apprehended an order, which has objectively two aspects, the statement of what is to be done which is a cognitive aspect, and the imperative mode (do this) in which the action was couched, which is a volitional aspect. From the subsequent act of carrying out the command we can assume that the cognitive aspect perdured in the mind and the volitional aspect was accepted by an act of compliance. When the hypnotic state is lifted, and the person performs the act, we find he is conscious only of the idea of the action to be performed, and of an impulse to perform it and of the actual performance. He is unconscious of the original apprehension of the order under either aspect, of its perdurance, of the translation of command into compliance, and of the origin of the impulse to action from the act of compliance. There are thus three unconscious psychic causalities—the connection of cognitive aspect of the command in the mind with the execution of the command, the translation of its imperative mode into an attitude of compliance, and the instigation of the impulse to act from the unconscious attitude of compliance with a command. The justification of the assertion that these are nevertheless causally connected is established by the fact that the action eventually executed is identical with the action originally commanded.

Freud's second case comprises the descriptions of the actions of hysterical patients. A hysterical woman having a typically patterned 'fit' is unaware that the movement and gestures she is executing are a dramatization of ideas which are actually present in her mind. By psychoanalysis these ideas can be made conscious, and their appropriateness to the actions demonstrated, but during the attacks of hysteria, the woman is not only unable to make any connections between her gestures and the unconscious mental patterns which direct them, but does not even represent to herself any conscious pattern of these motions. She is also unaware of the impulses which supply the dynamic force behind her ideas and actions.

Freud adds evidence from daily life—from wit and parapraxes, from the interpretation of dreams, but with the two examples outlined above, the establishment of the unconscious mind seems to be complete.

Nevertheless, there remains a nagging doubt which historically manifests itself as a reluctance on the part of many psychologists to accept the idea of unconscious mental activity. And it seems that, in the terms of the demonstration we have set up above, we can point precisely to the source of this doubt.

We have defined psychic events as intentional events, and unconscious psychic processess as series of causally connected psychic events which are outside the area of mental awareness. But intentional events have a two-fold polarity—they not only make reference to objects outside the self, they point also to the subject of the intentional act. When I know something I concomitantly know that I know it, at least in an oblique and virtual way. Now it is possible, as we have shown above, to demonstrate causal series of mental events whose causality depends on psychic or intentional modes of operation, i.e., by representative and significative modes of connection, and this is what Freud has called unconscious mind. But is it possible to argue that you do not have true mental operations if the subjective, reflexive aspect of an intentional act i.e., the reference of object to subject through the act, is eliminated? And, by definition, unconsciousness eliminates the reflexive part of a mental act since unconscious events are precisely those which we cannot refer to ourselves as subjects.

In his justification of the concept, Freud does not try to solve this last doubt, beyond the assertions which we have mentioned, that conscious and mental cannot be equivalated. Nevertheless, by the time he has finished his paper on the unconscious, Freud has taken a long step towards solving even this last difficulty.

III. *The Definition of the Unconscious*

After establishing the existence of the unconscious, Freud takes the next logical step, in the light of Thomistic procedure, and begins to work out its definition. A precise definition which expresses a core reality should be couched in terms of a proximate genus and a specific difference, and the proximate genus in turn is a class which is divided from a broader genus by a specific difference, and this process continues until it reaches the simplest genus in any order, which can only be intuited and not properly defined. *(Post. Anal.,* Bk. II, L. xiv, 537ff) The material elements which enter into the composition of any object are expressed in the generic part of the definition, the precise formal element which constitutes it is expressed in the last specific difference. (*Post. Anal.,* Bk. II, L. xiii, 530ff) A complete definition adds the agent and purpose of the object defined.

In the last six of the seven sections comprising the paper on *The Unconscious,* Freud assembles the materials for a definition, which he finally proposes in the last section. In the second section he proposes two alternative hypotheses which might express the formal difference between the conscious and the unconscious. His first hypothesis is that the conscious differs from the unconscious insofar as there is a fresh registration of an idea in a different psychological place in the mind. The second hypothesis is that the idea remains in the same place but undergoes a functional change of state. The first concept would require that one and the same idea exist simultaneously in two places in the mental apparatus; the second hypothesis requires only a difference in the energy affecting an idea. The first hypothesis involves a topographical distinction between the unconscious and conscious minds, and suggests that there are two minds in man; the second hypothesis involves only a functional difference, and leans towards the idea of one mind with two distinct levels and manners of operation.

In the third section, Freud deals with the material elements which make up conscious and unconscious mind. He notes that instinctual urges cannot be called unconscious in the strict sense because in any case they only attain consciousness through the ideas to which they attach themselves. It is thus meaningless to speak of them as conscious or unconscious in themselves. The case for emotions and affects is not quite so simple, but Freud leans to the opinion that they too do not properly belong to the unconscious. His overall conclusion is that ideas and imagery are the elements of the mind which can meaningfully be termed either conscious or unconscious.

In the fourth section Freud considers the efficient and final causes of unconscious mind, namely the process of repression which produces it, and the reasons repression operates.

In the fifth section he considers the kinds of processes which operate in the unconscious as opposed to the conscious mind; how ideas in the unconscious follow laws different from those of conscious thought.

In the sixth section he outlines various relations between the two kinds of mind, and finally, in the last section, he moves into a definitive statement of the specific difference between the conscious and the unconscious. "We now seem to know all at once what the difference is between a conscious and an unconscious presentation." (p. 201)

On the basis of the factors Freud describes as pertinent to the operations of conscious and unconscious minds, we can draw up definitions in terms of the Thomistic four causes. Taking Freud's first hypothesis, we can define the unconscious mind as a part of the mind whose ideas and images are prevented from being re-registered in the part of the mind of which we are aware by a process called repression, which operates in order to prevent the occurrence of unpleasure. On the basis of the second hypothesis, this definition would only have to be changed to read: the unconscious comprises those ideas and images which are deprived of the energy needed for a reflective mode of functioning, by a process called repression, which operates to prevent the occurrence of unpleasure.

In these two definitions, and also in the third definition which he will eventually propose, the material, efficient and final causes are the same. The material cause, or elements of the unconscious mind, are ideas and images, which, with their associative processes of various kinds, constitute a mind or part of a mind separated from the conscious mind. The proximate genus, therefore, of unconscious mind, is mind taken broadly enough to include both conscious and unconscious ideas. Mind itself is a category of vital operations which differs from other categories by having the characteristic property of intentionality.

The efficient cause which renders ideas unconscious is the process of repression. In an earlier paper, (*Repression,* Standard Works, p. 146) Freud defines repression: ". . . the essence of repression lies simply in turning something away, and keeping it at a distance, from the conscious." (p. 147) It is "essentially a process affecting ideas on the border between the system Ucs. and Pcs. (CS.) . . . " (*The Unconscious,* p. 180) It operates in a two-fold manner. The cathexis, or charge of libidinal energy attaching to the idea is withdrawn, reducing its energic level to something below the level required for preconscious operation, and an anticathexis, or counter-energy, is applied, to keep it from re-emerging on the conscious level. (Ibid., pp. 180-181)

The final cause, or purpose of repression, is to prevent unpleasure. Although Freud did not generally think in teleological terms, there is no doubt that at least in the case of repression, he was willing to posit a final cause. He calls unpleasure a condition of repression, but not a pre-condition; he rather describes it as a condition which operates as a motive force. "It has consequently become a condition for repression that the motive force of unpleasure shall have acquired more strength than the pleasure obtained from satisfaction" (*Repression,* p. 147). Later in the same paper he expresses himself even more unequivocally: "We recall the fact that the motive and purpose of repression was nothing else than the avoidance of unpleasure." (Ibid., p. 153)

With the final, efficient and material causes in hand, the question still remains: precisely what is the formal difference between conscious and unconscious mind? Is the unconscious mind those ideas whose re-registration is being prevented, or is it, as the theorizing on the nature of the force of repression suggests, those ideas which have a diminished cathexis and an anticathexis working against them? Before he has settled on either of these alternatives, Freud lights on a third which to his mind is in every way more satisfactory.

In analyzing the peculiarities of speech exhibited by schizophrenics, Freud comes to the conclusion that schizophrenia uses words the way the transference neuroses use things. In the transference neuroses, thoughts proliferate in the unconscious mind by associative processes connecting repressed images and memories of objects or experiences with similar images and memories; in schizophrenia the associations are not between images and memories of similar objects, but by similarities of words associated with objects. Then Freud comes to the threshold of his new definition of the unconscious: "What we have permissibly called the conscious presentation (or idea) of the object can now be split up into the presentation of the word and the presentation of the thing; the latter consists in the cathexis, if not of the direct memory-images of the thing, at least of remoter memory-traces derived from these." Then comes the 'eureka'; "We now seem to know all at once what the difference is between a conscious and an unconscious presentation. The two are not, as we supposed, different registrations of the same content in different psychical localities, nor yet different functional states of cathexis in the same locality; but the conscious presentation comprises the presentation of the thing plus the presentation of the word belonging to it, while the unconscious presentation is the presentation of the thing alone. The system Unc. contains the thing-cathexes of the objects, the first and true object-cathexes; the system Pcs. comes about by this thing-presentation being hyper-cathected through being linked with the word-presentations corresponding to it. . . . Now, too, we are in a position to state precisely what it is that repression denies to the rejected presentation in the transference neuroses; what it denies to the presentation is translation into words which shall remain attached to the object." (Standard Works, pp. 201-202)

Thus the final definition of the unconscious is: the ideas and images of the mind whose verbalization is prevented by the process of repression in order to prevent the occurrence of unpleasure.

The validity of this definition can be established if we can show that this precise difference will achieve the end which the cause of the unconscious mind intends, and if we can also show that this difference will account for the diversities of process which distinguish the conscious from the unconscious mind, for a precise definition purports to express the core reality which is directly related to the end of a thing and which underlies all the properties it exhibits.

The cause of the unconscious mind, repression, has, as its goal, the prevention of unpleasure. The characteristic differences which distinguish conscious from unconscious mental processes are summed up by Freud in the fifth section of his paper.

In the unconscious, ideas are exempt from mutual contradiction, that is, wishful impulses whose aims would appear to us to be incompatible continue side by side without diminishing or cancelling each other out. Again, in the unconscious, the primary processes rule, and these are characterized by displacement and condensation. In displacement, the whole cathexis or energizing force behind one idea may be lost to another, with the effect that the second idea takes on the role of gratifying the instinctual impulses previously attached to the first. In condensation, one idea may appropriate the whole cathexis of several other ideas, with the effect that it functions thereafter to represent in global form all that the other ideas represented. In secondary processes, by contrast, ideas are distinguished by definition from each other and cannot serve to represent or function in place of each other. Again, unconscious ideas are unaffected by the concept of time—they are not altered by time, nor ordered among themselves by time, nor affected in any way by the exigencies of time. In the unconscious, ideas demand satisfaction without reference to timeliness. And, as a matter of fact, unconscious ideas are unaffected by any regard to reality—they press for satisfaction without reference to the objective possibilities, the circumstances which make a thing opportune or inopportune, the social acceptibility of the desire, etc. Their urge to fulfillment depends only on their intrinsic force, and the direction they take only on the pleasure or unpleasure they promise.

The precise question here is: is it the non-verbalization of ideas which enables the mind to avoid unpleasure and causes the differences between mental processes in the conscious and unconscious?

An adequate answer to this question would entail another study at least as long as the present one, and would probably involve us in a number of controversial issues. Nevertheless, a preliminary survey of the various elements in the question suggest that the final results will be affirmative.

The first problem which would have to be settled is the relationship between things and memories or images of things on the one hand, and words, on the other. Somewhere along the way the question would have to be raised concerning the relation of ideas to images; are they synonymous terms, or do they express different aspects of one thing, or two different things? Freud never worked out a detailed theory on these points; in Thomistic psychology on the other hand these interrelationships have traditionally formed one of the major areas of interest.

In Thomistic psychology, a sharp distinction is made between the idea and the image of a thing; the idea is an abstract and universal representation, the image is a concrete and singular representation. Between the idea and image is the complex process of abstraction, operating in a variety of ways for a variety of purposes. Images themselves are never directly connected with words; only the ideas derived from images are expressed in words. (Unless perhaps a man learns to connect words with images and things as a parrot does; this however is not considered specifically human verbalization.) Verbalization, then, is a process which presupposes the acts which might be summarily termed 'ideation,' and these include classification, definition, division, judgment, reasoning, etc., all of which depend on the preliminary act of abstraction.

If we accept this psychology, Freud's definition of the difference between conscious and unconscious becomes highly significant. In Thomistic terms, the unconscious includes all the memories, imagery, fantasy, etc., which for some reason or another are not or cannot be subjected to the process of abstraction and ideation, and therefore cannot be translated into words. On this supposition, the processes which rule in the unconscious are all easily explicable; they are all processes which proceed in accord with laws of imagination as opposed to or lacking the influence imposed by laws of ideation.

The laws of imagination include the laws of association by similarity, contrast and contiguity, the appearance of images in the mind in virtue of their clarity, repetition and recency, and the emergence of images on the basis of emotional or drive force behind them. The laws of ideation are basically definition, division and reasoning. Definitions are abstractions of the essences of things, operating at various levels more or less removed from the concrete nature, all the way from the specific essence to the most remote genus. Divisions follow on definitions, breaking concepts into their component parts. Judgments couple concepts by affirmation and disassociate them by negation; reasoning, into which enter all the prescriptions of logic, leads from judgments to new judgments.

On the strength of ideation, man investigates an outside reality in which he tries to find the meaning, locates his own meaning and orients it towards reality, and attempts a synthesis of his internal and external experiences in terms of what he can accept as true, good, logical, realistic, and so on. All is expressed to himself and to others in words.

In the light of this concept of ideation, can we say that the prevention of unpleasure is a result of the prevention of ideation and verbalization of some images and memories? The unpleasures the mind can avoid by repressing memories into the unconscious are those which arise from the becoming conscious of id drives which are hard to contain, and which therefore threaten either to compel behavior which will be punished (in reality or by the superego), or to raise the level of psychic tension to an unbearable degree. When threatened by the eruption of such unconscious mental contents, the mind reacts with anxiety. Earlier in his writings, Freud theorized that anxiety was produced directly from the id by the conversion of undischargeable libido. Later he modified this opinion, and made ego organization and the function of judgment key steps in the generation of important forms of anxiety (signal anxiety), and even in those cases in which anxiety seems to arise directly from interference with libido discharge, it is the threat of reducing the ego to a state of helplessness in the face of rising tension which generates anxiety. (See: *Inhibitions, Symptoms and Anxiety,* Standard Works, Vol. XX, pp. 140-1)

In either case, ego organization and the capacity to form judgments about threatening situations are prerequisites for triggering off anxiety. Anxiety in turn must operate to fend off some more extreme form of unpleasure. We may assume that this ultimate unpleasure is simply the ego's disintegration under the onslaught of id forces, but this will not explain how the prevention of the verbalization of the unconscious contents assists the ego. If it undergoes disintegration, it undergoes disintegration whether or not it recognizes clearly the forces to which it is succumbing. But if we assume that the ultimate unpleasure is not simply the disintegration of the ego but the ego's recognition of its own disintegration and the traumatizing memories which will effect it, its verbalization, i.e., acknowledgement and articulation, of the on-going process of its own destruction which it is being obliged to predict, witness, and record, then the prevention of verbalization enables the ego to escape the last turn of the screw. Then, to non-verbalize is to fend off the ultimate unpleasure, to refuse to allow consciousness to a memory is to escape the unbearable. The distinction between conscious and unconscious as verbalized and non-verbalized is directly related to the goal of repression, the prevention of the maximum unpleasure.

We can also understand the diversity of characteristic modes of operation in the unconscious as compared to the conscious in virtue of the same distinction.

Unconscious ideas are exempt from mutual contradiction because ideas only attain the status of being able to contradict each other when they are formulated in judgments. Unconscious ideas are devoid of a time sense, and of any meaningful relationship with external reality because they have not been incorporated into the context of meaningful ideas which is the specific product of ideation. Hence unconscious ideas are more subject to processes of condensation and displacement, since these are the proper modes of operation at the level of concrete imagery and affect, and are only disestablished when ideation moves in to organize and hierarchize images and memories on the basis of conceptual laws.

A brief survey, therefore, of the effects introduced into mental operations by the processes of ideation and hence of verbalization tend to confirm Freud's judgment that the precise effect of repression is to prevent the translation of memory and image presentations in the mind into image plus word presentations. (Cf. *New Introductory Lectures on Psychoanalysis: Syntopicon,* Freud, p. 837ff.)

If this analysis is correct, it should also serve to remove the last vestiges of reluctance to accept the concept of an unconscious mind. We conjectured above that this trace of hesitation might come because a mind that has no subjective reference at all seems to be no mind at all. Ideas and images which in no way appear as related to oneself seem to be somewhat bizarre mental furnishings. But if it is true that the unconscious mind contains not necessarily ideas and images which are totally lacking in subjective reference, but ideas and images whose subjective reference is highly diminished because they cannot be woven into the peculiarly human state of consciousness which is consciousness of the world as meaningful, we have an unconscious mind with a minimal but still operative reference to the subject. Even concrete images and memory traces can reveal themselves as 'of the subject,' however dim the reference might be. In fact, this minimal reference is reported as having existed in some cases after repression has been removed.

In summary, then, it seems that we can offer, as a final suggestion, the notion that the contents of the mind are capable of operation at four or five distinct levels. The highest degree of consciousness is attained in those ideas which are abstracted from the phenomenal level, are meaningful, integrated with each other and governed by logical coherence and reference to reality, and are to the highest extent capable of articulation, and in close unity with these ideas are the images and memories which constitute their bases, and which are subject to ordering and manipulation in terms of these ideas. These contents constitute the focus of that awareness we call consciousness. Within this sphere of awareness are also emotions and affects, and the control of motility, which in any event do not pertain strictly to unconsciousness. This conscious part of the mind can be divided into conscious and preconscious, depending on whether it is actually operating or only habitually present, subject to easy recall. At a somewhat different level, but still pertaining to awareness, are the contents of imagery or fantasy which emerge into view in virtue of the force of instinctual drives, but which are not part of the context of meaningful ideas. The contents of dreams, the fantasy which sometimes accompanies parapraxes and wit, the fantasies symptomatic of psychological stress in the neuroses and psychoses, etc., belong to this level. Since the 'meaning' of these contents is a matter of their unconscious motivation, they are not intelligible or capable of being integrated into the higher structure of ideational contents in the form in which they present themselves. If they become objects of ideation, submitted to the abstractive processes, they are not generally understood, since their manifest appearances give few clues to their real significance, and thus there are parts of the mind which strike us as strange, fearful, bizarre and irrational. Finally, at the deepest level, there are the fantasies which operate either entirely or at least for all practical purposes, outside the influence of ideation, and are thus outside the area of awareness, and constitute the unconscious mind proper.

Saint Stephen's Priory
Dominican House of Philosophy
Dover, Massachusetts

BIBLIOGRAPHY

[1] Aquinas, Thomas, *Posterior Analytics.* (1955, Marietti Editori Ltd., Turin, Italy)

[2] Aristotle, *Topics* (The Works of Aristotle, Trans. W. D. Ross, Vol. I) (1928, Oxford University Press, London)

[3] Freud, Sigmund, *The Unconscious—A Note on the Unconscious in Psychoanalysis—Repression—Inhibitions, Symptoms and Anxiety.* The Complete Psychological Works of Sigmund Freud, Standard Edition, Ed. James Strachey, Vols. XII, XIV, XX. (Hogarth Press, London.) *New Introductory Lectures on Psychoanalysis,* as printed in *Syntopicon, Freud.*

[4] Jones, Ernest, *The Life and Works of Sigmund Freud,* 3 vols. (1961, Basic Books, New York)

[5] Nagel, Ernest, *The Structure of Science.* (1960, Harcourt Brace, New York)

[6] Rappaport, David, *The Structure of Psychoanalytic Theory* in *Psychology: A Study of a Science,* Ed. Sigmund Koch, Vol. 3. (1959, McGraw-Hill, New York)

COMMENTS

Victor Gourevitch

COMMENTS ON THE PAPER OF FATHER MICHAEL STOCK

Father Stock's thesis is remarkably bold. He proposes to show that Freud's doctrine of the unconscious is, in all decisive respects, perfectly compatible with what he calls "the Aristotelian-Thomistic line of thought." Freud's teaching regarding the unconscious is at the very center of his more comprehensive teaching regarding man as a whole. The question inevitably arises whether Father Stock would also contend that Freud's more comprehensive teaching is, in all decisive respects, compatible with the comprehensive "Aristotelian-Thomistic" teaching regarding man. He does not address himself to this question. He implies that it is possible to consider the problem of the unconscious in isolation from its broader context. This is, no doubt, one reason why his argument is not only technical but also extremely formal. Even in its own terms, however, it seems to me to fail to prove the main point which it sets out to prove.

Father Stock's first major point is that the tradition[1] does not equate the mental with the conscious. There is, therefore, no reason why it needs to reject or even to suspect references to unconscious mental phenomena. This is certainly true. Perhaps no important thinker ever has equated the two. We are nevertheless indebted to Freud for his insistent warnings against the commonsense temptation to do so. Habits, skills, dispositions, memories, appetitions, to say nothing of so-called vegetative and other latent processes have always been recognized to belong to the twilight and nether regions of consciousness. Disagreements arise only once we ask how the relationship between conscious and unconscious mental phenomena is to be understood.

Father Stock tries to show that non-conscious mental phenomena are also not somatic. He thinks that the Freudian view requires that this be proven, and that the traditional view allows it to be. He nowhere indicates why he thinks that Freudian view requires it. As a matter of fact, it does not seem to. If we confine ourselves to the papers by Freud on which Father Stock draws most heavily, we do not find Freud making any such claim. All that he argues is that, in the present state of our knowledge, an account of psychological phenomena cannot be derived from an account of physiological phenomena.[2] In other words, Freud merely rejects what has since come to be known as "reductionism." But it is one thing to say, with Freud, that unconscious phenomena—or mental phenomena generally—are not usefully described in somatic terms, and another thing entirely to say, with Father Stock, that unconscious phenomena are not somatic. If we look beyond Freud's early and provisional formulations to his mature doctrine, the discrepancy between what he says and what Father Stock says he says is only increased. What Freud came to call the id is certainly unconscious. It is by no means clear what might be meant by saying that it is not somatic.[3] On Father Stock's version it would even be difficult to understand Freud's speculations about the localization of mental processes. Yet such speculations abound in his work.[4] The sharp distinction between psychic and somatic which Father Stock attributes to Freud is not to be found in Freud.

Granted that the tradition allows for mental phenomena which are not conscious. How does it view their relationship to the somatic? For Aristotle the question of whether a "somatic" or a "psychic" account of a psychological phenomenon is the more scientific, is a question of deciding whether what he calls a "material" or a "dialectical" account is the more scientific account of the phenomenon. Is, for example, anger better characterized as a boiling of the blood or as a desire to return pain for pain? His answer is unequivocal: an adequate scientific account must allow for both.[5] Thomas Aquinas is, if anything, still more emphatic on this point. In commenting on these remarks of Aristotle's he substitutes "formal" for Aristotle's "dialectic."[6] In other words boiling of the blood is the material definition or cause of anger, and the desire to return pain for pain is its formal definition or cause. Both Aristotle and Aquinas, when they look for the causes of such psychological phenomena as anger, that is to say of feelings or emotions, situate the material cause in the body. In contrast to the tradition, Father Stock looks for all four of the causes in what he calls "the psychic sphere" (p. 411). Not that the tradition and Father Stock differ about the meaning of "material cause." They do not. Both mean by it "that out of which." They differ about what the actual material cause of psychological phenomena is. If Aristotle had been told that the material cause of anger is not blood boiling up around the heart, he would presumably have asked for a more adequate description of what happens in the body when we are angry. If he had been told that we simply do not have any adequate description of this sort, he would presumably have said that in that case we do not know the material cause of anger. But Father Stock argues differently. He says that we can and do know the material cause of the unconscious without ever leaving "the psychic sphere." The question at issue is, then, whether the tradition would allow Father Stock's claim that we can and do know psychological phenomena in terms of all four causes without ever making any reference to the body. It seems to me very doubtful that it would, except possibly in a very derivative sense. Father Stock altogether distinguishes much more sharply than does the tradition between the psychic and the physical. He seems to think of them as spheres standing in a casual relationship to one another. Thus he says, for example, that ". . . the influence of affects on psychological process and [of] psychological process on affects do not belong purely to the psychic sphere. . . ." (p. 411). There is no way in which such talk about psychic spheres interacting with bodily spheres can conceivably be understood in traditional terms. Unlike Father Stock, the tradition does not, in speaking about psychological phenomena, distinguish between spheres. It distinguishes, rather, between different but complementary accounts of one and the same being. To put the difference between Father Stock's formulations and the tradition's in traditional terms, Father Stock implies an efficient cause between the material and the formal.[7] To my knowledge there is no basis for such a suggestion in Aristotle.

We are thus led to the following conclusion. Freud and the tradition do agree that psychological phenomena require distinct but complementary accounts in material as well as in formal terms, "material" here referring to actual operations of the body.

To the extent to which Freud and the tradition agree on this, they disagree with Father Stock's statement of the problem.

According to Father Stock *the* efficient cause of the unconscious is repression (pp. 419, 420). This interpretation is central to his entire argument. But even in the comparatively early paper on "The Unconscious" upon which Father Stock so heavily relies for his interpretation, Freud says at the very outset ". . . the repressed does not comprise the whole unconscious. The unconscious has the greater compass: the repressed is a part of the unconscious."[8] Freud did not change his views on this point.[9] Once again, Father Stock's version of Freud's view leaves no room for what Freud later came to call the id, although toward the end of his paper he himself refers to "id drives" (p. 424).

According to Father Stock the formal cause of the unconscious is ideas and images which are not or cannot be verbalized (pp. 421-6). While this is the view expressed by Freud in the 1915 paper on "The Unconscious," he subsequently modified it. He came to hold that, as far as feelings are concerned, the specific difference between conscious and unconscious is no longer to be sought in the distinction verbalized and non-verbalized. This last distinction is now said to apply only to ideas.[10] Father Stock's final definition of the unconscious refers to ideas but not to feelings: ". . . the ideas and images of the mind whose verbalization is prevented by the process of repression . . ." (p. 421). This definition is thus manifestly at odds with Freud's mature views. Father Stock might conceivably choose to defend his definition in either of two ways. He might contend that his definition agrees with Freud's claim that unconscious ideas are non-verbalized ideas. But on this argument his definition can be shown to be incomplete because it fails to allow for unconscious feelings in constrast to unconscious ideas. Alternatively, he might contend that his definition agrees with Freud's views by locating the distinction between conscious and unconscious in the distinction between the verbalized and the non-verbalized. But on this argument his definition can be shown to be incomplete because it fails to allow for feelings that are conscious without being "attached to verbal images."[11] On any interpretation of it, Father Stock's statement of the formal cause of the unconscious does not represent Freud's mature views faithfully. Father Stock's most elaborate structure of analogies between Freudian and Thomistic psychology rests on the claim that the distinction verbalized non-verbalized is the specific difference between the conscious and the unconscious (pp. 422 ff.). The foundation crumbles if, with Freud, we admit feelings which are conscious without being verbalized. Without the foundation the structure collapses.

We must now turn to Father Stock's discussion of what he calls the final cause of the unconscious. This discussion will reveal the fatal flaws in his argument. Very generally speaking, the difficulties we will encounter will be due to Father Stock's conception of final causality. Father Stock claims to be speaking from an "Aristotelian-Thomistic" point of view. This expression is, of course, very popular. It is also very imprecise. It is doubtful whether Aristotle and Thomas Aquinas agree on a sufficient number of important questions to allow it to be used meaningfully.[12] Tacitly Father Stock seems to acknowledge as much. He never refers to Aristotle directly. He refers directly only to Aquinas and to Aquinas' commentaries on Aristotle. Regarding final causality there are significant differences between Aristotle and Aquinas, even if we confine ourselves to Aquinas the commentator. Aquinas is much more categorical and sweeping in what he says about final causality than is Aristotle. Hence modern thought clashes much more directly with Aquinas than it does with Aristotle. It is, therefore, also much more difficult to resist the onslaught of modern thought against final causality if one understands it in Thomistic terms. For all practical purposes Father Stock seems to have reinterpreted final causality in a way that brings it into line with the modern criticisms of it. Modern thought does not deny a regulative function to analysis in terms of final causes. But it understands the relationship between final causality, chance, and nature in a radically different way from the way in which that relationship is understood by the tradition, and in particular by Aristotle. As a result, "final causality" can be used equivocally. The traditional term may be given a modern meaning. This equivocation is the nerve of Father Stock's purported proof of the compatibility, in all decisive respects, between the tradition and Freud's doctrine of the unconscious.

Father Stock understands the final cause on the model of human action or production. Final causality so understood is "true final causality". Hence "final cause" has only a diluted meaning where there is no deliberation. In non-human contexts "final cause" is said to be intelligible only by analogy. This entire conception of final causality goes directly counter to the Aristotelian conception of it. Aristotle had, in fact, explicitly warned against the temptation of understanding final causality on the model of deliberation.[13] By disregarding this warning, and together with it Aristotle's understanding of final causes, Father Stock fatally weakens the meaning of "final cause". By his criterion, analysis in terms of final causes will be justifiable only on epistemological grounds, that is to say on grounds wholly alien to the tradition. Father Stock's account does not allow for the traditional view ac-according to which art imitates nature.[14] On the model which he proposes, to speak of final causality with respect to nature would be to imply an agent in nature or of nature. Although Aristotle explitictly rejects this way of understanding final causality,[15] Thomas Aquinas does speak in such terms.[16] Hence to reject Thomas' interpretation does not entail rejecting Aristotle's conception. Father Stock seems to think that it does when he says that we speak of final causality with respect to nature by analogy only.

By understanding final causality on the model of action and production, Father Stock greatly narrows the traditional scope of final causality. At the same time his model leads him to conceive of final causality as much more efficacious than Aristotle had conceived of it. He says that the final cause causes the efficient cause directly, and the other causes indirectly. "The four causes are interconnected. The final cause causes the efficient cause, since it establishes the purpose for which the agent acts, the efficient cause causes the formal cause, since the latter is the precise effect it produces, and the formal cause determines the material cause, since it is related to it as determinant to derminable" (p. 412). He illustrates the relationship between the final and the efficient causes as he sees it in the following terms. A carpenter's purpose–to make a table–". . . had to crystallize at some point into the resolve to make this table" (p. 412). By saying that the final cause is the cause of the causes he implies that it is the most powerful of the causes. Although Thomas Aquinas speaks in such terms about the final cause,[17] the passage in Aristotle upon which these remarks are supposed to be a gloss[18] does not allow anything like the same privilege to the final cause. On one occasion Father Stock indirectly refers to one of Aristotle's many discussions of the inter-connectedness of the causes by referring to Aquinas' commentary on it.[19] In these passages neither the original nor the commentary provide the least warrant for speaking about the final cause as causing any of the other causes. Indeed, it would be very puzzling to know what to make of the claim that the causes are causally related to one another.

Aristotle did not conceive of final causality on the model of doing or making. A better clue to the way in which he conceived of it is provided by the fact that he argued for it, by and large, by reductio ad absurdum of its denial in given cases or sets of cases. Or rather, Aristotle points to the fact that, in a great variety of cases, human as well as non-human, to deny final causality leads to greater absurdities than to affirm it. In such cases final causality must be taken into account alongside of the other kinds of causality, and in such cases to know means to know in terms of all four causes. The fact that we may not be able to account for final causality in a rigorous way does not eliminate the need to invoke it. Its irreducibility is more evident than is its justification. We must, as the tradition has it, explain the less well known in the light of the better known. That is why it is absurd and promiscuous to claim final causality for all phenomena,[20] or for nature as a whole. The necessity attaching to final causes is properly called "hypothetical."[21]

Father Stock reinterprets the traditional conception of final causality in a number of other ways. When he illustrates the four causes by the example of a table he speaks of the final cause of the table as if it were one and the same as the final cause of the carpenter's making of the table (cp. p. 412). But the final cause of the table is its function as a table, regardless of whether the carpenter made it for money or for any other end.[22] The same objection applies to what Father Stock says about the final cause of the unconscious. After having asserted that repression is the efficient cause of the unconscious, he goes on to speak about the final cause of repression as "the avoidance of unpleasure" (p. 420), as if the final cause of repression were necessarily one and the same as the final cause of the unconscious. But they no more need to be the same than the final cause of the carpenter's making a table needs to be the final cause of the table. Now, whether the unconscious can in any intelligible sense be said to have a final cause or end in the way in which it is appropriate to talk about the function of the table is open to question. That is not to imply that only actions or artifacts can properly be said to have functions, final causes, or ends. The sense organs can clearly be spoken of in terms of functions. Similarly, one can make sense of the statement that the end of man is virtue, regardless of whether one chooses to agree with it or not. But it is not at all clear that one can speak of the unconscious, as Freud described it, in any of these ways.

Many other difficulties stand in the way of speaking about a final cause as "the avoidance of unpleasure." Properly speaking, final causes are, or would be, choice-worthy for their own sakes. Hence avoidance would not, properly speaking, satisfy the criterion for a final cause. It would be a final cause only incidentally. According to the tradition "final cause" can, generally speaking, be used interchangeably with "good." Unpleasure cannot, generally speaking, be used in that sense. Nor can "the avoidance of unpleasure," even if the double negation is translated into an affirmation. For pleasure is not the good. It may be good, and perhaps even a good. To grant that the tradition accepts pleasure as a good is still not to grant that it is compatible with Freud on this point. It would first have to be shown that Freud understands pleasure in the same way as the tradition understands it. But this is not the case. Father Stock uses the traditional term without regard for its traditional meaning. There is nothing inherently objectionable in this procedure. He might use the term in ways that could be shown to be compatible with its traditional meaning. He does not use it in such ways, however. He uses it in ways that have explicitly been rejected by the tradition.

Let us, for the sake of the argument, nevertheless, assume with Father Stock that the unconscious does have a final cause, and that it is the same as the final cause of repression, namely "the avoidance of unpleasure" (pp. 420, 422, 425). In other words, we are assuming that the unconscious is subject to the pleasure principle. According to Freud "the pleasure principle is . . . a tendency operating in the service of a function whose business it is to free the mental apparatus entirely from excitation or to keep the amount of excitation in it constant or to keep it as low as possible. We cannot yet decide with certainty in favor of any of these ways of putting it; but it is clear that the function thus described would be concerned with the most universal endeavour of all living substances—namely to return to the quiescence of the inorganic world."[23] Hence "the pleasure principle seems actually to serve the death instincts,"[24] and ". . . we shall be compelled to say that *'the aim of all life is death'* "[25]. If we accept Father Stock's account of the final cause of the unconscious as the avoidance of unpleasure, then the final cause of the unconscious must be said to be death. Aristotle anticipated just such a misunderstanding of what he meant by final cause. He calls it an absurdity.[26]

The absurdity arises from the failure clearly to distinguish between final causes and final results; and hence between final causality and chance or spontaneity. Father Stock characterizes the difference between chance and final causality in the following terms: "Chance effect, for example, occurs according to determinable patterns of distribution. Where these patterns are [not] found, where the effect is, according to Aquinas' words, something which occurs all the time or in the majority of cases, we can suppose that some influence is working which is not chance, but allied rather to purposive action, which achieves its end all the time or for the most part." (p. 413). In such cases we "postulate a form of final causality." This statement is not altogether clear. It makes sense only if the [not] I have bracketed is deleted. It then amounts to saying that final causality is "postulated" when occurrences exhibit regularity by happening always or for the most part. Once again, Father Stock's formulation is much weaker than a traditional statement would have been. We have come to expect that. But it is also very much stronger than a traditional statement would have been. Indeed, Father Stock seems to regard regularity of occurrence as a sufficient condition for speaking about final causality. The traditional would at most regard it to be a necessary condition.[27] Since not everything that happens even in nature happens according to final causality, we need clues before asking, in a given case, whether it did happen according to final causality.

One such clue is provided by the fact that something happens always or for the most part.[28] Even if we were to claim that the fact that something happens always or for the most part is a necessary condition for speaking of it in terms of final causality, it would still not be a sufficient condition for doing so. Although it always or for the most part requires rain for corn to grow, Aristotle explicitly rejects the inference that it rains for the sake of the corn's growing.[29]

Father Stock would have us talk about chance whenever there is not a regularity in the pattern of occurrence, that is to say whenever, on his account of it, considerations of final causality would not apply. This directly contrasts with the tradition which would speak of chance or spontaneity only in cases where considerations of final causality would apply.[30]

Father Stock's departures from the view he claims to be speaking for are very great. He does not reconcile that view with the Freudian view. He accommodates it to a modified Freudian view. As a result he also has no way of escaping the difficulties attendant upon the Freudian view. A conception more consonant with the tradition's than his is would have helped him to avoid a number of the more obvious difficulties which his concessions to the Freudian view get him involved in. If Father Stock postulates a final cause "for the sake of which" in the light of what happens always or for the most part, then he cannot avoid the Freudian conclusion that the final cause of life is death. Father Stock may wish nevertheless to reject this conclusion. Let us assume that he would argue that what happens always or for the most part is not death, but life, ordinary men leading ordinary lives. Freud never tired of pointing out that most men lead lives of quiet desperation, while even the noblest lives and the greatest achievements are the result of frustration of the quest for pleasure, not its fulfillment. According to Freud's well-known view all that we regard as most distinctive about man's humanity, although in the light of hindsight it proves worthy of having been striven for all along, is actually the consequence of the striving for another end altogether, namely pleasure.[31] If, with Father Stock, we regard pleasure as the final cause of unconscious striving, then the Freudian account of the relationship between the quest for pleasure and its results, between the final cause and the final outcome, satisfies almost literally Aristotle's definition of chance.[32] In an important sense Aristotle's doctrine about final causes may be understood as an attempt to escape the conclusion that men's lives are altogether the playthings of chance.[33] Father Stock seems to have left us no way in which to escape that conclusion.

Aristotle, in his teaching about final causality, does not attempt to prove or to guarantee that men's lives are altogether beyond the sway of chance. He does not even trace its bounds with any real precision. He merely shows that there must be such bounds. Some of his interpreters and commentators have made much more sweeping claims on behalf of final causality than did he. Their claims have, rightly, evoked very sharp criticism. Father Stock seems to think that such criticisms of the interpretations apply with equal force to the original. It is, however, by no means evident that they do. In any case, the vitality of the tradition is not demonstrated by attempting to show it is so spineless as to be adaptable to any end. The tradition is not merely a set of empty categories and formal concepts. It is a body of living thought, and it speaks in the most direct terms to our present problems. The vitality of the tradition can only be shown by attempting to show that it can account for the phenomena. It is possible, of course, that in the last analysis it cannot do so. But nothing is gained—and much is lost—by avoiding the confrontation. For then the problems themselves become obscured.

Dept. of Philosophy
Wellesley College
Wellesley, Mass.

NOTES

[1] I prefer to speak about the tradition rather than about the Aristotelian-Thomistic view; cp. nn. 12 and 29 *infra.*

[2] Freud, "The Unconscious," *Collected Papers* (Hogarth Press), vol. IV, p. 107.

[3] Freud, *The Ego and the Id* (Hogarth Press), p. 28.

[4] *E.g. ibid.*, p. 31.

[5] Aristotle, *De anima* 403a 5-403b 8; cp. *Physics* 194b 10, ff.

[6] Thomas Aquinas, *In de anima* I, L. 2, 24-27.

[7] At one point he explicitly says ". . . the formal cause determines the material cause . . ." (p. 412).

[8] *op. cit.*, p. 98.

[9] "We recognize that the Ucs does not coincide with what is repressed; it is still true that all that is repressed is Ucs, but not that the whole Ucs is repressed.' *The Ego and the Id,* p. 17.

[10] *The Ego and the Id,* pp. 21-26.

[11] *Ibid.,* p. 26.

[12] ". . . St. Thomas created the very Aristotelianism he gives the impression of merely interpreting." ". . . St. Thomas' Aristotle is a Thomistic creation." Pegis, A. C., *At the Origins of the Thomistic Notion of Man,* The Saint Augustine Lecture, 1962 (Villanova University), Macmillan, 1963; p. 4.

[13] *Physics* 199b 26, ff. Cp. 196b 16-20.

[14] *Physics* 199a 8-199b 33. Consider especially the argument that even when it completes nature, art is guided by nature: *Physics* 199a 15, ff., 199b 26, ff.

[15] *Metaphysics* 1032a 26, ff.

[16] *in Met.* VII. L. 6, 1394.

[17] "haec species causae potissima est inter alias causas; est enim causa finalis aliarum causarum causa," in *Phys.* II, L. 5, xi—186.

[18] *Physics* 195a, 23, ff.

[19] in *Post. Anal.* II, 9. Cp. Aristotle, *Post. Anal.* 94a 20-95a 9.

[20] *Physics* 196b 17, contra Thomas Aquinas, in *Phys.* II, L. 8, 211.

[21] Aristotle, *Physics* II, 9.

[22] The classical refutation of Father Stock's claim is, of course, Socrates' refutation of Thrasymachus. Plato, *Republic* 345e-347a.

[23] Freud, *Beyond the Pleasure Principle* (Bantam Books), pp. 107, ff.

[24] *Ibid,* p. 109.

[25] *Ibid,* p. 70.

[26] *Physics* 194a 30, ff.

[27] Actually it does not even go that far. For in that event, *e.g.* the fact that virtue is rare would entail that it is not the end of man.

[28] That for Aristotle regularity is not a conclusive consideration is further evident from the fact that he speaks of it in an hypothetical disjunction; he does not claim that this disjunction is necessary or that it is exhaustive. *Physics* 199a 3, ff.

[29] *Physics* 198b 16, ff. In Aquinas the question of where to draw the limits of final causality is complicated by the problem of providence, *in Phys.* II, L. 12, 250.

[30] Aristotle, *Physics* 196b 17, ff., 197a 4 ff., 197b 18, ff.

[31] See especially *Beyond the Pleasure Principle,* pp. 76-78. See also the text quoted by Rieff, *Freud: The Mind of the Moralist* (Anchor), p. 130.

[32] See the passages referred to in n.30 *supra;* cp. also Aristotle, *Poetics* 1460a 19-26.

[33] Aristotle, *Nicomachean Ethics* 1094a 17, ff., cp. *Metaphysics* 994a 8, ff.

13

MILIČ ČAPEK

THE MYTH OF FROZEN PASSAGE: THE STATUS OF BECOMING IN THE PHYSICAL WORLD

Presented May 14, 1964

I think it is a duty of every speaker to explain first the title of his paper. In 1951 Professor Donald Williams of Harvard wrote what Professor Nicholas Rescher of Pittsburgh University recently called a "passionate paper"; its title was "The Myth of Passage."[1] In this paper the idea of passage or becoming was dismissed as a piece of superfluous, confused and romantically tinged metaphysics. But it is a part of the human situation that one man's fact may be another man's myth; and this is especially true of the philosophical situation as long as philosophy remains a free dialogue unrestricted by any kind of external administrative control. Since I am an addict to the "myth of passage," the title of the above mentioned paper suggested the title or rather the counter-title of mine; for the purpose of my paper is to show that if any metaphysics deserves to be called a myth, it is the metaphysics of Being rather than that of Becoming; and, more specifically, that the discoveries of twentieth century physics, far from weakening the ontological status of becoming, strengthen it. But I would like to emphasize that the provocative title—"The Myth of Frozen Passage"—which I am using with Professor Williams' kind permission—is to some extent deceptive as it does not do justice to the fact that there is a considerable area of agreement between his thesis and mine. Our dialogue today may be regarded, especially by those who are acquainted with the history of ideas, as another and most recent phase of the dialogue between Heraclitus and Parmenides which started so many centuries ago; while those who superciliously dismiss any kind of ontology may regard it as a dialogue between two addicts or two myths—the myth of passage and the myth of frozen passage.

Although this paper is not primarily intended as a polemic against Professor Williams', it will be convenient to quote one characteristic passage of his paper, the analysis of which will serve us as the basis for a far broader discussion of the problem of becoming in general. On the very first page of his paper Professor Williams writes:

> the homely realm of natural existence, the total of world history, is a spatio-temporal volume, of somewhat uncertain magnitude, chockablock with things and events. Logic, with its law of excluded middle and its tenseless operators, and natural science, with its secular world charts, concur inexorably with the vision of metaphysics and religion that truth and *fact* are thus eternal. I believe that the universe consists, without residue, of the spread of events in space-time, and that if we accept realistically the four-dimensional fabric of *juxtaposed* actualities we can dispense with all those dim nonfactual categories which have so bedevilled our race: the potential, the subsistential, and the influential, the noumenal, the numinous, and the non-natural. (italics mine.)[2]

In this passage we have all the salient features of what Nicholas Rescher aptly characterized as "the revolt against process"[3] of which Professor Williams is not the only representative; besides him Rescher names also Professors Quine, Nelson Goodman and Strawson, and in our discussion we shall come across some other names. Nor should we forget the very ancient and respectable tradition of which Donald Williams is fully aware when he refers to the concurring visions of the timelessness of truth and reality in both the traditional metaphysics and traditional religion. (I knew he would object against my using the words "timeless" and "eternal" as synonymous, but we shall certainly face the question to what extent the difference between these two terms is not merely verbal.) In any case, there is no disagreement about the fact that the recent rejection of becoming is merely the most recent phase of the persistent *eternism,*[4] which pervades the whole history of Western thought without being confined to it, as various forms of Oriental mysticism show. Thus we should not be surprised to find in Williams' thought the same features as in the traditional forms of Eleatism and eternism. They are as follows:

1. The coeternity of truth and fact.
2. An unlimited confidence in the traditional, i.e. two-valued logic; more specifically, the applicability of the law of the excluded middle to *all situations,* whether past, present or future: consequently a rejection of any kind of potentiality, including the potentiality of the future.
3. Geometrization of time, i.e. the tendency to regard time as one dimension of the four-dimensional manifold, not *essentially* different from the remaining three spatial dimensions.

It is clear that these three features are not logically independent; this is especially true of points 1 and 2. The last feature was for obvious reasons stressed only in the post-Renaissance era when the Eleatic tradition found a powerful ally in classical analytical mechanics and Laplacean determinism which treated time as the fourth dimension of space symbolized by the t-axis of independent variables: but *implicitly* the geometrization of time was present at the very dawn of Greek thought: Zeno's paradoxical denial of becoming was based on his confusion of time with a geometrical line.

On the other hand, Professor Williams departs from the Eleatic and Spinozistic tradition in two important points:

1. He does not accept the Parmenidean and Plotinian myth of "Absolute One" for he unhesitatingly accepts *the reality of diversity.* But his diversity—or "manifold" to use his language—is essentially of static type. In this sense, Donald Williams belongs to the tradition of *static pluralism* of the Herbart and McTaggart type to which Bertrand Russell was very close, at least in some of his writings.
2. Unlike Spinoza, unlike Laplace, unlike Calvin, Zwin-

gli and all other predestinationists Professor Williams does not believe that the timelessness of truth entails determinism. Whether this claim can be consistently upheld, especially if truth and reality are made coextensive and coeternal, is another question the discussion of which cannot be entirely avoided, even though it is almost beyond the scope of this paper.

My primary concern in this paper will be the frequently made claim that the static, i.e. passageless, becomingless view is supported by the relativity theory, in particular, by the relativistic fusion of space and time. I intend to show that a static interpretation of space-time not only creates an unsolvable epistemological problem, but also blissfully ignores some very important features of the proposed union of space with time which suggest a *very opposite* i.e. a *dynamic* interpretation.

While in Professor Williams' essay there is no specific reference to the relativistic space-time, but only to the four-dimensional manifold in general, there is such reference in Professor Quine's recent book *Word and Object* where it is claimed (p.172) that the discovery of the principle of relativity "leaves no reasonable alternative to treating time as spacelike."[5] No argument is given in support of this view except a bare statement that "according to Einstein's relativity principle space and time are distinguishable only relative to a velocity." What Professor Quine here means is that the simultaneity of events is relative to observer's frame of reference; in other words, the observers in different inertial systems split the four-dimensional manifold differently into its spatial and temporal components. While in classical Newtonian physics the class of objectively simultaneous events unambiguously separated the past from the future or, more generally, the events anterior to a certain date from the events posterior to it, there are apparently no such objective and unambiguous boundaries in relativistic physics. From this there seems to be only a small step to the conclusion that the relativization of simultaneity entails a *relativization of succession;* the objectivity of the temporal order seems to be lost, at least on the cosmic scale and the only reality which seems to be left is that of the four-dimensional Minkowskian manifold containing the static relations exhaustibly describable in tenseless sentences.

I tried to point out the fallacious character of the static, spatializing interpretation of space-time several times; thus I am inevitably running the risk of repeating myself.[6] But since the same misinterpretation occurs again and again without being substantiated by any new argument, a certain amount of repetition in criticizing it is perhaps excusable. The source of the misunderstanding may be traced, I think, to Minkowski's famous address in 1908 and, more specifically, to his unfortunate, but highly characteristic choice of the word by which he characterized the merging of space and time into a single four-dimensional continuum of points-events. He characterized this continuum as "four-dimensional world" (*die Welt* in German)[7] and the creeping static connotation of this word clearly suggests that he, unconsciously or semi-consciously, understood the union of space and time as an operation in which the temporal component was, if not absorbed, then at least subordinated to the spatial one. Minkowski's choice of the word was hardly accidental; otherwise it would be difficult to explain why so many physicists followed him in using the same or a similar static terminology. Is it accidental that nearly all treatises dealing with the relativity theory use the term "space-time" and almost none "time-space"? In the whole literature about relativity I found only one exception to this rule: Professor G. Giorgi who in his article in *Scientia* in 1934 used the term "chronotope" (from the Greek *chronos*-time and *topos*-space) which in English means "time-space."[8] This unconscious tendency to give a priority to the spatial component would in itself be innocent if it were confined to the semantics only; unfortunately, it was not so. Numerous more explicit statements of both physicists and philosophers indicate that the relativistic space-time was regarded as *a sort of hyperspace* with time as an additional fourth dimension. According to Einstein, "becoming" in the three-dimensional space has been transformed into "being" in the world of four dimensions;[9] according to Herman Weyl, "the objective world *is,* it does not become"; it appears to become only to our "blindfolded consciousness" (*abgeblendete Bewusstsein*) which creeps along its world-line into the future.[10] Ebenezer Cunningham wrote in 1914 that "if all motional phenomena are looked at from this point of view, they become timeless phenomena in four-dimensional space. The whole history of the physical system is laid out as a changeless whole."[11] Nor is it surprising to find that Ludwig Silberstein, the author of *The Theory of Relativity,* claims that the theory of relativity was anticipated by H. G. Wells' famous novel *The Time Machine* in which a fictitious traveler moves along the "fourth dimension" as freely as along the the three spatial dimensions.[12] In Schilpp's volume on Einstein, Kurt Gödel seriously considers the Wellsian idea of a round trip "into any region of the past, present and future and back again" and in a footnote he even computes or rather makes an estimate of the quantity of fuel which a rocket ship would consume to make this fantastic trip.[13] A similar idea is found in Ernst Cassirer who claimed that in the theory of relativity the anisotropy of time disappears as the distinction between the "past" and the "future" is purely conventional, comparable to the difference between "plus" and "minus" sign in space.[14] Both Weyl and Cassirer regarded the relativity theory as a confirmation of Kant's view about the ideality of time; time, being only a form of our perception, cannot be applied to "things-in-themselves." (It is true that this was a peculiar adaptation of Kant for the purposes of the static view of physical reality; Kant certainly did not exclude time from the *physical* world which, being phenomenal, is necessarily perceived through the form of time. Both Cassirer and Weyl confused the physical world with the realm of things-in-themselves; Kant excluded succession from the latter, not from the former.) Among contemporary philosophers of science the most recent defenders of Weyl's view that the four-dimensional space-time is devoid of becoming are Adolf Grünbaum and O. Costa de Beauregard.[15]

But there were some important dissenting voices both among physicists and philosophers who explicitly warned against a facile spatializing interpretation of Minkowski's proposed fusion of space with time. Emile Meyerson in his beautiful work, unfortunately little known in this country, *La déduction relativiste,* quotes Paul Langevin, to whose writings I shall probably refer again, saying that "to call time the fourth dimension of space is altogether meaningless."[16] It is even more significant that Einstein himself apparently hesitated on this point. He was clearly aware that the asymmetry of time is preserved even in its relativistic union with space for, as he said, "we cannot send wire-messages into the past." On April 6, 1922 Einstein attended the meeting of French Philosophical Society at Paris where Meyerson stressed that the asymmetry of time is preserved even in the relativity theory; Einstein quickly agreed, but when we read his one-sentence answer, we cannot help feeling that this point was not particularly interesting to him, and that, despite his agreement with Meyerson, he apparently did not give too much thought to this problem.[17] His reaction was noticeably different six years later. Meyerson developed his argument against spatialization of time more explicitly in the book, referred to above, *La déduction relativiste* (Paris, 1925) and Einstein wrote a highly favorable and extensive comment on it in *Revue philosophique* (53e année. mars-avril 1928, pp. 161-166). Einstein especially praised Meyerson's warnings against the spatializing interpretation of time and he stressed that this misinterpretation is present not only in a number of popular expositions of relativity, but even in some technical works written by physicists. Although later Einstein seemed to lean again toward a static view as can be seen from his positive interest in Gödel's idea of the observer visiting the past and returning to the present,[18] it is clear that his view about the status of time and becoming was more hesitating and more ambiguous than is generally accepted,

Similar ambiguities exist in the thought of Herman Weyl. He emphasized that it is more accurate to say that the spatio-temporal continuum has three plus one dimensions than to call it four-dimensional;[19] in other words, the temporal dimension preserves its distinct character. It is true that he repeated his statement about the becomingless character of the objective world as late as in 1934 in his book *Mind and Nature* (p. 76); but in the same book and almost on the same page he explicitly says that

> The immediately experienced is subjective but absolute; no matter how cloudy it may be, in this cloudiness it is something given and not otherwise. To the contrary, the objective world which we continually take into account in our practical life and which science tries to

> crystallize into clarity is necessarily relative . . . Who desires the absolute, must take subjectivity, the ego for which things exist, into the bargain; who is urged toward the objective cannot escape from the problem of relativity![20]

I am sure Professor Grünbaum is not aware of this passage; it clearly shows that Weyl was not as resolutely opposed to the ontological status of becoming as he sometimes sounded. Even in his early book *Was ist Materie?* (1924) he raised the question whether the irreversibility of subjective time is not based on the unidirectionality of the physical time; in other words, he did not seem to confine becoming to the introspective realm only.[21] Eight years later he wrote a book whose title must sound ominously to Professor Williams' ears—*The Open World* (1932); the title has a suspiciously Bergsonian ring and the bugaboo of passage seems to lurk behind it. Two years later the book which I quoted—*Mind and Matter*—appeared; and from Weyl's equating subjectivity with absoluteness it would or it should follow that the subjectively intuited becoming must have an objective ontological counterpart. I do not claim that Weyl made his view clear; but there can be hardly any question that his thought on this point was more complex and more ambiguous than some defenders of *one aspect* of his thought realize.

Much less ambiguous warnings against the spatializing misinterpretations of the relativistic union of space with time were issued by A. E. Eddington, P. W. Bridgman, G. J. Whitrow among scientists, by A. N. Whitehead, H. Bergson and H. Reichenbach among philosophers.[22] The last two thinkers, who in other respects had very little in common, recognized, correctly in my view, that the static interpretation of Minkowski's world is merely the most recent phase of the perennial Eleatic tradition. In his posthumous book *The Direction of Time* (p. 11) Reichenbach writes almost in a Bergsonian way:

> This timeless universe is a four-dimensional Parmenidean Being, in which nothing happens, "complete, immovable, without end . . . ; it *is* all at once, a continuous *one*." Time flow is an illusion, Becoming is an illusion; it is the way we human beings experience time, but

> there is nothing in nature which corresponds to this experience.[23]

Reichenbach rejected this Eleatic interpretation in his posthumous work as much as in his earlier books; he even went so far as to regard any metaphysical elimination of becoming as almost a neurotic symptom, a manifestation of the feeling of insecurity generated by the depressing picture of restless change:

> The coming of death is the inescapable result of the irreversible flow of time. If we could stop time, we could escape death—the fact that we cannot makes us ultimately impotent, makes us equals of the piece of lumber drifting in the river current. The fear of death is thus transformed into a fear of time, the flow of time appearing as the expression of superhuman forces from which there is no escape. The phase "passing away," by means of which we evasively speak of death without using its name, reveals our emotional identification of time flow with death.[24]

This is, of course, an interesting, but controversial piece of psychoanalyzing; there are other possible emotional attitudes toward time, and Reichenbach is aware of them. The title of Joseph Needham's book *Time, the Refreshing River*,[25] (which, I was told, Professor Robert Cohen likes—I do too)—shows an emotional attitude altogether different from that described by Reichenbach. In other words, using the language of the same metaphor: there are certainly some pieces of wood which would like to stop their drifting by freezing the river of time; but there are other pieces of wood, and even more, some living beings which enjoy navigating and swimming in that river and which would certainly resent being eternally and motionlessly caught and frozen in a bloc of ice into which the Parmenidian magic would like to convert the restless surge of existence. All that I am trying to say is that all such emotional attitudes are philosophically irrelevant; it is not important whether you prefer to have the universe as a giant refrigerator or as a giant swimming pool; it does not really matter or at least it *should not* matter whether you suffer from metaphysical sea-sickness and vertigo—as the enemies of passage seem to suffer—or whether you suffer from metaphysical claustrophobia

—I would say a refrigerator-phobia—as the friends of passage seem to. What is relevant are the rational and empirical reasons which the opponents adduce in favor of their preferences. In other words, to base the judgment of any view, whether scientific or philosophical, on psychological, psychiatrical or psychoanalytical speculations about its origin, is questionable and unfair; to dismiss "the myth of frozen passage" because of its alleged origin in the neurotic fear of passage is as unfair as to dismiss the alleged myth of passage on the spurious ground that its acceptance betrays the romantic twist of mind. Reichenbach himself had far better reasons for his rejection of the static interpretation of space-time, even though he did not always draw all the consequences from them and even did not state all the reasons which he could have used. It is with reasons of this kind—both rational and empirical—that we are now going to deal, after this preliminary, though incomplete survey of the divided opinion. Let us first consider the most frequently and apparently most plausible argument for the static interpretation of the spatio-temporal manifold—the argument to which I already briefly referred and which probably convinced Professor Quine: the relativization of simultaneity. Its plausibility is undeniable: if there is no objective cosmic Now which would be a substratum for objectively simultaneous events, what is more natural than to infer that the objective distinction between the past and present loses its meaning? In other words, that succession itself is relative, dependent on the choice of frame of reference and not constituting an objective feature of reality? Yet, this conclusion is false and its apparently convincing power is due to the fact that it logically follows from the premise which, unfortunately, is itself false. To claim without qualification that simultaneity and succession of events are relative is simply not true. In truth we can make such a claim only when we completely overlook the mathematical implications of Minkowski's formula for the constancy of the spatio-temporal interval. Even a cursory inspection of this formula tells us that there are five different possible relations between two events;

a) The simultaneity of the isotopic (equipositional) events, i.e. events occurring at the same place.
b) The succession of the isotopic events.

c) The simultaneity of *heterotopic* events, i.e. of events occurring at different places.

d) The succession of *heterotopic* events which are *causally related.*

e) The succession of events which *are not* causally related.

Each of these categories of events requires a separate analysis.

a) Let us first consider the problem of the simultaneity of the events occurring at the same place. In the language of Minkowski's diagram this case is represented by the intersection of two world-lines. For instance, the clash of two trains, of two stars or the clash of an electron with a photon is represented by the point of intersection of two corresponding world-lines. Now although the velocities of two stars, two trains or an electron (not that of a photon because of the constancy of the velocity of light!) will vary according to our choice of the frame of reference, the corresponding world-lines will have the *same* point of intersection in *every* system. In other words, the simultaneity of isotopic events is a *topological invariant;* such events simultaneous within one system will remain simultaneous within any other system. It is hardly surprising that it is so: for if the isotopic events would be simultaneous in some systems and not in other, this would result in very strange causal anomalies: for instance, a bullet would hit its target in one system and not in another, Compton effect would take and would not take place according to our choice of the frame of reference. But the absolute character of the simultaneity of the isotopic events clearly is incompatible with the sweeping claim that the simultaneity of *all* events is purely and unqualifiedly relative.

b) The succession of the events occurring at the same place has an equally absolute character. Like their simultaneity, the temporal order of such events constitutes a topological invariant; two isotopic events which are successive in one frame of reference, remain successive in any other system. This was established by Paul Langevin as early as 1911 and it follows as another mathematical consequence of the formula of Minkowski for the invariance of the world-interval. In Langevin's words, the *world-lines* which by definition are constituted by the succession of isotopic events, are *irreversible* in all frames of reference.[26] The irreversibility of the world-lines is thus in the full sense of the word *absolute.* This implies that in no system of reference can an effect precede its cause. In classical physics, in which no upper limit was imposed for mechanical velocities, it was, at least in principle, possible *to perceive* an effect before its cause; it would be sufficient to move with the velocity greater than light from the Earth and look backwards: then one would see the past history reversed and if you would go sufficiently fast and sufficiently far and have a sufficiently powerful telescope, you could watch your own birth. In the relativistic physics not only the temporal order of cause and effect is irreversible, but even a *reversal of their visual perceptions* is not possible.[27] The constancy of the velocity of light in all systems of reference is the reason for this impossibility.

c) The situation is quite different for the simultaneity of distant, i.e. heterotopic events: since the advent of Einstein's theory the simultaneity of such events became fully and without qualification relative. In other words, a couple of events simultaneous within one frame of reference will appear to succeed each other in other groups of systems, and will appear in a reversed order in still other systems. Only within those systems which are at rest with respect to the original system, will the same two events appear simultaneous as well. In the terms of Minkowski's diagram, the observers moving with different velocities will necessarily disagree what the cosmic Now is; they will split the four dimensional world—I very much prefer to say the four-dimensional world-process—differently, and their instantaneous three-dimensional cuts will *not* coincide. In other words, there is no such a thing as "nature at an instant" or "space at an instant"; which is only another way of saying that there is *no absolute space, no world-wide instant, no absolute juxtaposition* which would serve as a substratum for absolutely simultaneous events. If we realize the logical correlation of the concepts of simultaneity and juxtaposition; if we realize that the class of objectively simultaneous events at a given instant is, according to Newtonian physics, constituted by the three-dimensional instantaneous cuts across the four-dimensional becoming; if we further realize that such instantaneous cross-sections are excluded by the relativization of the simultaneity of the heterotopic events, then we arrive inevitably at the conclusion that the relativity of simultaneity is nothing but *the relativity of juxtaposition.*

d) In considering the succession of heterotopic events, i.e. those events which occur at different places, we should sharply differentiate those couples of events which are causally related from those which are not. Two events are causally related when their distance in space is in any system *smaller* than their separation in time multiplied by the velocity of the fastest causal action. Again, an inspection of Minkowski's formula for the constancy of the world-interval shows that the temporal order of such events is independent of the choice of frame of reference; their succession is also a topological invariant. This provides us with the possibility of generalizing our previous result; not only the world-lines of material "particles," but all *causal lines* are irreversible in every system of reference. In truth the world-lines are merely special instances of causal lines: every world-line of material "particle" is really an isotopic causal line or what Whitehead called a "route of occasions"; it represents a life career of a particle stationary within a certain system. By a suitable change of the frame of reference such a stationary particle may be converted into a moving particle; in other words, its world-line may become a causal line in a wider sense when it becomes a vehicle of some causal influence, i.e. a link between two world lines. Similarly, there is no difference whether we speak of causal lines of light or of the world-lines of photons; there is one significant difference, however, that no change in our frame of reference can convert a photon into a stationary particle. This is one difference between matter and radiant energy which the relativity theory did not remove; on the contrary, it is one of the cornerstones not only of the relativity theory, but also of contemporary physics in general: the notion of stationary photon is devoid of meaning.

e) The last category concerns the couples of events which are causally unrelated and unrelatable: their distance in space is in any frame of reference larger than their separation in time multiplied by the velocity of fastest causal action. Again a brief inspection of the formula for the world interval shows that the events succeeding each other in one system may appear simultaneous in another system or even appear in a reversed order in still another system. This case was really covered when we considered the second category (b); this is only natural because the relativization of the simultaneity of distant events implies a relativization of their succession and vice versa. In truth, no events can ever be judged simultaneous unless they are causally unrelated for the one simple reason that there are no instantaneous connections and interactions in nature; and a purely spatial line joining two juxtaposed simultaneous events is a paragon of instantaneous connection. *But no such instantaneous connections exist in an objective sense;* what is a geometrical line, i.e. a set of juxtaposed points for one observer will appear as a succession of events—of course of causally unrelated events—for another observer.

Philosophically, the most significant result is that while *there is no juxtaposition of events which would be a juxtaposition for all observers, there are certain types of succession which remain such in all frames of reference.* These types of succession are represented by the causal series, i.e. by the world-lines, including the world-lines of photons. In other words, unlike spatial juxtaposition, the irreversibility of the world-lines has an *absolute* significance, possessing genuine and objective reality independently of the conventional choice of the system of reference. The ontological priority of time over space or, more specifically, of succession over juxtaposition could hardly have found a more convincing illustration. In relativistic space-time—I prefer to say time-space—there are only two kinds of relations which have an objective significance: *successive causal connections and contemporary causal independence.* The latter relation replaced the classical concept of simultaneity as the boundary between the past and the future. In the Newtonian universe the past was separated from the future by an instantaneous three-dimensional cross-section; this was precisely the function of the Newtonian absolute space as a substratum of absolutely simultaneous events. In the relativistic universe the past is separated from the future by the wedge-shaped four-dimensional area symbolized in our two-dimensional diagram by the wedge-shaped area which Eddington called "elsewhere." In other words, the past is separated from the future *even more effectively* than in the classical scheme![28] We are thus more than justified in concluding that the relativistic union of space with time instead of being a spatialization of time is a *dynamization of space;* this conclusion is further strengthened by considering the general theory of relativity. Time—or should I say the mythical entity of passage?—prevents me from discussing it in detail; let me mention only one or two facts. The merging of time-space with the changing physical content in the general theory as well as the astronomical evidence for the expansion of the universe clearly indicate that the dynamization of space is *far more pronounced* than the special theory of relativity suggests.

Why was there and why is there still so much misunderstanding about the correct meaning of the relativistic fusion of space and time? The answer is simple as long as we are acquainted with the history of ideas: there is a Parmenides at the bottom of our mind, always ready to convert becoming into being, process into substance, time into space. Because of our Eleatic subconscious we tend to misinterpret and distort the relativistic time-space in a static fashion. Moreover, the four-dimensional manifold, which Professor Williams discussed, is certainly *not* the relativistic manifold. Let us listen to what he wrote on p. 468 of his article:

> An easy interpretation would be that the world content is uniquely organized in the time direction because the time direction itself is aboriginally unique. Modern philosophical wisdom, however, consists mostly of trying to put the cart before the horse, and I find myself more than half convinced by the oddly repellent hypothesis that the peculiarity of the time dimension is not thus primitive but is wholly a resultant of those differences in the mere *de facto* run and order of the world's filling. It is then conceivable, though doubtless physically impossible, that one four-dimensional area of the manifold be slewed around at right angles to the rest, so that the time order of that area, as composed by its interior lines of strain and structure, runs parallel with a spatial order in its environment. It is conceivable, indeed, that a single whole human line should lie thwartwise of the

> manifold, with its belly plump in time, its birth at the east and its death in the west, and its conscious stream perhaps running alongside somebody's garden path.

This in an interesting piece of science fiction or rather philosophy fiction which, however, under no circumstances can claim to be supported by the relativistic physics; on the contrary it is clearly and hopelessly incompatible with it. What Professor Williams claims is that it is possible to cut the four-dimensional manifold in such a way that the world-line of one human body will stretch in a space-like fashion—"along somebody's garden path," for instance—in a suitably chosen frame of reference. This is clearly impossible according to our discussion above: for the succession of causally related events retains its temporal order in *every* frame of reference; for this reason the birth and the death which represent the initial and the terminal event of the world-line of one human body can *never* appear simultaneous to any observer. The degeneration of the temporal order of causally related events into a simultaneity was in principle possible as long as the infinite speeds were allowed; an observer moving with an infinite velocity would have a simultaneous panoramic vision of the whole of history—but only of *past* history, not of future events. (Future history is a contradiction in terms, anyway.) But as long as we take relativistic physics seriously, we must dismiss any instantaneous velocity, any orthogonal crossing of the world-lines as a piece of the outdated pre-relativistic reasoning.

But Professor Williams went even further than the Newtonian physics permitted in claiming that by an appropriate change of the system of reference it is possible to make a certain world-line *running backward in time, from the future to the past*:

> It is conceivable too then that a human life be twisted, not 90° but 180° from the normal grain of the world. F. Scott Fitzgerald tells the story of Benjamin Button who was born in the last stages of senility and got younger all his life till he died a dwindling embryo. Fitzgerald imagined the reversal to be so imperfect that Benjamin's stream of consciousness ran, not backward with his body's gross development, but in the common clockwise

> manner. We might better conceive a reversal of every cell twitch and electron whirl, and hence suppose that he experienced his own life as we do ours, but that he observed everyone around him moving backward from the grave to cradle. I may be overbold to unveil such speculations, since to some they will seem a warning of the dangers of any dimensional view.[29]

Not only are such speculations overbold, but they are clearly imcompatible with relativistic physics. The assumption of instantaneous velocity would imply the flattening of the surfaces of the frontward causal cone; the world-line of the corresponding observer would be orthogonal to the "normal" course of time and the observer itself would be omnipresent in all its points. As Williams correctly sees, the backward running local time would require that the corresponding world line would be twisted by an angle greater than 90°; in other words, the corresponding causal cone would be turned backwards as an upturned umbrella. It is not necessary to dwell on all causal anomalies and plain absurdities which would result from such situations; the essential point is that as long as we respect the basic principle of relativity theory according to which all possible causal lines emanating from the point Here-Now must be included in the frontward causal cone; then no world-line starting from Here-Now can ever reach the region of Elsewhere; *a fortiori,* none of them can be bent backwards to be included in the rearward causal cone; consequently, the notion of the backward or pastward moving world-line does not have any place in the relativistic universe. The phantasies of H. G. Wells, Scott Fitzgerald and other reveal indeed the dangers of the "dimensional" view of time, i.e. of the traditional spatialization of time.

I already mentioned a number of physicists and philosophers dissenting from the static interpretation of space-time. I should not have omitted Professor Philipp Frank who as early as 1938 severely criticized Sir James Jeans' static and fatalistic interpretation of the "world of Minkowski."[30] Needless to say I wholeheartedly agree with Frank's criticism, even though I think that some of his arguments could be made even more cogent. But he certainly was right in calling Jeans' interpretation "metaphysical"; he failed, however, to specify by *which kind of metaphysics* Jeans was inspired. The reading of Jeans' *The New Background of Science,* would have supplied him with this information; on p. 109 Jeans wrote:

> From the time of Plato onwards, philosophical thought has repeatedly returned to the idea that temporal changes and the flux of events belong to the world of appearances only and do not form a part of reality . . . For this reason philosophers have insisted that reality must be timeless, and time merely, in Plato's phrase, "a moving image of eternity."

Then after quoting approvingly the passage from Bradley's *Appearance and Reality,* Jeans concludes:

> We may notice how the absorption of space and time into a higher unity, the space-time continuum, which transcends both and is changeless, satisfies the requirements of the philosophers, although only at the expense of relegating evolution to the realm of appearance.[31]

This is clear enough. But if we are severe to Jeans, we have to be equally severe to Kurt Gödel who ties his interpretation to the tradition of timeless idealism as explicitly as Jeans; only instead of Plato he mentions Parmenides, instead of Bradley, McTaggart, especially his article "The Unreality of Time."[32] What is more surprising is to find some physicalistically oriented philosophers in the company of timeless idealists invoking the heritage of Plato, Bradley and McTaggart! Evidently, the common aversion to "the myth of passage" produces some strange bedfellows.

After showing that the myth of frozen pasage belongs essentially to the prerelativistic modes of thought, the only possible way to save it would be to claim that it is really independent of any particular physical theory, not excepting the physics of relativity. This is what Professor Quine explicitly claims.[33] I shall not discuss the alleged *positive* arguments for this claim the most important of which seems to be the alleged timelessness of truth; what I am going to point out are the unusual difficulties which this view implies. As the quoted Jeans' passage says, in the passage-less view "time is relegated into the realm of appearances." In other words, time still exists, though only on a subjective, introspective, psychological level. But does not this create intolerable dualism of "the realm of reality" and "the realm of appearances," a metaphysical dualism far worse than the dualism of Descartes? For even the Cartesian dualism was not as sharp as one created by degrading time to a mere appearance: no matter how sharply the realm of unextended spirits contrasted with the spatial realm of matter, they both shared the *temporal* character; both minds and bodies were, for Descartes, in time. On the other hand, in excluding succession from the objective physical world we are adding to the duality of subjective and objective another duality of timeless and temporal: how are these two realms intelligibly related? Jeans in the passage impugned by Philipp Frank answered it in the same way as Weyl did:

> time is interwoven with space [so] that it is impossible to divide it into past, present, and future in any absolute manner. This being so, the tapestry cannot consistently be divided into those parts which are already woven and those which are still to be woven. Such a distinction can have no objective reality behind it . . . The shortest cut to logical consistency was to suppose that the tapestry is already woven throughout its full extent, so that the whole picture exists, although we only become conscious of it bit by bit—like separate flies crawling over a tapestry.[34]

Jeans' metaphor of the flies crawling over the already existing tapestry is strikingly similar to Weyl's metaphor of consciousness crawling up the world-line of its body. Professor Grünbaum accused me of taking Weyl's metaphor too literally and wrongly inferring from it that

> we are already dead without realizing it now; but our consciousness creeping along the world-line of its own body will certainly reach any preexisting and nominally future event which in its completeness *waits* to be finally reached by our awareness. But our future awareness will create this event as little as Columbus's discovery created the American continent.[35]

Grünbaum calls it a "careless" and "gross distortion," although in the very same paragraph he does not say anything different in a slightly less metaphorical way when he characterizes Weyl's thesis "that coming into being is only *coming into present awareness.*" (Grünbaum's italics)[36] Now "coming into present awareness" means, it it means anything at all, that what I experience as a new, fresh introspective and sensory quality, existed timelessly *prior* to and *independently of* this particular experience of mine; then death—as any other physical event—in Grünbaum's own language, will not come into being, but will only come into my own awareness. His irritation seems to indicate that he feels vaguely the absurdity of his own predicament into which he got by excluding becoming from the physical world.

I am not sure whether I interpret correctly Professor Williams' and Professor J. J. C. Smart's thought; but I have the impression, gained mainly from my conversations with both of them, that they go even farther than Jeans, Weyl and Grünbaum, since they seem to deny becoming even on the *phenomenal* level. To me such a view is entirely unintelligible; even Plato, even Parmenides, did not seem to go so far. If even such all-pervading experience as that of temporality is denied, what should one say about it? Should I recommend reading, if not William James's chapter on "stream of thought," at least Marcel Proust and Virginia Woolf to convince my opponents that becoming is the feature which pervades our whole sensory and introspective experience? Or should I simply suggest to them that they just listen to music—music of any kind? The only meaningful way to defend the exclusion of becoming on the phenomenal level would be to claim with Descartes, C. A. Strong and Alexius Meinong that our present psychological moment is a *simple mathematical instant,* devoid of becoming, and that succession is never experienced, but always only constructed. Even Bertrand Russell did not go so far.[37] But even if we concede the impossible claim that our psychological moment is devoid of temporal thickness, we inevitably have to admit the succession of such moments; but in admitting this, the despised bugaboo of passage creeps again into our picture. It seems that, no matter how ingenious our refrigerating technique is, it never succeeds in freezing the flow of existence; the artificially produced ice always breaks somewhere under pressure of the irresistible and unfreezable becoming.

It is true that some passages of Williams seem to suggest that his disagreement with me is merely semantic. He claims that he does not deny the reality of time; that he does not substitute "a safe and static world, a block universe, a petrified *fait accompli*, a *totum simul*, for the actuality of risk and change." He claims that his theory "does not assert . . . that future things 'already' exist or exist 'forever.' " Emphatically also it does not, as is frequently charged, "make time a dimension of space," any more than it makes space a dimension of time." (p. 470). What is even more important, Professor Williams denies that his four-dimensional manifold entails any kind of Calvinistic or Laplacean predetermination. All this sounds like an enthusiastic espousal of the open universe and the reality of becoming. Unfortunately, there are other passages directly contradicting the passages just quoted. If he really does accept the reality of time, why does he believe that the universe is "the four-dimensional fabric of *juxtaposed actualities*"? (p. 458). Juxtaposition and succession are mutually exclusive; so are actuality and futurity. If he denies that future things "forever" exist, why does he explicitly defend the idea that there is "an eternal world total in which past and future events are as determinately located, characterized, and truly describable, as are southern and western events." ("Sea Fight Tomorrow" in: *Structure, Method and Meaning*, p. 306.) I do not doubt that Professor Williams honestly believes that he believes in the reality of time; but all his geographical metaphors and, especially the highly revealing term "juxtaposition" which he uses, indicate that his time, like the time of Grünbaum, is merely nominal, being a result of the confusion of genuine temporality with its spatial symbol.

Yet, there is a certain area of positive and unambiguous agreement. I agree with Professor Williams that passage is nothing apart from events;[38] concrete events do constitute passage, although I certainly do not regard them in the same atomistic fashion as he does. Like him, I do not believe that the occurrence of any event can be separated from its own very nature; although unlike him, I do not believe, for the very reason he invokes, in the specificity of future events. (I think he is getting in serious difficulties when he claims that the future events may be *determinate* without being *determined*. Although I do not accept Professor Richard Taylor's fatalism, it at least has the virtue of consistency.)[39]

Finally, I heartily agree with him that all kinematic metaphors like flow, passage, movement, crawling, advancing, etc. are inadequate; but they are inadequate because of misleading spatial associations, not because they suggest the incomplete, unfinished character of reality. What the alleged myth of passage asserts is not the existence of some metaphysical uniformly flowing river, carrying everything, with it, because such an image, borrowed from our limited macroscopic surroundings, cannot adequately convey the all-pervasive and imageless character of becoming. The denial of the temporal incompleteness of the world, as I tried to show, cannot be substantiated by relativistic physics; on the contrary, it is clearly incompatible with it. Moreover, it generates enormous epistemological difficulties while it does not succeed in eliminating becoming on the phenomenal level. If it tries to do even this, it becomes a weird metaphysical myth, completely divorced from our most obvious and undeniable experience.

Dept. of Philosophy
Boston University
Boston, Mass.

NOTES

[1] *The Journal of Philosophy*, 48 (July 19, 1951), 457-472;
[2] *Ibid*, p. 457
[3] Nicolas Rescher, "The Revolt against Process", *The Journal of Philosophy, 59* No. 15 (July 19, 1962), 410-417.
[4] The term coined by J. A. Leighton.
[5] W. Quine, *Word and Object* (Cambridge, Mass., 1960), 172.
[6] Cf. my articles: "The Doctrine of Necessity Re-examined", *The Review of Metaphysics* 5 (1951), 11-54, esp. 40-47; "Relativity and the Status of Space", the same journal, *9* (1955), 169-199; *The Philosophical Impact of Contemporary Physics* (Princeton, 1961), Chapter XI.
[7] H. Minkowski, "Space and Time". In: *The Principle of Relativity*, (London, 1923) 73-91.
[8] G. Giorgi, "L'Evolution de la notion du temps", *Scientia 38* (1934), suppl. p. 36.
[9] E. Meyerson, *La Déduction relativiste* (Paris, 1925), p. 100.
[10] H. Weyl, *Was ist Materie?* (Berlin, 1924), pp. 82, 87; *Philosophy of Mathematics and Natural Science*, (Princeton, 1949), p. 116.
[11] E. Cunningham, *The Principle of Relativity* (Cambridge, 1914), p. 191.
[12] L. Silberstein, *The Theory of Relativity* (London, 1914), p. 134.
[13] K.Gödel, "A Remark about the Relationship between Relativity Theory and Idealistic Philosophy" in: *Albert Einstein, Philosopher-Scientist, ed.* by P. Schillp (Evanston, 1949), pp. 560-561.
[14] E. Cassirer, *Zur Einsteinschen Relativitätstheorie* (Berlin, 1921), p. 119. The

English translation appeared as a supplement of the translation of Cassirer's book *Substanzbegriff und Funktionbegriff* (New York, 1953).

15 A. Grünbaum, *Philosophical Problems of Space and Time* (New York, 1963), esp. Chapter 10; O. Costa de Beauregard, *La notion da temps. Equivalence avec l'espace.* (Paris, 1963). Cf. my discussion with Costa de Beauregard in *Actes du Xe Congres de sociétés de philosophie de la langue française: Discussions* (Paris, 1959), pp. 66f.

16 E. Meyerson, *op. cit.*

17 *Bulletin de la societé française de philosophie,* 6 avril 1922, p. 112.

18 *Albert Einstein, Philosopher-Scientist,* pp. 687-88.

19 H. Weyl, *Space-Time-Matter,* tr. by H. L. Brose (London, 1922), p. 283.

20 H. Weyl, *Mind and Nature* (Philadelphia, 1934), p. 76. The same passage occurs already in his *Philosophy of Mathematics and Natural Science,* p. 116.

21 *Was ist Materie?* (Berlin, 1924), p. 84.

22 cf. A. E. Eddington, *Space, Time, Gravitation,* (Cambridge, 1920), pp. 51-52; *The Nature of the Physical world* New York, 1933), ch. III, esp. pp. 50-52; 57-58; P. W. Bridgman, *The Logic of Modern Physics* (New York, 1948), p. 74; G. N. Whitrow, *The Natural Philosophy of Time* (London, 1961), *passim; A. N.* Whitehead, *The Concept of Nature,* pp. 53-54; 178; H. Bergson, *Durée et simultaneité,* 2nd ed. (Paris, 1923), esp. ch. VI; H. Reichenbach, *The Philosophy of Space and Time,* tr. by Maria Reichenbach (New York, 1957), pp. 112-113.

23 *The Direction of Time* (Berkeley California, 1956), p. 11.

24 *Ibid.,* p. 4.

25 *Time: the Refreshing River* (London and New York, 1943).

26 P. Langevin, "Le temps, l'espace et la causalité dans la physique moderne" *Bulletin de la Societé francaise de philosophie,* Séance du 19 octobre 1911, p. 37.

27 Still more accurately: "of their visual perceptions in the vacuum or in the medium of negligible index of refraction."

28 Cf. Eddington, *The Nature of the Physical World,* pp. 57-58: "The limit to the velocity of signals is our bulwark against that topsy-turvydom of past and future, *of which Einstein's theory is sometimes wrongly accused*" (italics mine).

29 D. Williams, *loc. cit., pp. 468-69.*

30 Philipp Frank, *Interpretations and Misinterpretations of Modern Physics* (Paris, 1938), pp. 46-55.

31 Sir James Jeans, *The New Background of Science,* (New York, 1933), p.110.

32 *The Mind,* N. S., 17 (1908).

33 *Op. cit.* p. 172.

34 The passage quoted is from Jeans's Sir Halley Stewart Lecture of 1935 *Man and the Universe.* Cf. nearly an identical passage in *The New Background of Science,* p. 108.

35 M. Capek, *The Philosophical Impact of Contemporary Physics* (Princeton, 1961), p. 165. The passage is reproduced, though not completely, by Grünbaum in his *Philosophical Problems of Space and Time,* p. 328.

36 Grünbaum, *op. cit.,* p. 329.

37 B. Russell "On the Experience of Time", *The Monist 25* (1915), p. 217: "We shall thus be forced to conclude that one experience cannot last for more than one mathematical instant *which is absurd.*"

38 D. Williams, *Op. cit.* pp. 463.

39 R. Taylor "Fatalism", *The Philosophical Review 72* (January 1962), pp. 56-66.

PHYSICS AND FLUX: COMMENT ON PROFESSOR ČAPEK'S ESSAY

DONALD WILLIAMS

Professor Capek and his essay are of a very good sort, and it is a privilege to try conclusions with them. They deal with a subject of the first importance not only for the *Naturwissenschaften* which, I gather, it is the main purpose of this Society to philosophize about, but for the *Geisteswissenschaften,* for traditional and literary philosophy, and for common life. They do it, moreover, with good humor and erudition and a fine concern for the concrete facts and not just the frivolities of grammatical and conceptual schemes. Though we both have expressed ourselves warmly on the issues, it is the warmth of curiosity rather than religious commitment. Each of us cares very much what the nature of time is, in the sense that we should very much like to ascertain it, whatever it may be, but neither of us cares much what it is in the sense of *wishing* time were one way rather than another. I have felt the thrill of creative-evolutionism, and Mr. Capek, I am sure, has had his visions, physical and metaphysical, *sub specie aeternitatis.*

This very amity suggests that it is more entertaining than instructive to portray our differences by assimilating them either to emotional attitudes, to ancient philosophical prototypes, or to particular theses about fate and freedom. That the so-called eternalist theory can't have much to do with the fear of death is witnessed by two of the authorities Mr. Capek quotes, one deeming that the theory guarantees we'll never die, the other implying that we are dead already but don't know it yet. The fact is, of course, that I am no more and no less desirous of having a long world line than Mr. Capek is that he keep on becoming. Similarly, the particular analogy of the idea of the four-dimensional manifold with the Eleatic and Spinozistic philosophies of the Big Being, which Professor Capek handles with discretion, has often been overdrawn. The Eleatics, and in less degree Spinoza and Bradley, were infatuated with Oneness, and hostile to time principally because it is a principle of division, of distinctness, diversity, and disconnection. Space, however, is all these in a higher degree. A chief target of the Eleatics seems to have been the geometrizing Pythagoreans, whereas Bergson, when he repudiated the spatialization of time, felt that he restored a quite Spinozistic unity and continuity. Hence I depose that it is not at all surprising that I should combine the doctrine of the manifold with the cavalier attitude toward determinism and fatalism which Mr. Capek cites, for in the four-dimensional room of the universe everything hangs loose from everything else, and the most fateful connection can be no more than a *de facto* regularity of design in the wallpaper.

What I advocate as 'the doctrine of the manifold,' I should now perhaps more specifically state, is simply a philosophical acceptance, as an ultimate literal truth about the way things are in themselves, of the conception that nature, all there is, was, or will be, 'is' (tenselessly) spread out in a four-dimensional scheme of location relations which intrinsically are exactly the same, and hence in principle commensurate, in all directions, but which happen to be differentiated, in our neighborhood at least, by the *de facto* pattern of the things and events in them—by the lie of the land, so to speak. We are all perfectly familiar with the fact that the prodigious difference of the vertical dimension of space, with its terrifying asymmetry of up and down, above and below, from all those comparatively indifferent directions we call horizontal, is not due to any intrinsic difference between vertical and horizontal distances but only to a certain characteristic complex of matter and force in our vicinity whose 'grain,' so to speak, runs one way and not the other. Just so, I argue, there is a somewhat more pervasive pattern of physical qualities and relations which constitutes the even more momentous oddity of the temporal direction, with its even more striking asymmetry of earlier and later, in contrast with all the so-called spatial directions. Very much as the singularities of arrangement which distinguish the vertical from the horizontal were explained by Descartes and Newton, so recent scientific attention to 'the direction of time' has begun the description of the arrangements which distinguish the temporal from the merely spatial. George Santayana, repudiating this account, said it analyzed time as 'a frieze of arrows,' obviously intending that we think of all the arrows as headed the same way. This sort of image of asymmetry *en masse* is implicit in most scientific as well as philosophical accounts of our affair. Such a design, i.e., the reality, whatever it be, which corresponds to this figure, I'll call the temporal motif, and any instance of it I'll call 'a time stream'—making sure, of course, that such a 'stream' is not thought of as *moving,* first, because it does not 'move' in the ordinary sense, and secondly, because what we do describe as 'moving' in the ordinary sense, in the merely spatial dimensions, we can only describe as *extended,* with more or less obliquity, in the four-dimensional scheme. We can well enough picture a time stream as like those complexes of printed arrows which represent ocean currents on a map.

I do not know the minimum possible size for a time stream, but I am confident there is one: an instance of temporality requires a certain volume for its enactment just as does an instance of green. On the other hand it does not seem at all necessary that our time stream should either be infinite or exhaust the universe, nor that there should be just one time stream, nor even that very much of the universe should be temporal at all, any more than much of the universe is characterized by up and down. It seems convenient, therefore, to call any extensive relation "spatial," and to call "temporal" just the temporal direction in a time stream. This is my answer to the question why it is proper to give the priority to the idea of space and is why I was wrong when I said that the manifold theory does not make time a dimension of space; it does. On the other hand, our theory declares that the world is 'timeless' only in those areas where the temporal motif is absent. It is not as a whole 'timeless' because it has at least some patches of time in it, but it might have been timeless altogether, in which case, of course, it would have no life or mind in it, since it is just because life and mind are so completely creatures of the temporal motif that time is so vertiginously unique for us.

Professor Capek's objections to this fairly commonplace philosophy of time are handicapped by his abstinence from presenting any alternative philosophy. No objections to any theory are of much force so long as it seems likely that there are worse objections to all the rival theories, and this is particularly the case with the philosophy of time. I cannot surely classify Professor Capek under any of the heads of time theory which I have been able to discover or invent. I don't think he rejects the manifold *in toto,* holding that the world is exhausted in the one present intuition, including its mere factitious image of a past and a future. On the other hand, he certainly denies any theory which accepts the whole manifold as a fundamental reality but adds the dynamism of a moving cosmic Now—most conspicuously such a now as a crawling spot of consciousness might define. He denies both the reality of the future and the intelligibility of any such conscious crawler (being right, I think, in the latter regard, though wrong in the former). His discourse seems most congenial to the theory of the truncated manifold—that the system of past and present is forever real, but that the manifold is endlessly growing or 'becoming' on the side of the future. With only these hints for background, then, let us see what is to be learned from his trenchant reflections on our pure theory of the manifold.

Some of his critique we may set respectfully aside as not really arguments against the theory but only expressions of his rejection of it. They are somewhat desultory complaints of features which will not seem faults except to persons already committed against the theory: for example, that my party speaks of things' or events' being 'juxtaposed' in any of the dimensions indifferently, and that we speak of the actuality and even 'the history' of the future. These usages are illicit only if the theory is wrong, and there is no way of showing them illicit except by some independent disproof of the theory. Let us remember, though, that the reference to future actualities is altogether crucial, its propriety turning essentially on the validity of logic and especially the law of excluded middle. Mr. Capek seems to recognize that the argument from logic is the eternalist's favorite, and he complains of it in the Aristotelian way that it means that the future events are all fated or necessary. Since, however, he neither suggests how the logical argument is to be obviated nor deals with the rejoinder that the 'fate' or 'necessity' which it yields is a triviality without moral, scientific, or even metaphysical importance, we may suppose that he wishes to avoid all such well-worn paths of argument for the fresher ones of his two principal attacks. These are, I take it, that the mere-manifold theory conflicts, first, with the physics of relativity, and secondly, with unmistakable traits of immediate experience.

With respect to relativity theory, Mr. Capek argues (a) that it does not require the pure theory of the manifold, as Professor Quine has asserted, and (b) that it does not permit the extravagant configurations I conjured up to illustrate the secondary and local nature of time. He takes note that I have never formulated or defended the doctrine of the manifold on Einsteinian grounds, a fact due mostly to my scientific incompetence, though in considerable part also to extreme philosophical uncertainty in every quarter, including Einstein's own cogitations, just how far the theory of relativity represents a cosmological state of affairs, how far it is positivistic or pragmatic epistemology, and how far it is pure convention. It is in a somewhat gingerly mood, therefore, that I make the following comments on his arguments from physics.

The contention he ascribes to Mr. Quine is that relativity theory makes mandatory the purely dimensional interpretation of space-time because by denying the possibility of defining any instantaneous cross-section of the manifold it precludes *a fortiori* any such one cosmic Present as (I gather) the theory of Absolute Becoming is supposed to entail. On this matter I must say that I am still quite with Mr. Quine. Surely the libration between temporal and spatial magnitudes, countenanced and indeed required by the physics of relativity, cannot be given even rudimentary expression without presupposing the full doctrine of the manifold. I am charmed rather than convinced by Mr. Capek's thesis that the Einsteinian or Minkowskian zones of silence, the wedges between the unequivocally past and the unequivocally future, only accentuate the difference between what has become and what hasn't and that instead of spatializing time they rather temporalize space. The total effect of the zone pattern (if we continue to talk so figuratively) is not that past and future are propped apart but that they overlap, and the very idea of the zones so requires the ultimacy of the idea of the manifold that the fortifying of space with time in which Mr. Capek so engagingly exults is conceivable only because the versatile sort of time it presupposes is (in his terms) already 'spatialized' and not 'dynamic.' Mr. Capek stresses, of course, that the physical theory keeps the relativization of temporal order within severe bounds, but it does not appear how Pure Becoming or an Absolute Now is any better reconcilable with a little libration than with a great deal. If his own theory is that of the truncated manifold, for example, what sort of metaphysical raggedness, adequate to the facts he cites, can he intelligibly attribute to the growing edge of the universe?

Whereas Mr. Quine is rebuked for believing that the amount of departure from absolute simultaneity and succession required by the theory of relativity entails the doctrine of the manifold, I am said to have discredited that doctrine by inferring from it the possibility of far greater departures than the theory of relativity allows. I hoped to have protected myself on this score by indicating, in the passage Mr. Capek quoted, that I am citing logical possibilities in relation to the abstract theory of the manifold; if they are contrary to specific principles of relativity theory, this should be sufficiently covered by my concession that they very well may be physically impossible. Even if the theory of relativity is taken as irrefragable gospel, it is none the less philosophically illuminating and imaginatively liberating to observe the consistency, not merely with itself and with the general scheme of the manifold, but with recent physical theses about the direction of time, of the notion that a major time stream, such as the one we parochially call the expanding universe, should embrace or embed temporal substreams which head across it at various angles. These substreams might curve, they might bend back on themselves, they might be of that N-shape which Professor Putnam has described as physically necessary and sufficient for so-called 'time travel,'[1] permitting, for example, that episodes in a man's life which are fifty years later than his birth, as reckoned by his life processes down the bore of his own time stream, occur long before he was born in the chronology of the main drift.

Though it suffices for our present philosophical purposes that it doesn't matter whether the more bizarre of the imagined deviations from normal spatiotemporal pattern could occur compatibly with the principles of relativity theory, it is not irrelevant to remark, first, that there is impressive reason to think that they are compatible with the theory, and secondly, that it is not quite fantastic to suggest that if the theory is indeed resistant it may be due for revision. To the first effect we can, of course, appeal to Gödel's celebrated suggestion, respectfully received by the relativists, that their science not only accommodates 'time travel' but gives the basis for calculating the amount of fuel needed for such excursion.[2] With regard to the second, we are up against the difficulty of interpreting the theory of relativity itself and the fact that in the staggering perspectives of recent scientific cosmology there hardly seems to be a definite belt of genuinely *physical* impossibilities between mere technological impracticability on one side and, on the other, mere violations of such terminological conventions as those by which an astrophysicist decides what objects deserve to be *called* "moving bodies," "causes," "predecessors," and so on.

Being frankly exploratory and speculative here I confess that Professor Capek has put the cap on my suspicion that the relations of the physical theory of relativity to the philosophy of the manifold are very equivocal and anomalous, if not downright contradictory, and that for all its particular successes the theory is logically in very unstable equilibrium. In certain essential connections, and to get stated at all, the theory needs the full doctrine of the isotropic manifold for its *Spielraum,* but in others it is profoundly, incongruously, and, let us say, 'romantically' devoted to the uniqueness and ubiquity of time. Thus it was that both our party and Mr. Capek could cite the theory, to opposite effect.

The doctrine of the manifold in its original inspiration was realistic and absolutistic, providing for an exactly determinate distance between any two point-instants, and a determinate ratio of such distance to the distance of any other two, irrespectively of the orientation of one pair to the other, and irrespectively, *a fortiori,* of any 'signalling' between the pairs, and regardless, finally, of that function of four-dimensional configuration called 'motion.'

Relativity theory, on the other hand, seems to impute a sort of logical magic to time and motion which sophisticates and contravenes the original features of the manifold. To this extent Mr. Capek may be justified in claiming Einstein for his party, but I wonder if this is not an extraneous and unfortunate aspect of the theory, due, perhaps, to two peculiar philosophical circumstances, first, that it is belatedly loyal to the metaphysical conceptions of enduring substances and causal transmission, instead of taking seriously the idea of four-dimensional complexes of properties, and second that it reconstrues those conceptions operationally, as if a thing could have no characters save those engendered by its affecting other things. If one defines one's terms and their relations in this way, by the characteristics of causation in the main local time stream, one's very vocabulary prevents contemplating reversals of time and cause and blots out the otherwise significant probability that the design of most of the universe is not temporal nor in the ordinary sense 'causal' at all. Operationalism, moreover, regarded in the large, is a pragmatic or romantic variety of the verification theory of meaning, and it doesn't need much argument to establish that since operations and verifications of any sort are temporal processes, any philosophy whose semantics resolves the world into them has given to so-called 'human time' a privileged status not wholly naturalizable in the manifold. This, I think, re-confirms one of the surest results of analytic philosophy in this century, viz., that operationalism and the verification theory of meaning won't do; but argument on that topic would take us too far afield.

Turning finally to Professor Capek's distinctively empirical argument, I fancy that he does not intend any considerable difference between what he calls the epistemological difficulties of the theory of the manifold and what he calls the theory's conflict with experience. I agree with him that this is the crux of the business, for I believe with him that there is such a thing as experience, unmistakable inspective and introspective evidence, which is the stuff of our personal lives, the only source of our concepts, and the only ground (though by no means the only object) of our beliefs. I think, moreover, nay, I find, that it is given, not made, and is authoritative for and not plastic to our linguistic and conceptual operations, which themselves, indeed, subsist as items in this same sphere of presence or presentation. It is a truism then to say that the philosopher fails wholly of his duty who does not accommodate to this experience the linguistic and conceptual schemes which are his maps of nature at large. Roughly, there are two modes in which this adjustment may be made. The more usual mode is that of unphilosophical science and common sense, content to affirm merely that there is some intelligible correlation between the phenomena and the conceived system of ulterior objects and events, as when we associate a heard musical note with an atmospheric vibration or with a characteristic neural discharge. For many purposes this is enough; the hypothesized physical structure, no matter how disparate it be from the datum, explains the latter, and in the extensionalistic systems of most of our contemporary philosophers of science would be said to explicate, analyze, and indeed be identical with it. These are queer thin senses of "explication" and "identity," however, in which we would have to say that squareness explicates and is identical with blueness if all and only blue things happened to be square. The less trumpery achievement would be to find an explication for our perceptual quality—felt passage, for example—which is identical with it in the sense in which *equilateral rectangle* explicates and is identical with *square*.

What then can we say at present about our spatiotemporal manifold on one hand and the unidirectional whoosh of felt time on the other? Mr. Capek is right that a person who said that this passage is 'mythical' as I did might fairly be taken to mean that there is no such experience, and he is right that the hardihood which thus denied the feeling of time flow would have to deny consciousness and its raw feels altogether. The academic world is full of behaviorists who have such hardihood, but I think it is as absurd as he does. What I meant was, first, that to call the moot character "passage" or "flow" is to indulge in myth because neither as a phenomenon nor as a pattern in the physical manifold can it resemble in any definitive or logically important respect such literal 'passage' as that of a ship through a canal, any more than a mind can be 'acute' in the sense in which an angle is acute, and secondly, that the feeling is so equivocal, and the doctrine of the manifold is so plain and firm, that we may be reasonably confident that to speak of felt passage is to speak allegorically of some function of distribution in the manifold.

Beyond this I do not see my way quite clearly, but Mr. Capek will be patient, I hope, when he reflects that if I have trouble explaining how it appears that there is passage or becoming although really there is only the manifold, he must have corresponding trouble to explain how it appears that there is a manifold though really there is passage and becoming. It is a sobering thought that not only is the physical nature and habitat of the simplest raw feel still mysterious, but even the correlation between the so-called private space of personal experience and the public space of the surveyor and astronomer is so elusive that it drove Leibniz, Russell, and Whitehead to the extravagances of monadism. Two proposals I think we should reject at once. One is what may be called pseudo-monism, the extensionalist trick of saying that the feel of passage 'is' the correlative dimensional structure when one means only that they are correlated. The other is the extreme dualism which affirms that the qualities intuitively identified as passage or becoming belong to one set of entities, 'appearances' in or to an immaterial spirit, for example, while the manifold and the time motif are traits of a fundamentally different set of entities, 'reality' or physical things in themselves. I do not renounce dualism for Mr. Capek's reason, that the bare idea that two sets of entities should be very dissimilar is scandalous, but because it pays no logical dividends in the present case. Time, as Mr. Capek says, is just the property which mind and matter are supposed to have in common, and the logical motives which impel us to think of physical things, past, present, and future, as disposed in a temporal dimension as well as in the spatial dimensions, impel us to think of our experiences, past, present, and future, even if they be immaterial, as extended at least in the temporal dimension, and the main problem of reconciling time as felt passage with time as an extension is rather intensified by dualism than solved.

It would be excusable but a little irresponsible if I now abandoned the issue with only the statement that it will be time to settle the special question of the relation of the time quality of experience to the time structure of nature when we have settled the general question of the relation of experienced qualities to physical structures. I think that already, however, we can suggest a useful variety of attacks on Mr. Capek's and my problem. We do not have to wait on a general theory of qualities to notice a phenomenological contrast between the mere raw feel of sweet or warm and the diffuse unique sense of passage or becoming (so-called), nor need we delay to distinguish between the latter mere sense of becoming, which can be had in any specious present, and that other observation, the least congenial with the manifold, that each specious present is succeeded by or flows into another. The mere feel of passage, with its inherent forward strain and imbalance, seems at worst more *appropriate* to the physical time motif generally than most perceptual qualities are to their physical conditions, and in my article to which Mr. Capek mainly referred I mentioned more special traits of vital and mnemic process which must accentuate the conscious effect. What is now left of the final problem of succession or supersession, I think, resolves itself into the question why our specious presents, though overlapping one by one, are limited, so that we never experience our whole conscious lives at once, each as a single field. Though the concept is strained and awkward I am confident that this characteristic of the fields, which seems in principle a fairly plain topological one, is enough to explain in the most thorough sense, i.e., strictly to analyze or explicate, that experience of succession which constitutes the so-called movement of time, and hence the experiences of 'novelty' and 'becoming,' and since nothing could be more primitive and familiar, nor more integral to our notion of conscious biographies, than just this structural peculiarity of experience, we might be said now at last to have solved the problem. The question why experience is thus composed I do not think is meaningless nor even futile. Being the question, however, what constitutes the peculiar unity of a conscious field, which is one degree more difficult than Hume's question what constitutes a lifetime's personal identity, it is indistinguishable from the question what constitutes a conscious field at all, i.e., what makes it conscious. And this is more than even an unblushing student of psychophysics can be expected to settle at the end of a string of comments on comments on theories of time.

In conclusion, let us charge ourselves with a couple of cautions. First, it is not clear that the conscious field is fundamentally different in the presently most relevant respect from other fields in nature. Secondly, even if it be so, we must not think of the conscious sequence in Weyl's way, as a metaphysical lightning bug crawling up a time line. It, like everything else, a planet or a storm, for example, is rather a four-dimensional worm laid out once for all in the manifold, and can no more establish one absolute or moving Now than a clock or a stone can. On the other hand, just as Mr. Capek reminds us that the so-called movement of time, even in his theory, is not really a *movement,* so the so-called staticity of the manifold is not really staticity. We are not denying either genuine movement or the quality which Mr. Capek calls becoming, but trying to explain them, in appropriately different ways. Our explanations, moreover, are such that instead of curtailing either movement or becoming they greatly increase their incidence. Mr. Capek, for all that he is so eloquent about it, thinks that there is no becoming except in the one thin skin of the present; we say, however, for what it may be worth, that there (tenselessly) 'is' (real) becoming wherever in the manifold the temporal motif is.

Dept. of Philosophy
Harvard University
Cambridge, Mass.

1 Hilary Putnam, "It Ain't Necessarily So," *Journal of Philosophy 59* (Oct. 25, 1962), pp. 658-671.

2 Kurt Gödel, "A Remark About the Relationship Between Relativity Theory and Idealistic Philosophy," in *Albert Einstein: Philosopher-Scientist,* edited by Paul A. Schilpp (New York, 1955), pp. 555-562.